Coastal Zones

Coastal Zones

Solutions for the 21st Century

Juan Baztan
Omer Chouinard
Bethany Jorgensen
Paul Tett
Jean-Paul Vanderlinden
Liette Vasseur

ELSEVIER

AMSTERDAM • BOSTON • HEIDELBERG • LONDON • NEW YORK • OXFORD
PARIS • SAN DIEGO • SAN FRANCISCO • SINGAPORE • SYDNEY • TOKYO

Elsevier
Radarweg 29, PO Box 211, 1000 AE Amsterdam, Netherlands
The Boulevard, Langford Lane, Kidlington, Oxford OX5 1GB, UK
225 Wyman Street, Waltham, MA 02451, USA

ISBN: 978-0-12-802748-6

British Library Cataloguing in Publication Data
A catalogue record for this book is available from the British Library

Library of Congress Cataloging-in-Publication Data
A catalog record for this book is available from the Library of Congress

For information on all Elsevier publications
visit our website at http://store.elsevier.com/

Contents

Part II
Developing Solutions: Challenges for Communities in the Context of Global Change

Contributors

Joseph Onwona Ansong Intergovernmental Oceanographic Commission of UNESCO, Paris, France

Francisco Miranda Avalos Foro Hispano Americano de Intercambio de Información sobre Temas del Mar (ONG OANNES)

Alioune Ba Departement de Geographie, Université Cheikh Anta Diop, master GIDEL, Boulevard Martin Luther King, Dakar, Senegal

Juan Baztan Université de Versailles Saint-Quentin-en-Yvelines, OVSQ, CEARC, Guyancourt, France; Marine Sciences for Society, www.marine-sciences-for-society.org

Anastasie Beye Mendy Departement de Geographie, Université Cheikh Anta Diop, master GIDEL, Boulevard Martin Luther King, Dakar, Senegal

Kenny Black Scottish Association for Marine Science, Scottish Marine Institute, Argyll, Oban, UK

Anne Blanchard Center for the Study of the Sciences and the Humanities, University of Bergen, Bergen, Norway

Kieran Bowen Bowen Marine Ltd, Bishopstown, Cork, Ireland

Scott Bremer Center for the Study of the Sciences and the Humanities, University of Bergen, Bergen, Norway

Ruth Brennan Scottish Association for Marine Science, Scottish Marine Institute, Argyll, Oban, UK

Elisabetta Broglio Institut de Ciències del Mar, CSIC, Barcelona, Spain; Marine Sciences for Society, www.marine-sciences-for-society.org

Ana Carrasco Observatorio Reserva de Biosfera, Cabildo de Lanzarote, Arrecife, Spain

H. Caymaris Intendencia de Rocha, Rocha, Uruguay

Omer Chouinard Université de Moncton, Moncton, New Brunswick, Canada; Marine Sciences for Society, www.marine-sciences-for-society.org

C. Chreties IMFIA, Facultad de Ingeniería, Universidad de la República, Montevideo, Uruguay

Colleen Mercer Clarke Telfer School of Management, University of Ottawa, Ottawa, ON, Canada

John D. Clarke Telfer School of Management, University of Ottawa, Ottawa, ON, Canada

D. Conde Centro Universitario Regional Este, Universidad de la República, Rocha, Uruguay; Seccion Limnología, Facultad de Ciencias, Instituto de Ecología y Ciencias Ambientales, Universidad de la República, Montevideo, Uruguay; Espacio Interdisciplinario, Universidad de la República, Montevideo, Uruguay

Elizabeth Cook Scottish Association for Marine Science, Scottish Marine Institute, Argyll, Oban, UK

Loreta Cornacchia ISPRA - Istituto Superiore per la Protezione e Ricerca Ambientale, Rome, Italy; Royal Netherlands Institute for Sea Research (NIOZ), Yerseke, The Netherlands

Charlotte Da Cunha Université de Versailles Saint-Quentin-en-Yvelines, OVSQ, CEARC, Guyancourt, France

Keith Davidson Scottish Association for Marine Science, Scottish Marine Institute, Argyll, Oban, UK

D. de Álava Centro Universitario Regional Este, Universidad de la República, Rocha, Uruguay

Raimonds Ernsteins UNESCO Chair in Sustainable Coastal Development, University of Latvia, Riga, Latvia

Awa Fall Niang Departement de Geographie, Université Cheikh Anta Diop, master GIDEL, Boulevard Martin Luther King, Dakar, Senegal

François Galgani IFREMER, Centre de Corse, Laboratoire Environnement Ressources PAC/Corse Imm Agostini, ZI Furiani, Bastia, France

M. García-León Universitat Politècnica de Catalunya, Laboratori d'Enginyeria Maritima LIM/UPC, International Centre Coastal Resources Research, CIIRC, Barcelona, Spain

Joaquim Garrabou Institut de Ciències del Mar, CSIC, Barcelona, Spain

Ndickou Gaye Departement de Geographie, Université Cheikh Anta Diop, master GIDEL, Boulevard Martin Luther King, Dakar, Senegal

Judith Gobin The University of the West Indies, St. Augustine Campus, Trinidad and Tobago

Sathya Gopalakrishnan Department of Agricultural, Environmental and Development Economics, The Ohio State University, Columbus, OH, USA

V. Gràcia Universitat Politècnica de Catalunya, Laboratori d'Enginyeria Maritima LIM/UPC, International Centre Coastal Resources Research, CIIRC, Barcelona, Spain

Thierry Huck UBO-CNRS-LPO, UFR Sciences F308, Brest, France; Marine Sciences for Society, www.marine-sciences-for-society.org

Arnaud Huvet IFREMER, Centre de Brest, Laboratoire Physiologie des Invertébrés, Plouzané, France

Alejandro Iglesias-Campos Intergovernmental Oceanographic Commission of UNESCO, Paris, France

Bethany Jorgensen The University of Maine, Orono, Maine, USA; Marine Sciences for Society, www.marine-sciences-for-society.org

Mélanie Jouitteau CNRS, UMR 5478, Université de Pau et des Pays de l'Adour, Pau, France; Université Bordeaux III, Pessac, France

Matthias Kaiser Center for the Study of the Sciences and the Humanities, University of Bergen, Bergen, Norway

Alioune Kane Departement de Geographie, Université Cheikh Anta Diop, master GIDEL, Boulevard Martin Luther King, Dakar, Senegal

Janis Kaulins UNESCO Chair in Sustainable Coastal Development, University of Latvia, Riga, Latvia

Andrew G. Keeler University of North Carolina Coastal Studies Institute, Wanchese, NC, USA

X. Lagos Dirección Nacional de Medio Ambiente, Ministerio de Vivenda, Ordenamiento Territorial y Medio Ambiente, Uruguay

Craig E. Landry Department of Agricultural & Applied Economics, University of Georgia, Athens, GA, USA

Daniel E. Lane Telfer School of Management, University of Ottawa, Ottawa, ON, Canada

Anita Lontone UNESCO Chair in Sustainable Coastal Development, University of Latvia, Riga, Latvia

Dylan McNamara Department of Physics & Physical Oceanography, University of North Carolina, Wilmington, NC, USA

Andrus Meiner European Environment Agency, Copenhagen, Denmark

Aquilino Miguelez Observatorio Reserva de Biosfera, Cabildo de Lanzarote, Arrecife, Spain

Laura J. Moore Department of Geological Sciences, University of North Carolina, Chapel Hill, NC, USA

A. Brad Murray Nicholas School of the Environment, Duke University, Durham, NC, USA

Michelle Mycoo The University of the West Indies, St. Augustine Campus, Trinidad and Tobago

Sabine Pahl Plymouth University, Drake Circus, Plymouth, UK

D. Panario UNCIEP, Instituto de Ecología y Ciencias Ambientales, Facultad de Ciencias, Universidad de la República, Montevideo, Uruguay

Ika Paul-Pont IUEM, CNRS/UBO, Laboratoire des Sciences de l'Environnement Marin, Plouzané, France

G. Piñeiro Departamento de Evolución de Cuencas, Instituto de Ciencias Geológicas, Facultad de Ciencias, Universidad de la República, Montevideo, Uruguay

Steve Plante Departement Sociétés, Territoires et Développement, Université du Québec à Rimouski, Rimouski, Québec, Canada

Gregory Quenet Université de Versailles Saint-Quentin-en-Yvelines, Guyancourt, France

Jacques Quensière IRD, UMI RESILIENCES Bondy Cedex, France

Tiavina Rivoarivola Rabeniaina Université de Moncton, Moncton, New Brunswick, Canada

L. Rodríguez-Gallego Centro Universitario Regional Este, Universidad de la República, Rocha, Uruguay

A. Sánchez-Arcilla Universitat Politècnica de Catalunya, Laboratori d'Enginyeria Maritima LIM/UPC, International Centre Coastal Resources Research, CIIRC, Barcelona, Spain

Aichetou Seck Departement de Geographie, Université Cheikh Anta Diop, master GIDEL, Boulevard Martin Luther King, Dakar, Senegal

L. Seijo Oficina de Planeamiento y Presupuesto, Presidencia de la República, Montevideo, Uruguay

Martin D. Smith Nicholas School of the Environment, Duke University, Durham, NC, USA

S. Solari IMFIA, Facultad de Ingeniería, Universidad de la República, Montevideo, Uruguay

Philippe Soudant IUEM, CNRS/UBO, Laboratoire des Sciences de l'Environnement Marin, Plouzané, France

Céline Surette Université de Moncton, Moncton, New Brunswick, Canada; Marine Sciences for Society, www.marine-sciences-for-society.org

Andrea Taramelli ISPRA - Istituto Superiore per la Protezione e Ricerca Ambientale, Rome, Italy; IUSS, Istituto Universitario di Studi Superiori, Pavia, Italy

L. Teixeira IMFIA, Facultad de Ingeniería, Universidad de la República, Montevideo, Uruguay

Paul Tett Scottish Association for Marine Science, Scottish Marine Institute, Oban, Argyll, UK

Diatou Thiaw Departement de Geographie, Université Cheikh Anta Diop, master GIDEL, Boulevard Martin Luther King, Dakar, Senegal

Richard Thompson Plymouth University, Drake Circus, Plymouth, UK

Mariano Gutiérrez Torero Universidad Nacional Federico Villarreal, Lima, Peru

Emiliana Valentini ISPRA - Istituto Superiore per la Protezione e Ricerca Ambientale, Rome, Italy

Jean-Paul Vanderlinden Université de Versailles Saint-Quentin-en-Yvelines, OVSQ, CEARC, Guyancourt, France; Marine Sciences for Society, www.marine-sciences-for-society.org

Liette Vasseur UNESCO Chair in Community Sustainability: from local to global, Department of Biological Sciences, Brock University, St. Catharines, Ontario, Canada

N. Verrastro Centro Universitario Regional Este, Universidad de la República, Rocha, Uruguay

J. Vitancurt Centro Universitario Regional Este, Universidad de la República, Rocha, Uruguay; Dirección Nacional de Medio Ambiente, Ministerio de Vivenda, Ordenamiento Territorial y Medio Ambiente, Uruguay

Sebastian Weissenberger UQAM, Montreal, QC, Canada

Ilga Zilniece UNESCO Chair in Sustainable Coastal Development, University of Latvia, Riga, Latvia

Foreword

The coast is an edgy place. Living on the coast presents certain stark realities and a wild, rare beauty…It's a place of tide and tantrum…of tense negotiations with an ocean that gives much but demands more…the coast remains…uncertain about tomorrow.

Carl Safina, The View from Lazy Point: A Natural Year in an Unnatural World

This description of the coast is certainly compelling. But what is it about the shoreline that so powerfully draws us to it? And why should any of us care about coastal zones?

Probably the greatest reason is that the ocean and coasts provide ecosystem services that permit each of us to live on this planet, whether we are situated near the ocean or not. Carbon absorption, oxygen production, habitat for myriad creatures, and biodiversity are just a few of the services. But perhaps more important is the fact that coastal populations are growing dramatically. Already, more than 40% of the world's population lives within 100 km of the coast, with projections that within the next few decades, this will rise to 75%. In much of the developing world, coastal populations are exploding. This of course puts tremendous pressure on coastal systems, elevating potential losses due to natural hazards or extreme climate events, such as coastal erosion, sea-level rise, storm surges and tsunamis, and at the same time greatly increasing the likelihood of amplified anthropogenic impacts, intensified competition among the growing numbers and types of users, and enhanced conflicts between natural processes and human development.

As Executive Secretary of the Intergovernmental Oceanographic Commission (IOC) of the United Nations Educational, Scientific and Cultural Organization (UNESCO) from 2010 to 2015, I gained first-hand understanding of the importance of coastal systems and the critical issues they are facing, having personally received requests from developed, developing, and emerging coastal nations alike for assistance in undertaking and/or obtaining the necessary science to underpin effective policies, strategies, and regulations. The IOC, of course, has been dealing with these topics for more than half a century. Created in 1960 to promote international cooperation and to coordinate programs in ocean research, services, and capacity development, the IOC continues to address both deep ocean and coastal science concerns, having produced guidelines for Marine Spatial Planning (MSP) that have been used by more than 40 countries as well as a handbook for Integrated Coastal Area Management

(ICAM), coordinating tsunami warning systems around the globe, and working on the ground to help member states in various regions deal with coastal zone hazards and issues.

This book, *Coastal Zones: Solutions for the 21st Century,* is both timely and essential. Its multinational authors have for many years demonstrated their concern for coastal communities as well as their commitment to addressing coastal issues through integrated science, and here bring forward not only challenges, but more importantly, opportunities and possible solutions. They elucidate the importance of coastal zones for ecological, social, and economic reasons, while pointing out the severe anthropogenically induced environmental degradation that is occurring against a backdrop of risks and alterations due to climate change. Throughout the chapters, the authors propose holistic approaches, arguing that in order to be effective, initiatives designed to promote sustainability must be co-constructed with affected communities. They stress the need for coordination and cooperation to overcome strong intersectoral competition as well as transdisciplinary, community-centered adaptation strategies. Certain chapters speak to sustainability of industries such as fishing and aquaculture, and the need to adopt conservation strategies that are socially acceptable for all stakeholders. Others caution that human-designed solutions to coastal hazards such as erosion and storm protection need to consider trade-offs over the longer term, since in fact they can produce risks more dangerous than the original. Overall, the book provides innovative approaches by which coastal communities around the world may address their coastal zone management issues through inclusive governance that is inspired by multidisciplinary science and active, meaningful intersectoral stakeholder engagement.

Wendy Watson-Wright, PhD
Former Executive Secretary
Intergovernmental Oceanographic Commission of UNESCO

Introduction

Juan Baztan[1,2], Omer Chouinard[3,2], Bethany Jorgensen[4,2], Paul Tett[5], Jean-Paul Vanderlinden[1,2], Liette Vasseur[6]

[1]Université de Versailles Saint-Quentin-en-Yvelines, OVSQ, CEARC, Guyancourt, France; [2]Marine Sciences For Society, www.marine-sciences-for-society.org; [3]Université de Moncton, Moncton, New Brunswick, Canada; [4]The University of Maine, Orono, Maine, USA; [5]Scottish Association for Marine Science, Scottish Marine Institute, Oban, Argyll, UK; [6]Department of Biological Sciences, Brock University, St. Catharines, Ontario, Canada

Coastal zones, the narrow transition areas that connect terrestrial and marine environments, are our planet's most productive and valued ecosystems (Crossland et al., 2005). Sixty percent of the world's major cities are located in coastal zones, and 40% of the all the people on the planet live within 100 km of a coastal zone (Nicholls et al., 2007). Within coastal areas, we see the tightly intertwined relationships between humans and coastal resources amplifying the most urgent questions of limits and equilibrium, sustainability, and development in our world today.

Over the past 25 years, efforts have been made to understand and improve the relationships between our societies and our coastal ecosystems. They have led to more than 100 national and transnational coastal zone plans, protocols, and conventions. Nevertheless, we realize the balance between development and stewardship still tilts toward development. Many more efforts are needed to restore harmony between use and conservation of coastal zones. Furthermore, most advances have been driven top-down, often with scant regard for grass-roots interests.

In November 2011, the "Coastal Zones: 21st Century Challenges" working group, a consortium of academics and members of research centers across the globe, collected 115 points of view and synthesized them into one document addressed to delegates attending the Rio+20 Conference held in Brazil in June 2012. This baseline document represented the interdisciplinary collaborative work of more than 200 coastal zone researchers from all continents (see Appendix). The idea for this book grew from seeds planted by the baseline document, and it was nourished through subsequent workshops organized by the working group.

Rooted in the baseline document, our objectives for this book are to: (1) highlight the looming challenges facing coastal zones around the world and

(2) explore potential solutions from the perspective of the scientific and technological community, as part of the effort to construct and achieve the Rio+20 goals and soon to come, the proposed Sustainable Development Goals (SDGs). Indeed, among the proposed SDGs, Goal 14 is of particular importance to us: Conserve and sustainably use the oceans, seas, and marine resources for sustainable development. Several others are also directly or indirectly connected to our work and dialog.

This book links perspectives from regional, national, and international efforts with local needs for actions in communities where coastal zone challenges are faced daily. It is designed for a diverse audience that encompasses academics and "on the ground" practitioners and community stakeholders. Stakeholders and practitioners need to know how to reach their groups or communities, how to involve them in finding long-term solutions, how to identify underlying issues and understand how problems are integrated in order to determine a path forward, and so forth. We hope you will gain new insight from our unique effort to compile and connect present challenges and possible solutions from different locations around the world, as opposed to focusing on one single region.

For students, this book provides an invaluable reference to better understand the steps of Integrated Coastal Zone Management, from problem description to potential approaches to solutions, and to see how these steps can be implemented in communities. From this book, students will learn the pros and cons of various approaches, understand the issues from an interdisciplinary point of view, and find new ideas for projects and research.

In our experiences as professors, researchers, and practitioners, there are few textbooks on Integrated Coastal Zone Management that adopt a transdisciplinary approach—by which we mean one that draws on stakeholder knowledge and interpolates it with perspectives from the natural and social sciences to provide a basis for the co-development of an effective understanding of socioecological systems in the coastal zone. We aim to provide a broad perspective and to consider not only problems but also approaches that may lead to solutions. With "wicked" problems like those facing coastal zones, it is easy to find literature that highlights the intractability of the challenges we face. It is much harder to find research concerning potential solutions to help communities and stakeholders. This work aims to help fill that crucial gap.

We would like to take a moment to acknowledge the many people who have brought this book to fruition. It is a truly interdisciplinary collaborative work, and we sincerely thank those whose input, encouragement, and effort have made it possible. From the "Coastal Zones: 21st Century Challenges" working group members, to the contributing authors for their inspiring work, to the communities who have worked with and supported us, as well as the publishing team at Elsevier, especially Candice Janco and Marisa LaFleur, with their unfailingly polite attempts to keep us on track, and Mohanapriyan Rajendran and his production team. Thank you all for your contributions to and patience with the process.

By presenting a wide range of approaches to the challenges coastal zones face in the twenty-first century, this book will expand your "toolkit" for collaboratively transforming coastal communities and ecosystems to achieve a more sustainable future. We hope it inspires you to engage these challenges in new ways, with renewed vigor. As a colleague recently said, "Let's do the work—that's the only solution."

REFERENCES

Crossland, C.J., Kremer, H.H., Lindeboom, H.J., Marshall Crossland, J.I., Le Tissier, M.D.A. (Eds.), 2005. Coastal Fluxes in the Antropocene: The Land–Ocean Interactions in the Coastal Zone, Project on the International Geosphere-Biosphere Programme Series. Global Change—The IGBP Series. Springer-Verlag, Berlin, 232 pp.

Nicholls, R.J., Wong, P.P., Burkett, V., Codignotto, J., Hay, J., McLean, R., Saito, Y., 2007. Coastal systems and low-lying areas. In: Parry, M.L., Canziani, O.F., Palutikof, J.P., van der Linden, J.P., Hanson, C.E. (Eds.), Climate Change 2007: Impacts, Adaptation and Vulnerability: Contribution of Working Group II to the Fourth Assessment Report of the Intergovernmental Panel on Climate Change. Cambridge University Press, Cambridge, UK, pp. 315–356.

Part I

Facing the Challenges

Chapter 1

Paradigm Shifts, Coastal Zones, and Adaptation to Fast-Paced Change: Moving Toward Transdisciplinary Community-Centered Approaches

Jean-Paul Vanderlinden[1,2], Gregory Quenet[1], Charlotte Da Cunha[1], Juan Baztan[1,2]

[1]*Université de Versailles Saint-Quentin-en-Yvelines, OVSQ, CEARC, Guyancourt, France;*
[2]*Marine Sciences for Society, www.marine-sciences-for-society.org*

Chapter Outline

INTRODUCTION

If one projects oneself further into the twenty-first century, taking stock of what is known today about the future, one quickly recognizes the need to develop new strategies to face the rapid changes that coastal areas will be going through. Adaptation, at a pace rarely known to humankind, may very well be the most challenging endeavor for coastal communities.

Using adaptation to climate change as a case study, we argue that a paradigm shift must occur. It is now necessary that knowledge creation transcends the traditional organization of science, and that this transcendence must be locally driven, implemented, and translated into policies.

In order to achieve such a shift, the humanities in general, and environmental humanities in particular, should move to the forefront of adaptation science with, and for coastal communities. This leads to an apparently contradictory situation where the traditional organization of science should be fading into the background while simultaneously the very categories associated with this organization must be acknowledged in order to mobilize the conceptual tools that were not sufficiently used in the past (e.g., environmental history, eco-philosophy, literature, performance studies, and ethics).

ADAPTATION TO CLIMATE CHANGE AS A CASE STUDY

The fight against climate change and its consequences has long focused on reducing greenhouse gas emissions—that is to say, mitigation. This may have further increased the pressure on coastal areas (e.g., off- or on-shore wind farms, shifts in transportation modes; see chapter 15 from Bremer et al.). However, with the realization that climate change and its impacts are inevitable, adaptation policies are occupying an ever-increasing space both in the science and policy spheres. This raises three challenges: adaptation is (1) a concept of uncertain form (Tubiana et al., 2010), (2) which deals with uncertainty, and (3) which calls for transdisciplinary analysis.

Why is adaptation to climate change still "a concept of uncertain form"? The definition proposed by the Intergovernmental Panel on Climate Change, the more consensual, is very generic and offers neither methodological nor political content: an "adjustment of natural systems or human systems when facing a new environment or a changing environment" (McTeggart et al., 1990), or an "adjustment in natural or human systems in response to actual or expected climatic stimuli or their effects, which moderates harm or exploits beneficial opportunities" (Adger et al., 2007).

Adaptation policies are difficult to define for three main reasons. The first refers to the temporal dimension of adaptation, thought to be a long process but which needs to be anticipated by a proactive approach. The second reason refers to the uncertainties surrounding the impacts of climate change, particularly at the local level on the coasts. The third reason refers to the evolutionary nature of the adaptation concept, which assumes a constant evolution, a continual readjustment of knowledge and choices. Beyond these difficulties, the conceptualization of adaptation itself remains unclear: it builds itself following the elaboration of public policies in the framework and modalities of action they put in place, and this empirical definition is often used to avoid a theoretical definition that would clarify its long-term goals. This leads to challenges in planning, and a quagmire of implementation procedures (Simonet, 2011).

Not only is adaptation a concept of uncertain form, it is a concept geared at dealing with uncertainty. One of the key challenges of adaptation is that adaptive actions are rooted in foresight exercises conducted under high levels of uncertainty. As such, these must be, but unfortunately seldom are, framed as part of coastal climate risk governance processes[1] (Renn et al., 2011). When

1. Risk governance may be defined as a systemic approach to the decision-making processes associated with risk (uncertain events associated with potential beneficial or harmful consequences), which seeks to reduce risk exposure and vulnerability by filling gaps in policy.

considered, adaptation becomes a unique locus for negotiating future pathways in order to act under uncertainty. Action under uncertainty involves, for the affected parties, resolving conflicting claims. These claims pertain to the communities' understanding of causal chains, to their assessment of pertinence, and to the expression of their values and norms (Renn, 2008; Touili et al., 2014). The ontology of adaptation is therefore closely linked with the uncertainty associated with the foresight-related content of the concept, and leads to the existence of plural perspectives—all equally legitimate, but none in a position to grasp the concept in its entirety. This reinforces the need for the development of robust conceptual foundations.

Finally, adaptation calls for transdisciplinary analysis mobilizing environmental humanities. Adaptation's theoretical grounding has revolved essentially around the analysis of potential hindrances to adaptive processes. These challenges include lack of precise knowledge on the future of local and regional climate regimes, lack of understanding of these future climate regimes by local and regional populations, misunderstanding of the economic cost and benefit of adaptation strategies, discrepancies between national governance cultures and local collective action, and "social" limits and values including "fairness." The length of this list, which is by no means exhaustive, may be explained by the fact that almost every discipline may contribute to the understanding of adaptation, thus leading to a constant redefinition of adaptation's conceptual content. This clearly indicates that adaptation as a concept transcends disciplines and calls for transdisciplinary analysis (Blanchard and Vanderlinden, 2010).

These three characteristics indicate that some of the knowledge base necessary to domesticate adaptation is not limited to the natural and social sciences. Philosophy (as the art of inventing concepts and giving them meanings), environmental history (as the source for understanding the historical embeddedness of the dialog between nature and culture), and performance studies (as the study of the deep equivalence of words and actions), to name a few, seem all to be needed in order for adaptation to be thought of in a way that is attuned to its characteristics.

Yet, adaptation to climate and environmental changes has only recently become a research topic relevant to many of the humanities. From the late nineteenth century, the term has primarily referred to biological theories of evolution, meaning the modification of a living organism according to its environment, or its situation (Darwin, 1859). Since then, this concept remains one of the most discussed and complex in biology (Pradeu, 2011). The term, like "resilience," remains borrowed from the natural sciences, and only made a late appearance in the social sciences (late twentieth century). More recent works put climate change in relation to social change, but they rarely use the concept of adaptation, except in the particular case of natural disasters. As a result, hazards are treated as an external feature of societies, avoiding genuine political and social thought about the real meaning of climatic threats in everyday life.

Today is about simultaneously seizing environmental change and social change in their evolutions and multiple interactions. However, the humanities have largely acted with reluctance to this proposal and to the concept of adaptation. Durkheimian sociology was built on the rejection of *circumfusa* and relies on the deeply entrenched belief that only social processes and structures can produce social processes and structures (Durkheim, 1893, 1894). Vidalian geographers and Annales historians avoided anything that might resemble a form of environmental determinism (Friedman, 2004). Yet we argue that now the humanities must be involved.

MOVING TOWARD THIS PARADIGM SHIFT: A GAP ANALYSIS AND ASSOCIATED RATIONALE

In our research, we have identified several key gaps that need to be addressed. The first gap lies within the need to redefine our understanding of how words, speech acts, and a particular category of these—the words and discourse produced by science—change reality upon their utterance. If we want scientific discourses on change (i.e., climate change), to change the way one sees the planet, we must find a way to convey that the planet has changed. We want words to act. How does one "do things with words"? From Austin's initial lectures (Austin, 1962) and his associated definition of performatives, where "the issuing of the utterance is the performing of an action," the concept of performativity has been explored in various fields: identity (the most famous instance lies in the works on gender by Butler, 1990, 1997), contextualization of speech acts (Austin, 1962; Searle, 1979), performance studies (Parker and Sedgwick, 1995), and science's capabilities to influence the world (MacKenzie et al., 2008). If we want the knowledge on changes to convey the changing nature of coastal areas and to act on coastal futures, we need to mobilize the progress made on these fronts within our specific context. The gap we need to fill is thus associated with taking fully into account how knowledge as "embodied performance" (Gil, 2008) may, through the analysis of change, past, present, and future, lead to a dynamic co-evolution of the planet we inhabit and the societies in which we live.

The second gap, closely associated with the first one, is the need to re-explore the way science-based scenarios, as "narratives giving a memory of the future" (Rasmussen, 2008) may be framed as stories and turned into performance, thereby modifying the very fabric of our world. Scenario planning has a long history, and has focused through time on building narratives geared at giving people and institutions a window on the unforeseeable future (Vanderlinden, 2015). Scenarios are hybrid forms; they are rooted in science, but are carried by hypotheses that do not need to be scientific per se, and conveyed by stories, or narratives. Some scenario exercises lead to genuine changes in policies, changes in daily practices, or enhanced adaptive capacities; others fail to do so. How can a science-based—but not only science-based—scenario push people to move today? How can a science-based scenario enable people to grasp the need to change and act? How can a science-based scenario change the world? What are the characteristics of "successful" narratives of change?

A third gap lies in the mobilization of environmental history. One of the central shortcomings of most approaches dealing with the analysis of change and human society's responses to change may lie in the great nature–culture divide. Environmental history is one field of the humanities closing this divide through extensive bridge-building between natural sciences and the humanities (e.g., Quenet, 2015). The systematic mobilization of environmental history will lead us to collectively see adaptation as a capacity of translation and enrollment, of connecting environmental changes and social changes thanks to multilateral negotiations in a material field of constraints. This will give us a window on the past that is precisely attuned to the challenges of adaptation for the future.

Finally, combining the three elements presented above leads us to a final gap, which lies in the empirical application of the promises associated with Latour's "Compositionist manifesto" (Latour, 2010, 2011). Rooted in the fact that the divide between nature and society (matters of fact and matters of concern) cannot be taken for granted anymore, compositionism stresses that things have to be put together while retaining their heterogeneity. If nature is not already assembled, the scientific facts of the matter have to be constructed and an assembly is necessary to compose a common world through arts and politics.

IMPLEMENTING SUCH A PARADIGM SHIFT: THE ARTisticc PROJECT

This gap analysis led the authors and several other colleagues to develop a project geared at initiating the process: the ARTisticc[2] (Adaptation Research: A Transdisciplinary Community and Policy-Centered Approach) project. ARTisticc has been designed to experiment with the paradigm shift described above in seven coastal communities: Uummannaq (Greenland), Tiksi (Sakha Republic/Russian Federation), Wainwright (Alaska, USA), Cocagne-Grande-Digue (New Bunswick, Canada), Bay of Brest (Brittany, France), Mbour (Senegal), and the Kanyakumari district and Nagapattinam regions (India). At these sites, the communities themselves will assess their science-based scenarios of the future. For each field setting, current adaptation will be identified with the participating coastal communities, and the analytical focus will be on the evolution, or lack thereof, of local institutions intertwined with noninstitutional or nonlocal forcing. Within each community, a past adaptation will be analyzed through the lens of environmental history. Depending on local specificities, a local artist or craftsperson will translate the scientific results into meaningful local artwork (through storytelling, playwriting, photography, local crafts, film, etc.). All of

2. ARTisticc is funded through the participation of the Belmont Forum International Opportunity Fund, and national funding agencies from France, USA, Canada, Russia, and India. Their contributions are gratefully acknowledged.

the project's outputs will thereafter be assembled, and composed, in order to confront policymakers with the results achieved.

CONCLUSIONS

The pace of change that coastal communities will be facing later in the twenty-first century calls for a shift in the way we construe, scientifically and socially, the adaption discourse. While natural and social sciences have been used in the past, the humanities and arts have been somehow less present. We argue that this is a major shortcoming of adaption as a scientific subject. Solutions to twenty-first century coastal challenges should, and hopefully will, include the arts and humanities.

REFERENCES

Adger, W.N., Agrawala, S., Mirza, M.M.Q., Conde, C., O'brien, K., Pulhin, J., Pulwarty, R., Smit, B., Takahashi, K., 2007. Assessment of adaptation practices, options, constraints and capacity. In: Parry, M.L., Canziani, O.F., Palutikof, J.P., Van Der Linden, P.J., Hanson, C.E. (Eds.), Climate Change 2007: Impacts, Adaptation and Vulnerability. Contribution of Working Group II to the Fourth Assessment Report of the Intergovernmental Panel on Climate Change. Cambridge University Press, Cambridge, pp. 717–743.

Austin, J.L., 1962. How to Do Things with Words: The William James Lectures Delivered at Harvard University in 1955. Oxford, Clarendon.

Blanchard, A., Vanderlinden, J.-P., 2010. Dissipating the fuzziness around interdisciplinarity: the case of climate change research. Surv. Perspect. Integr. Environ. Soc. (S.A.P.I.EN.S) 3 (1). Available from: <http://sapiens.revues.org/990> (03.02.15).

Butler, J., 1990. Gender Trouble: Feminism and the Subversion of Identity. Routledge, London.

Butler, J., 1997. Excitable Speech: A Politics of the Performative. Routledge, London.

Darwin, C., 1859. On the Origin of Species by Means of Natural Selection. John Murray, London.

Durkheim, E., 1893. De la Division du Travail Social. Félix Alcan, Paris.

Durkheim, É., 1894. Les Règles de la méthode sociologique. Rev. Philos. 37 (38), 465–498.

Friedman, S.W., 2004. Marc Bloch, Sociology and Geography. Encountering Changing Disciplines. Cambridge University Press, Cambridge.

Gil, S.P., 2008. Knowledge as embodied performance. In: Gill, S.P. (Ed.), Cognition, Communication and Interaction. Springer, Berlin, pp. 3–30.

Latour, B., 2010. An attempt at a "compositionist manifesto". New Lit. Hist. 41 (3), 471–490.

Latour, B., 2011. Waiting for Gaia. Composing the Common World through Art and Politics. Lecture given for the launching of SPEAP. French Institute, London.

Mackenzie, D., Muniesa, F., Siu, L. (Eds.), 2008. Do Economists Make Markets? On the Performativity of Economics. Princeton University Press, Princeton.

Mctegart, W.J., Sheldon, G.W., Griffiths, D.C. (Eds.), 1990. Climate Change: The IPCC Impacts Assessment. Australian Government Publishing Service, Canbera.

Pradeu, T., 2011. Philosophie de la biologie. In: Barberousse, A., Bonnay, D., Cozic, M. (Eds.), Précis de philosophie des sciences. Vuibert, Paris, pp. 378–403.

Parker, A., Sedgwick, E. (Eds.), 1995. Performativity and Performance. Routledge, London.

Quenet, G., 2015. Versailles une Histoire Environnementale. La Découverte, Paris.

Rasmussen, L.B., 2008. The narrative aspect of scenario building–how story telling may give people a memory of the future. In: Gill, S.P. (Ed.), Cognition, Communication and Interaction. Springer, Berlin, pp. 174–194.

Renn, O., 2008. Risk Governance: Coping with Uncertainty in a Complex World. Earthscan, London.

Renn, O., Klinke, A., Van Asselt, M., 2011. Coping with complexity, uncertainty and ambiguity in risk governance: a synthesis', AMBIO. J. Hum. Environ. 40 (2), 231–246.

Searle, J., 1979. Expression and Meaning: Studies in the Theory of Speech Acts. Cambridge University Press, Cambridge.

Simonet, G., 2011. L'atelier « H » ou la représentation de l'adaptation dans l'élaboration du plan climat de Paris. VertigO 11 (2).

Touili, N., Baztan, J., Vanderlinden, J.-P., Kane, I.-O., Diaz-Simal, P., Pietrantoni, L., 2014. Public perception of engineering-based coastal flooding and erosion risk mitigation options: lessons from three European coastal settings. Coastal Eng. 87, 205–209.

Tubiana, L., Gemenne, F., Magnan, A., 2010. Anticiper pour s'adapter. Le nouvel enjeu du changement climatique. Pearson, Montreuil.

Vanderlinden, J.-P., 2015. 'Prévoir l'imprévu' Ceriscope Environnement et relations internationales. Available from: <http://ceriscope.sciences-po.fr/environnement/sommaire> (28.02.15).

Making the Link

By Liette Vasseur

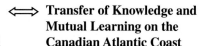

Paradigm Shifts, Coastal Zones, and Adaptation to Fast-Paced Change: Moving Toward Transdisciplinary Community-Centered Approaches ⟺ **Transfer of Knowledge and Mutual Learning on the Canadian Atlantic Coast**

Jean-Paul Vanderlinden, Gregory Quenet, Charlotte Da Cunha and Juan Baztan

Omer Chouinard, Tiavina Rivoarivola Rabeniaina and Sebastian Weissenberger

Vanderlinden et al. have demonstrated very well the importance of a holistic approach to climate change adaptation. Climate change is affecting all parts of social-ecological systems. This is especially true in coastal zones, where human lives have been for a long time coupled with their natural resources. How these systems will change over time is difficult to know. We have to note that they always evolve and will continue to do so as species adapt to changes. The problem is that climate change adds a new element in this evolutive complexity.

For human communities, memories and creativities distinguish us from other species. As is well illustrated in the ARTisticc project, the arts in all their forms can greatly contribute to better understanding the challenges and identifying opportunities to adapt to climate change.

In the next chapter, Chouinard et al. further explore, with several communities of New Brunswick, Atlantic Canada, how the history and stories of local people can contribute to the understanding of their social-ecological coastal system. Their narratives demonstrate how concerned they can be about climate change, especially regarding sea-level rise and storm surges. But issues are easily forgotten with the summer; this also shows the bias in the perceptions of people as a function of their immediate experience.

Both chapters underline the importance of connecting to local communities as they are the ones who experience first hand what is happening. Keeping their stories alive by recording them is a great opportunity for future generations. Lessons from the past can certainly serve in the future.

Chapter 2

Transfer of Knowledge and Mutual Learning on the Canadian Atlantic Coast

Omer Chouinard[1,3], Tiavina Rivoarivola Rabeniaina[1], Sebastian Weissenberger[2]

[1]*Université de Moncton, Moncton, New Brunswick, Canada;* [2]*UQAM, Montreal, QC, Canada;* [3]*Marine Sciences for Society, www.marine-sciences-for-society.org.* With the collaboration of **Mélinda Noblet,** *Université de Picardie Jules-Verne-Amiens, and* **Liette Vasseur,** *Brock University, Pays de Cocagne Sustainable Development Group (PCSDG).*

Chapter Outline

INTRODUCTION

New Brunswick is a Canadian province dominated by its coastline in its geography, demography, economy, and culture. It has 5500 km of coastline, 60% of the population lives near the coast, and fisheries and tourism contribute significantly to the economy. Adaptation to climate change (CC) in coastal areas is thus a major challenge for the province.

This chapter describes a partnership between academics and stakeholders from the Beaubassin (BPC) and Kent District (KDPC) Planning Commissions[1] on the eastern coast of New Brunswick, which aimed at a transfer of knowledge and mutual learning about adaptation practices in the communities belonging to these two planning commissions. The BPC distinguished itself in 2011 by proposing an innovative municipal bylaw that explicitly includes future sea level rise (SLR) due to CC. The legislation, which was first adopted in the rural community of Beaubassin East (CGVD28, bylaw 09-1B, February 2011), stipulates that the habitable part of any new construction must be at least 4.3 m above sea level. The figure is based on a 100-year storm surge (identical to that of 2000) to which 1.30 m of future SLR, in accordance with Ramhstorf et al. (2007), and isostatic adjustment, were added (Doiron, 2012). This regulation has emerged as a benchmark for adaptation practice in the Atlantic Region of Canada. We wanted to determine whether this might also be appropriate in the KDPC.

METHODS AND STUDY AREA

Four communities[2] in the two territories were part of the study:

- BPC: Beaubassin East (rural community, 6200 inhabitants) and Cap-Pelé (village, 2256 inhabitants), both in Westmorland County.
- KDPC: Cocagne (rural community since May 23, 2014, 2545 inhabitants) and Grande-Digue (Local Service District or LSD, 2182 inhabitants), both in Kent County.

The project included a mapping component and a sociological component. The mapping component was intended to provide baseline data on the natural ecosystem to better understand the issues facing the two territories. Scenario maps produced by Environment Canada (2006) were supplemented by a rapid environmental assessment in the field. The sociological component included semistructured interviews (Anadon and Savoie-Zajc, 2009) and focus groups (Kitzinger, 2004; Geoffrion, 2003):

1. Twelve semistructured interviews were conducted in the territory of BPC during the summer of 2013 on the following themes: the experience of environmental changes (winter storms of 2000 and 2010), changing behavior and habits, adaptation measures initiated individually or

1. There are 12 planning commissions in New Brunswick that provide shared services for municipalities, rural communities, and local service districts (LSDs) (NB, 2011).
2. New Brunswick has a three-layer local governance system: counties→parishes→municipalities/LSDs. Counties and parishes have more historic than actual significance. The 108 municipalities cover 8.0% of the province's land mass and are home to 64.5% of its population, There are five categories of municipalities: cities, regional municipalities, towns, villages, and rural communities. The 245 (as of July 1, 2014) local service districts (LSDs) are unincorporated areas covering most of the territory and are home to one-third of the population. They possess limited governance capacity.

collectively, and reactions and needs regarding these extreme events. In the KDPC, interviews conducted between 2010 and 2012 with about 20 residents from Grande-Digue and Cocagne (Chouinard et al., 2012) were analyzed.

2. Three focus groups were held: (1) KDPC (December 2012, six participants); (2) BPC (September 2013, seven participants); and (3) a cross-commission focus group with residents of BPC and KDPC (November 2013, 10 participants). The latter was aimed at promoting mutual learning on themes developed in the questionnaire, discussing and sharing best practices and experiences of each community in terms of adapting to CC, and taking into account the specificities of local governance.

RESULTS

Identification of Local Perceptions and Knowledge

Environmental

In general, the two communities had the same opinion regarding the changes that had taken place on the shores during the past 20 or 30 years. Interviewees cited, among other matters, the frequency of winter storms, rising sea levels, changes in the environment (erosion, loss of sand, and destruction of dunes), CC, and variability. The responses indicated a general understanding of the public on environmental issues. The majority of participants often referred to the past (childhood, family life, stories of neighbors) to illustrate this environmental change and clarify the "unusual" events. For example:

> *I spoke with a lady because her father was a fisherman and she was telling me… she was born in 1924; and when she was around 16 or 18, she worked in a fish plant. Then, when she was around 16 or 18 years old, she spoke about a mild winter when there was fishing in December and January, because she continued to work there. It must have been in 1930 or 1940. And she was saying that it was like the first year that she could see where they had worked in December and January in a plant. (Participant 4, Cap-Pelé[3])*

> *There is a lot of erosion in front of the house—30 feet since 1975, the grass is all gone. (Participant 14, Cocagne)*

> *What occurred to me in the beginning was the immensity, really, of the dunes in pictures of New Brunswick in 1938… where we can see what has happened to the beach in Cap-Pelé and in Barachois in about sixty years or so… for sure that there was a period of time where there has been much erosion…. (Participant 9, Barachois)*

---

3. All citations were translated from the French by the authors.

Participants in the southeast region all remembered the storms of 2000 and/or of 2010, which have been described in terms of the power of the rise in water level and the environmental and socioeconomic impacts. They mainly observed environmental changes such as land destruction, the disappearance of beaches, an increase in dune erosion, and the general effects of CC in the coastal zone. They also thought that extreme events were becoming more frequent and violent, even if, for some, extreme events experienced on the southeast coast of New Brunswick seem to not be as aggressive as compared with other regions:

I don't remember that one (storm of 2000) as much, but I remember seeing pictures and that was the worst one I think; I think the level in 2010 was a little bit the same but that there was less ice—10 cm less or something like that… but I believe that the damage was more visible in 2000 than in 2010. (Participant 3, Beaubassin)

Respondents in Kent who were considered key players in the community (having a wide influence within the community, long-time residents, etc.) recalled that 40 years ago, people were less or not at all concerned about environmental issues. One example mentioned by a participant was the Cocagne River bridge on Highway 11, which was found too low and vulnerable during the storm on December 21 and 22, 2010. Many respondents reported their observations regarding increase in sea-level, and its consequences. In Kent, people talked mainly about erosion problems and the accumulation of sediments in the river:

The trend is the erosion of the coast; however, I noticed the progress of the dune on the island of Cocagne. (Participant 11, KDPC)

Erosion is increasingly common. The cliff near my house has lost 2 feet over the past 6 years. Freezing and thawing are more frequent, this accentuates erosion and breaks stabilizing structures. (Participant 4, KDPC)

Psychosocial

All participants perceived a "problem" in coastal areas regarding natural events. Although the links between causes and effects were not always obvious, they expected to face more environmental disruptions and changes in the coming years. This perceived threat has strongly influenced their feelings of stress and anxiety. The general concept of CC seemed vague for many participants, and awareness of its consequences was mixed with uncertainty about the natural processes involved.

Analysis of the interviews conducted in the local service districts (LSDs) of Grande-Digue and Cocagne showed that the most knowledgeable people in these locations did understand the link between the evolution of atmospheric temperatures and increasing sea level. Others were more confused (Chouinard et al., 2012).

Respondents living in Cap-Pelé and Beaubassin East confirmed these conclusions: they were preoccupied by the environmental problems in their territories; their present situation and past experience strongly influenced their feelings of stress.

The residents of the communities of the BPC felt this anxiety only in winter, i.e., during periods of storm. In fact, a participant indicated that restored or enhanced protection measures alleviated her stress. She believed that the storm events of the last decade were either insufficiently significant (few casualties and damage less important than was anticipated), or else were not experienced by most participants. Therefore, she thought, people managed to easily forget.

On the one hand, there were fear and concern with the approach of spring ice-out. We can deduce that this reaction results from past experiences and apprehension of an uncertain future. On the other hand, we see that this awareness seems to last only during the period of storms and people forget quickly once summer arrives.

> *It stresses everyone out for a couple of days, but after that, it's fast forgotten! (Participant 8, KDPC)*

> *They (coastal residents) are stressed out and worried, especially when you say that your house… and you live in it! You know like, there are some that are just cottages by the sea, but you know some people live there year round! And you tell yourself that it only takes one storm that comes and floods my house, damages my house and like… a part of my land… and I can no longer live there… and where do I go after that! There is still this worry and they see how the change can come quickly… only with the storm of 2010. (Participant 8, Cap-Pelé)*

These responses can be explained by the fact that the complexity of the concept of CC and its causal link with the flooding and the erosion experienced daily. There are also feelings of uncertainty and apprehension about future events that do not facilitate planning and long-term actions.

Main Issues Raised by Respondents

Social: Behavior, Habits, and Associative Values

The new challenges facing the coastal zone seem to have had an effect on residents' self-reported lifestyles and habits. Our research suggests that such changes result from an awareness of environmental issues as well as an experience of harmful events.

In Kent County, environmental changes since the 1960s have led to stress for residents because storms eroded beaches, which continued to deteriorate in subsequent storms. These environmental issues have also had an impact on the community and on neighborhood relationships. But whereas some residents told of land-use conflicts, others argued that there had been a positive effect in strengthening community cohesion.

In the Cap-Pelé and Beaubassin East region, some said that extreme events had not affected relationships with the environment, nor resulted in conflicts. Others spoke of neighborhood conflicts regarding protection measures against erosion. For example, when a neighbor did not agree with rock fill, or had appropriated a few meters of additional land inland, to compensate for erosion losses. They added that neighbors helping neighbors was becoming rare, as is sharing awareness of problems with the municipality or among themselves.

> *When extreme events occur, people don't often help each other out and don't talk about it amongst themselves, nor in the community, but tend to close themselves off. (Participant 2, Cap-Pelé)*

Yet for other participants, extreme events had strengthened the cohesion between neighbors, including new as well as established residents, and had increased cooperation with local associations including those of Cap Bimet, Boisjoli, and Nicolas Beach. Moreover, they believed that citizens who were already members of an association were easier to mobilize for community activities, such as placing discarded Christmas trees in the breaches to protect the dunes or restoring access roads to the beach. It was easier for these communities to ask for help from the municipality in case of extreme events.

> *...during this period, we told ourselves that the neighbors were being kind and that we get along well; since we have to protect the dunes, because we live there and we can't only look at what happened behind (in front); so we said "well, from now on, we are going to protect the dunes" and everybody got involved. (Participant 9, Plage Nicholas)*

As for Cocagne and Grande-Digue, many respondents reported examples of land-use conflicts or tensions generated by erosion protection measures or new buildings, when those restrict access to the beach or obstruct the view of the ocean. These situations lead to frustration and tensions. Participants noted that people did not wish to tell a neighbor what to do, and inversely do not appreciate being told what to do by those around them, but are willing to listen to advice from a recognized community authority or group.

In the sector of Cocagne, two categories of respondents could be distinguished on the basis of their commitment. On the one hand, there were the concerned members of the Pays de Cocagne Sustainable Development Group (PCSDG). On the other hand, residents less involved in community activities stated that they felt fewer effects of CC. However, their lack of commitment did not mean that these residents were not affected by weather events associated with CC (Chouinard et al., 2012).

The majority of respondents understood the importance of solidarity in the community. They suggested involving summer as well as permanent residents in an association to undertake collective actions in order to develop a sense of attachment to the community.

We must find a way to involve seasonal residents in the community's projects. (Participant 3, Focus Group)

We must think about groups or platforms of associations. (Participant 5, Focus Group)

We need to find a collective solution, not an individual one because everyone is affected by coastal planning. (Participant 6, Focus Group)

Economic: Development of the Territories

Despite the vulnerability of coastal zones, residents noticed an increase in population in their regions, and in particular criticized the arrival of "rich" retirees and summer residents, who stay only six weeks of the year. Respondents did not understand how these people "from somewhere else" invested to buy a property on the beach with an ocean view without fear of losing their homes. These perceptions coincide with Le Bouëdec's (2002) analysis, according to which inland populations invent a new perception of coastal areas as exotic territories and places of emotion and of contact with nature, often detached from reality. With the decline of traditional activities like fishing and maritime transport, the coast is often seen as a luxury destination. But this is contradictory to traditional uses and has caused a deterioration of the relationship with local populations.

...twenty years ago, there had already been this acceleration in storms; it would have been noticed but less strongly because there were less people on the coast. (Participant 1, Focus Group)

Many retired people move along the coast, there is no more room on the coast and it's mostly big cottages; there are even people who live there permanently. (Participant 2, Focus Group)

Respondents also mentioned that insurance companies have begun to realize the impacts of CC on the coastal environment and that premiums for buildings along the coast were on the rise, in particular for water damage.

The future of the territory of Cocagne and Grande-Digue was much discussed in the focus groups. Some residents feared uncontrolled development, while others wanted economic growth. All, however, desired policies and regulations to ensure the viability and vitality of rural areas. They say that the challenge is to open the dialogue for a common vision and planning for CC adaptation, while taking into account the socioeconomic and demographic composition of the territory.

Assessment of Current and Future Adaptation Measures

All participants were aware of the risks facing coastal areas from flooding, storms, and rising sea levels. They admitted that adaptation measures initiated individually or by the community were not sustainable, while continuing to seek viable new measures. Our results allow us to make an assessment of the existing and proposed adaptation measures recounted by the respondents.

Protective Actions Undertaken

Protective actions were undertaken at the individual, collective, and municipal levels. Several residents protected their property through various means such as rock fill along the coast, dikes, or seawalls in front of houses. Collective actions carried out by the community and local associations aimed at protecting common goods such as dunes, beaches, and roads. Some examples are dune restoration, the use of recycled Christmas trees to protect dunes, the identification of risks and vulnerable areas, the implementation of emergency plans, and the adoption of a decree on the need for piles (beams, columns) for constructions along the sea front.

In general, respondents believed that small actions taken by communities had a long-term impact, particularly in terms of learning. For them, the goals of adaptation measures should be to mitigate negative impacts of natural disasters, to reduce exposure to hazards, and to maintain the territories and infrastructures of coastal communities.

Current Needs

First, the majority of respondents stressed the importance of solidarity in the local community. Citizens felt that there was too much laissez-faire at the individual level. Integration of summer residents in community projects was seen as highly desirable. Respondents considered access to information as well as a close collaboration with local authorities is important in order to achieve this.

Second, there was a certain feeling of incomprehension regarding the decisions and actions of the provincial government. Respondents considered that the laws and regulations put in place by the government were inadequate. Some decisions were hard to understand for them and some seemed contradictory. According to the majority of respondents, the level of involvement of provincial authorities was too low and the opinion of the population insufficiently considered.

> *I am surprised that the government gives permission to destroy beautiful dunes and it is thinking of building artificial dunes. (Participant 2, Focus Group, Beaubassin-Est)*

> *There is a contradiction in the measures taken by government: construction in buffer zone between properties and wetlands. (Participant 4, Focus Group, Grand Barachois)*

Respondents emphasized the weakness of the provincial regulations regarding private rock fill and the lack of communication between local governments, regional service commissions, and the province. Rural communities did not feel involved in the decision making process regarding construction in coastal areas.

> *...weakness at the level of provincial regulations: don't need to get a permit for rock fill. (Participant 4, Focus Group, Grand Barachois)*

> *Fredericton should not be giving permits directly, but should be going through the municipalities or rural communities. (Participant 2, Focus Group, Beaubassin-Est)*

Some respondents wanted the provincial government to establish a regulation on the authorization of building permits, enforceable (or supported) by municipalities.

> *...the need to modify planning regulations where it is necessary to get a permit for construction along the shore: whether it is for a structure, land development; the permit would be delivered by the municipality. (Participant 1, Cap-Pelé)*

> *The need for a building permit which should be obtained from the municipality or from the Rural Community (must be aware). (Participant 1, Cap-Pelé)*

Respondents questioned both the scrutiny of applications for environmental permits and the distribution of responsibilities among the various authorities. Better training of staff in local authorities and organizations was also seen as desirable.

> *...Delivery of environmental permits would be done at the regional level when involving small jobs, but big projects are in Fredericton. (Participant 3, Focus Group, Beaubassin East)*

> *If the municipality had some authority on its territory, there would be other eyes that could also analyze decisions, demand precisions for example, an informed person from the municipality linked to the association. (Participant 6, Focus Group, Cap-Pelé)*

> *...Reflecting on coastal planning. For example, the rural community and/or the municipality, in consultation with the Vision H20 Association who would have the power to intervene; an entity with the power to stop something. (Participant 5, Focus Group, Grand Barachois)*

Finally, respondents confirmed the need for a stronger cooperation between communities and the provincial government.

Perspectives

During the cross-commission focus group, participants noted that communities had progressed from awareness to action. While the steps were not all at the same levels of action, the objectives were common, namely land protection and the minimization of damage from winter storms. Representatives from each locality told us about their experiences.

The community of Bouctouche had already been produced a zoning plan requiring an elevation of the foundations of new buildings in risk areas. Despite this, the participants felt that the region remained vulnerable. Beaubassin East, Cap-Pelé, and Bouctouche had formulated an emergency plan as a response to frequent flooding caused by big storms and the feeling of menace that this had generated. This emergency plan made links among these communities in order to reinforce cooperation and break the isolation when land is "taken over by the ocean."

Finally, Kent County representatives stated that they did not have the means to deal with emergencies such as a long power outage, nor did they have a shelter with beds and running water. Unlike communities in the BCP, those in the KDPC had no rural plan, as both were LSDs[4] at the time (Cocagne became a rural community in May 2014).

Local, Regional, and Provincial Governance

The focus group participants thought that the province was beginning to take the situation into account. They recommended natural protection and higher foundations for homes. The new Regional Service Committee (RSC) was considering sharing and coordination of services between communities (LSDs and municipalities), building on the mutual aid that existed between associations, LSDs, municipalities, and rural communities.

At present, minimum limits have been fixed regarding the elevation over sea level of new construction for those communities that have planning regulations. In the case of the communities in Kent County, however, participants doubted that authorities would be able to enforce such limits without municipal plans or support of provincial regulations.

The weakness of self-governance of the LSD was of great concern. A project for creating a rural community in Cocagne and in Dundas was initiated, which could eventually lead to a joint rural plan, which would mark the cohesion between the actors on the coast and those inland. While Cocagne was incorporated as a rural community in 2014, the vote in the Dundas parish on May 12, 2014, ended in a tie (420 votes for, 420 against), thus lacking the number to win.

Adaptation Priorities

With regard to priority measures, there was little obvious change despite the perceived increase in risk. Several proposed changes have not been implemented due to lack of funding. One participant noted that the proposed renovation of the Cocagne Bridge could not be continued due to financial reasons. As a consequence, protective measures by individuals were increasing.

---

4. LSDs are communities that are not incorporated as a municipality or a city and depend directly on the provincial government for any decision making or planning.

The adherence of the population to associations was repeatedly emphasized as a priority item, since they facilitate the dissemination of information. Members in well-organized associations participate in activities, share ideas, and attend meetings (ranging from three or four times per year). Thus, associations can play the role of facilitators in society.

Among the changes desired by respondents was agricultural and economic development, favoring self-sufficiency through local businesses that are committed to the diversification of products and activities. The development of tourism and fishing activities should be pursued in relation to these goals.

Respondents want the government to look closely at new risks in the planning of the coastal zone, such as the erosion of beaches, the environmental and visual aspect of rock fill, and the increase in beachfront construction since the future of tourism depends on the attractiveness of the coastal environment.

> *We must make the government understand that coastal zone tourism is an industry (profit) and it should not take lightly what goes on in the coastal zone. (Participant 5, Focus Group, Grand Barachois)*

Respondents also stressed the importance of evoking the notion of land occupancy so that coastal residents realize the changes that have occurred: loss and abandonment of agricultural land, forest recovery on former agricultural land, etc. One solution proposed by some respondents is the purchase of land by municipalities to maintain and enhance environmental resources, and to ensure that these resources remain accessible to the community. One participant expressed concern about the growing population in the coastal zone and the fact that the conversion of coastal zone holiday cottages into permanent properties is legal because the owners already have the necessary licenses (septic tank installation for houses near the beach). Respondents also indicated that the regulations could easily be circumvented.

The KDRC informally advised researchers of a survey conducted in Kent County of the building permits issued in the past five years. Their inspectors identified 93 licenses, of which 37 should not have been issued. This indicates the absence of adequate regulation of new construction relative to climate risks.

The determination of respondents to protect their territory as a place of living, but also of identity-building, is evident in the comments made.

> *Loss of the coastal zone means loss of a coastal heritage, a habitat to be protected. (Participant 6, Focus Group, Cap-Pelé)*

The impact of the reform of planning commissions (NB, 2011) on local land management remained uncertain, according to participants. They could not say if it offered more efficient functioning of local institutions or whether it would benefit the development of communities or municipalities. Nevertheless, participants believed that there would be positive impacts during the next 10 months. The reform project seemed to start conversations in which people began to reflect on the organizational aspect. According to participants, residents feel the need to talk and express themselves on this subject, since they are aware of the decision-making process, regional management, and all services offered by the RSC.

CONCLUSIONS

This survey has led to several conclusions:

1. The study participants were aware of the vulnerability of coastal communities in southeastern New Brunswick and had adopted adaptation measures to cope with climate threats.
2. Participants stressed the importance to conduct collectively planned, long-term, proactive actions, instead of reactive actions at the individual level.
3. Participants felt a division between the people who built along the coast and those who lived inland. This discrepancy reinforced the lack of involvement of newcomers in the community.
4. The role of the provincial government was considered too weak in terms of legislation, guidance and decision-making. Furthermore, participants would like to be consulted more during the decision-making and the development of plans and strategies.
5. The importance of the decentralization of powers was often cited. Respondents wanted local authorities and the RSC to decide planning issues in order to make a recommendation to the province, so that there would be accountability and transparency at the local level.
6. The importance of local associations as a democratic body was stressed; they should be able to deliberate on collective activities for coastal development, particularly in the case of LSDs, which have less administrative and decision-making capacities.
7. There was a clear difference between the BPC and the KDPC, the first being composed of municipalities and the second of LSDs. The constitution of a municipality or rural community gave more authority to help communities, especially management plans and bylaws to manage the land. Residents of LSDs felt powerless at the institutional level and put more emphasis on the role of associations.
8. By contrast, the perception of the climate issue, the assessment of changes observed, and knowledge about CC did not differ between the two territories.

Studies in both territories have helped to identify the knowledge, concerns, and challenges identified by the participants. All agreed on the importance of CC for the development of the coastal zone and the need to address the topic in a better planned, better organized, and more efficient manner. Governance issues raised many questions among study participants, including the lack of support by the province, the inadequacy of the status of LSDs in order to implement local governance, and the yet-uncertain impact of the reform of local governance started in 2011 by the province (NB, 2011). At the same time, the role of associations and social cohesion in the process of developing adaptation strategies was strongly emphasized. It would be useful to extend the study performed here to other coastal territories of New Brunswick to obtain a more complete picture of the issue of CC and the challenges and solutions to adaptation and community needs as perceived by local citizens.

ACKNOWLEDGMENTS

We wish to thank Alice Kone for her participation in the elaboration, administration, and analysis of interviews and discussion groups. We also thank all the participants in the questionnaires and discussion groups. This project was made possible thanks to funds from the New Brunswick Environmental Trust Fund and Natural Resources Canada.

REFERENCES

Anadon, M., Savoie-Zajc, L., 2009. L'analyse qualitative des données. Rech. Qual. 28, 1–7.

Chouinard, O., Tartibu, N., with the collaboration of J. Gauvin, 2012. Adaptation réfléchie en milieu côtier: démarche d'engagement avec les résidents du milieu côtier des DSL de Cocagne et de Grande-Digue. Report Presented to the New Brunswick Environmental Trust Fund.

Doiron, S., Winter, 2012. From Climate Change Plans to By-laws: It's Time to Act. Plan Canada, pp. 30–34.

Environnement Canada, 2006. Impacts of Sea Level Rise and Climate Change on the Coastal Zone of Southeastern New Brunswick.

Geoffrion, P., 2003. Le groupe de discussion. In: Gauthier, B. (Ed.), Recherche sociale. De la problématique à la collecte des données, fourth ed. Québec: Presses de l'Université du Québec, pp. 333–356.

Kitzinger, J., 2004. The methodology of focus groups: the importance of interaction between research participants. Sociol. Health Illn. 16, 103–121.

Le Bouëdec, G., 2002. L'évolution de la perception de la zone côtière du XVe siècle, Actes de colloque à l'Université de Bretagne-Sud Lorient. (En ligne) http://assoc.lorient.ccsti.pagesperso-orange.fr/AIZC/lebouedec/lebouedec.htm (consulted on 09.12.13).

New Brunswick, 2011. Action Plan for a New Local Governance System in New Brunswick.

Rahmstorf, S., 2007. A semi-empirical approach to projecting future sea-level rise. Science 19, 369–370.

Making the Link

By Bethany Jorgensen

Transfer of Knowledge and Mutual Learning on the Canadian Atlantic Coast	⟺	Coastal Population and Land Use Changes in Europe: Challenges for a Sustainable Future
Omer Chouinard, Tiavina Rivoarivola Rabeniaina and Sebastian Weissenberger		*Alejandro Iglesias-Campos, Andrus Meiner, Kieran Bowen and Joseph Onwona Ansong*

From the local perspective of inhabitants along New Brunswick's Atlantic coast in Canada, we take flight across the Atlantic and pull back the lens for a bird's-eye view of efforts underway to improve policies at the regional level, spanning Europe's coastal communities.

These two chapters complement each other nicely. The first shows us the on-the-ground perceptions of the individuals who are impacted by coastal zone plans and policies, which are often decided at the regional level. The second gives us a regional perspective of Europe's coastal areas and offers suggestions for improving regional approaches to coastal zone management in an area that includes 26 different countries.

One of the main points from Chouinard et al.'s work is that the participants want to be more involved in the policymaking process, with the hope of shaping policies to be more practical and meaningful for their communities.

To follow this locally rooted study with a chapter about the challenges and potential for international policies and ecosystem-based management approaches at the regional level of Europe reminds us that successful solutions need to be coordinated across physical and temporal implementation scales as well as geopolitical boundaries.

As these two chapters illustrate, the way forward toward solutions demands that we maintain a telescopic understanding of the contexts within which we operate. We must find ways to work locally without losing sight of the global picture—and vice versa.

Chapter 3

Coastal Population and Land Use Changes in Europe: Challenges for a Sustainable Future

Alejandro Iglesias-Campos[1], Andrus Meiner[2], Kieran Bowen[3], Joseph Onwona Ansong[1]
[1]Intergovernmental Oceanographic Commission of UNESCO, Paris, France; [2]European Environment Agency, Copenhagen, Denmark; [3]Bowen Marine Ltd, Bishopstown, Cork, Ireland

Chapter Outline

INTRODUCTION

The coastal ecosystem serves as the habitat for living organisms such as plants, animals, and other organisms. Article 2 of the Convention on Biological Diversity defines the variability among living organisms from all sources including terrestrial, marine, and other aquatic ecosystems and the ecological complexes of which they are part; this includes diversity within species, between species, and of ecosystems as biological diversity. Ecosystems present in boundaries between major biomes typically rank among the most productive and diverse on the planet (McClain et al., 2003). Coastal ecosystems are part of this ecological complexes and rank among the most productive ecosystems in the world (Duarte and Cebrián, 1996). For this reason the coastal zone has historically been highly attractive and valuable for humans, and today it is often heavily urbanized with high population density and intensive agriculture.

Coastal ecosystems are shaped by the dynamic interactions between land and sea. The geology, movement of tides and currents, sediment deposition and erosion, extreme weather conditions, biological interactions, and in recent history, human activities, all contribute to formation and dynamics of the coastal ecosystem. Coastal ecosystems can be broadly defined as the part of land influenced by the sea, and the part of the sea influenced by the land. The constituent habitats include salt-adapted scrub and grasslands, wetlands, salt marshes, rocky shores, sandy beaches and dunes, tidal areas, estuaries, coastal lagoons, intertidal flats, reefs, sea grass meadows, and muddy sea beds, among others. These habitats act as breeding, spawning, and feeding grounds for migratory birds, fish, invertebrates, and mammals. Together they provide a range of ecosystem services such as providing food, erosion control, and atmospheric and climate regulation, and act as highly valued recreational areas (EEA, 2010). While the distribution, structure, and functioning of habitats within the coastal ecosystems always have been dynamic, the current rate of change is of concern. It has been estimated that the degradation of coastal habitats is exceeding the global loss of rainforests by four to ten times (Duarte et al., 2009).

A full holistic and integrated assessment of the current rate of change and how it affects the status of coastal ecosystems, biodiversity, and the services that coastal ecosystems provide to European coastal communities is not possible. European data on the status and trends of the biodiversity components of the coastal ecosystems are not comprehensive, and there are fundamental gaps, even for the small number of species and habitats targeted by the EU Nature Directives (EEA, 2010). With less data on the status and trends of biodiversity coupled with the complexity of the coastal environment, it is important that the interactions in the coastal areas are studied in order to ensure that the negative impacts on biodiversity are mitigated.

Dynamics of Coastal Environments in Europe

Coastal ecosystems—coastal lands, areas of transitional waters, and near-shore marine areas—are among the most productive, yet highly threatened, systems in the world. Valuable ecosystems, such as coastal dunes and seagrass beds, are continuously threatened in Europe's coastal zone. From 2000 to 2006, more coastal wetlands were lost, despite already a high conversion rate during the previous decades.

Population densities along the European coast are higher and continue to grow faster than inland, yet differences are less pronounced than globally. Populations tend to concentrate in areas most favorable for trade, marine industry, or recreation. And these activities sometimes become incompatible with the valuable coastal ecosystems (e.g., the Mediterranean).

European coasts are a natural environment that attracts socioeconomic development for various reasons. This attractiveness introduces multiple drivers of land use change, which can lead to increased stress on natural and human environments. The development-related loss of coastal land systems, related habitats, and ecosystem services continues to be most remarkable change of the coastal zone.

Artificial surfaces in coastal zones have been increasing in almost all European countries. Some 1347 km^2 of new urbanized areas were created from 2000 to 2006. The annual rate of urban sprawl at the coasts is higher than the European average, but it decreased compared to the previous period of 1990 to 2000. There are signs of urban development saturation at the coastline, and in 2000 to 2006 residential areas of development moved further inland.

The land development is driven mostly by residential sprawl and extended mainly at the expense of former arable land and pastures. The rest of newly urbanized areas from 2000 to 2006 were taken from forests and semi-natural vegetation. The highest increase in artificial surfaces (>10%) has been observed in coastal zones of Ireland, Southern and Eastern Spain, parts of Italy, and Cyprus.

Agricultural areas cover 31% of the European coast. Almost half is used as arable land, more than 25% are agricultural mosaics, and 20% are pastures. Despite that, the overall intensity of agricultural land cover exchange in European coastal zones slowed down significantly compared to the previous period of 1990–2000; 372 km^2 of arable land and permanent crops and 442 km^2 of pastures were still lost in from 2000 to 2006.

Human pressures on coastal resource use may compromise the ecosystem integrity. Recent patterns of overexploitation of key fish stocks in European regional seas have been altering the structure of marine ecosystems. Wind energy installations have vastly expanded on land and offshore, and projections show continued exponential growth. Despite numerous environmental impact studies, this new trend still contains uncertainties and needs careful planning to mitigate the potential threats to ecosystems. Appropriate stakeholder involvement will allow achieving most acceptable solutions for coastal communities.

There is growing evidence that Europe's coastal systems (including marine and terrestrial) are suffering widespread and significant degradation (e.g., loss of habitat, eutrophication, contamination, erosion, alien species) that poses a major challenge to policymakers and coastal managers. Land-based sources of pollutants, but also other indirect sources, play important roles in the formation of coastal pressures, and therefore linking the coastal zone and its water bodies with river basins is a priority.

Physical space (land and sea) is also a key resource that is needed to produce or sustain other natural resources and some ecosystem services. In this interpretation, the coastal zone is a finite "resource" because of its limited spatial extent, and ongoing conversion of coastal habitat to economic and urban use may prioritize short-term benefits at the expense of regulating and provisioning services that may be permanently lost.

Coastal areas can support only a certain amount of activity without entering into competition and environmental degradation. Due to the gradual expansion of different human activities, coastal zones have accommodated a number of different uses in space and time, often lacking long-term coordinated spatial planning, including marine areas.

TABLE 3.1 Natura 2000 Cover Within EU Coastal Zones in 2012

Zones Presented as Distance to Shoreline (km on Landward Side, Nautical Miles on Sea Side)	Area of Coastal Zone in EU (km^2)	Natura 2000 Coverage (km^2)	Proportion of Natura 2000
10–1 km	341,258	71,050	20.8
1–0 km	78,825	26,629	33.8
0–1 nautical miles	152,477	73,744	48.4
1–12 nautical miles	888,923	129,831	14.6

Source: European Commission, DG ENV, 2013.

As a reaction, the EU has been designating extensive coastal sites through its Natura 2000 network, both on land and at sea, to protect the coastal areas from further development (Table 3.1).

The state of the coast is often described through the management of coastal issues at local, regional, and also national scale. However, coastal issues are also recognized to be of relevance for Europe, because often they cannot be solved by the member states separately (e.g., common natural and cultural heritage, transfers of pollutants and sediments, tourist flow, maritime safety). Necessary trade-offs between different coastal uses require a long-term perspective and stakeholder involvement in the process, so far provided only by the Strategic Environmental Assessment directive, and for major projects, by the Environmental Impact Assessment directive.

Setting the Scene

Europe has a wide diversity of coastal landscapes resulting in a complex territory with regard to different aspects of environmental, social, cultural, and economic conditions. This variety of conditions requires a regionalized implementation of common principles. The challenge is to come out with policies that promote economic development without compromising ecological integrity. Coastal policies in Europe are discussed in the next paragraphs.

First, the European Commission is a signatory to the Regional Seas Convention (RSC). The RSC and its resulting Action Plans serve two major objectives in order to be the principal platform for implementing global conventions: (1) Multilateral Environmental Agreements and global programs or initiatives regionally, providing UN agencies or global programs with an existing mechanism to implement their activities on a regional scale; and (2) to be the regional platform for coordinating programs and projects that will contribute to the sustainable development of shared marine and coastal environments.

The following are the four European RSCs:

- The Convention for the Protection of the Marine Environment in the North-East Atlantic of 1992: The OSPAR Convention (OSPAR).
- The Convention on the Protection of the Marine Environment in the Baltic Sea Area of 1992: The Helsinki Convention (HELCOM).
- The Convention for the Protection of the Black Sea of 1992: The Bucharest Convention.
- The Convention for the Protection of the Marine Environment and the Coastal Region of the Mediterranean of 1995: The Barcelona Convention (UNEP-MAP).

The European Commission's Communication on Integrated Coastal Zone Management (ICZM) Strategy for Europe (COM(2000)547)[1] asked more than a decade ago for integrated management of the coastal zone that requires action at the local and regional level, guided and supported by a national vision and appropriate framework at the national level. The European Union has organized a stakeholder process and supported generation of factual information and knowledge about the coastal zone.

A European Union ICZM Recommendation in 2002 provided further guidance for implementation of the vision outlined by the ICZM Strategy. Along with setting the fundamental principles of sound coastal zone development, it triggered national stocktaking and preparation of national strategies for the coastal zone in most of the member states. Later, under the same initiative an activity to create indicators for monitoring the implementation of these coming strategies was started. These steps have paved the way for current policy initiative, based on stakeholder consultation, national stocktaking, and evidence collection (e.g., OURCOAST database[2]), and broadening the approach to climate change adaptation, erosion risk management, and maritime spatial planning.

Despite relying on ecosystem-based approaches, existing environmental policies for nature, water, and the marine environment only partly address the diverse and complex character of coastal zones, and territorial policies do not yet deliver the common spatial governance framework for the coasts. For example, Territorial Cohesion—a major European policy initiative to promote spatial planning and territorial dimensions of sustainable development—still needs to find its support so that coastal practitioners can counterbalance the development pressure from key economic sectors, such as tourism and transport infrastructure.

At the same time, coastal and transitional waters are recognized by the Water Framework Directive as integral part of river basin management districts. The

1. Communication from the Commission to the Council and the European parliament on integrated coastal zone management: A strategy for EuropeCOM (2000): http://ec.europa.eu/environment/iczm/comm2000.htm.

2. http://ec.europa.eu/environment/iczm/ourcoast.htm.

Marine Strategy Framework Directive is also addressing the coasts as part of Marine regions and subregions.

This direction has been enforced by the development of EU Integrated Maritime policy.[3] The European policies on governance and knowledge base through the European Directives establishing frameworks for community action in the field of marine environmental policy (2008/56/EU)[4] and for maritime spatial planning (2014/89/EU)[5] are implementing the systemic vision for the coastal areas. Most recent orientations provided by the European Commission's Blue growth strategy[6] recognize the need for innovative ways in using the sea and the coast.

All these policy developments are expected to create a growing demand for integrated spatial assessment of coastal areas, which is essential for creating a knowledge base for better regional management of coastal systems taking in to account land–sea interactions (Figure 3.1).

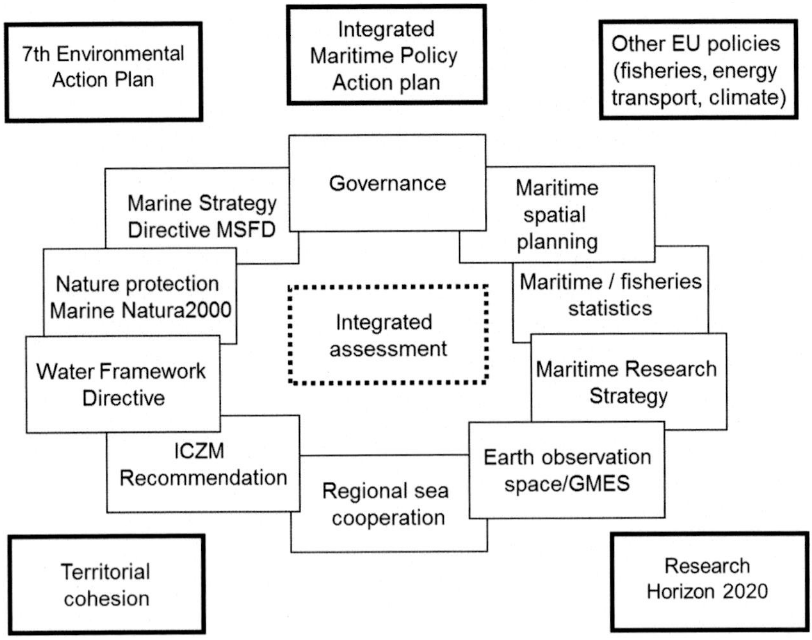

FIGURE 3.1 Context of European Union policy areas relevant to assessment of the coastal and marine environments. *Meiner, A. 2010. Integrated maritime policy for the European Union: consolidating coastal and marine information to support maritime spatial planning. J. Coast Conserv. 14 (1), 1–11.*

3. http://ec.europa.eu/maritimeaffairs/policy/index_en.htm.
4. http://eur-lex.europa.eu/legal-content/EN/TXT/?uri=uriserv:OJ.L_.2014.257.01.0135.01.ENG.
5. http://eur-lex.europa.eu/legal-content/EN/TXT/?uri=CELEX:32008L0056.
6. http://ec.europa.eu/maritimeaffairs/policy/blue_growth/index_en.htm.

Coastal Population in Europe

In 2013, approximately 200 million people lived in the 378 European coastal regions, comprising 41.74% of the total population of the 26 European coastal countries identified in Table 3.2. It is noteworthy that in almost all (96.7%) coastal regions the vast majority of the population lives by the sea, and population density at the coastline is much higher than in coastal regions as a whole. The EU's coastal region size is 1,772,768 km^2. The EU coastal regions are occupied by 199 million citizens, resulting in an average population density of 100 inhabitants per km^2 compared to 108 over the 22 EU countries with a seacoast. Six EU countries (Sweden, France, Finland, Italy, Spain, and Greece) have coastal regions representing more than 100,000 km^2 and occupying together 75% of the EU coastal area (Eurostat, 2009).

TABLE 3.2 Population of European Coastal Regions

	Population in Coastal Regions (Millions)				Share of Population in Coastal Regions Compared to National Population (%)
	2001	2006	2011	2013[a]	2013
Total population[b]	528.57	540.51	550.57	551.90	–
Coastal population[c]	193.55	194.12	197.57	199.38	41.74
Belgium	1.62	1.66	1.72	1.75	15.68
Bulgaria	1.09	1.08	1.08	1.07	14.69
Denmark	5.21	5.44	5.56	5.58	100
Germany	6.68	6.70	6.70	6.69	8.31
Estonia	1.00	0.99	0.99	0.99	74.73
Ireland	3.67	3.95	3.75	3.75	81.68
Greece	:	10.10	10.23	10.24	92.56
Spain	25.20	25.98	27.52	27.53	58.95
France	23.74	24.54	24.80	24.93	37.98
Croatia	1.31	1.33	1.46	1.46	34.26
Iceland	0.27	0.29	0.32	0.32	100

Continued

TABLE 3.2 Population of European Coastal Regions—cont'd

	Population in Coastal Regions (Millions)				Share of Population in Coastal Regions Compared to National Population (%)
Italy	29.89	30.51	32.96	32.96	55.22
Cyprus	0.70	0.76	0.80	0.82	100
Latvia	1.40	1.40	1.38	1.36	67.20
Lithuania	0.38	0.38	0.36	0.36	12.11
Malta	0.39	0.40	0.42	0.42	100
Netherlands	5.09	5.12	5.22	5.22	31.11
Norway	3.50	3.73	3.94	3.99	78.99
Poland	3.51	3.52	3.62	3.62	9.39
Portugal	7.11	7.31	7.43	7.41	70.66
Romania	0.97	0.96	0.96	0.96	4.79
Slovenia	0.10	0.10	0.10	0.11	5.34
Finland	3.00	3.63	3.72	3.73	68.73
Sweden	7.19	7.32	7.68	7.74	81.00
Turkey	:	38.02	39.96	40.61	53.70
United Kingdom	35.13	35.83	36.74	36.74	57.51

(1) 2011 instead of 2012 for MT.
[a]*Estimation made by Eurostat with national data reported in 2013.*
[b]*Including the national population of all the listed countries.*
[c]*Including the population living in coastal regions (NUTS3).*
Source: EUROSTAT (2014).

The share of the national population living in a coastal area, as well as the population density, depend on many factors, such as historical trade routes, economic development, climatic differences, and geographical characteristics such as the accessibility and configuration of the coastline (Map 3.1).

In the case of island states such as Malta or Cyprus, or peninsulas such as mainland Denmark, this share is 100% as most of the regions of these countries are coastal. Conversely, the share of inhabitants of coastal regions is only 5% of the total population in Romania and 9% in Germany.

Coastal regions in Northern Sweden and Finland are very large. A consequence of this is that the total population of these coastal regions is considered

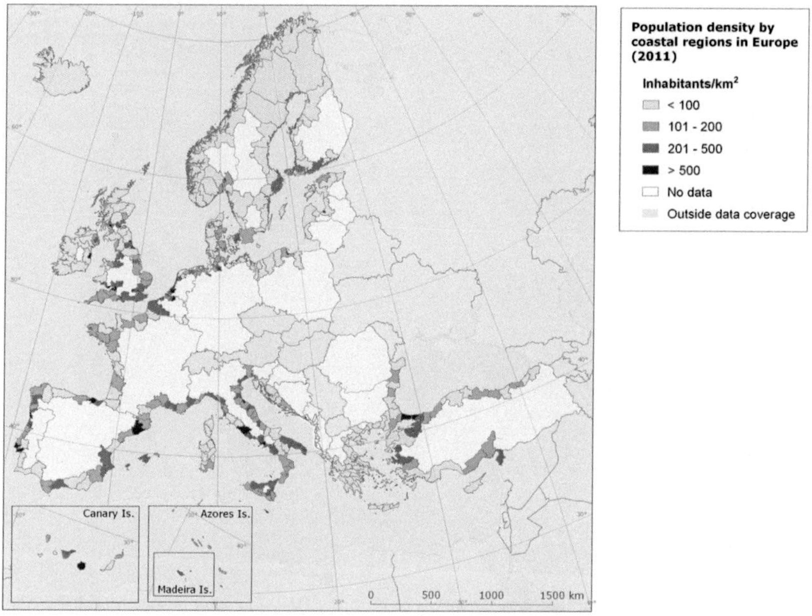

MAP 3.1 Population density in the European coastal regions, 2011. *Eurostat, 2013.*

coastal population, even if part of this population lives far away from the sea. However, this distortion is quite limited, as the population of these regions is small if compared to the national coastal population and a significant share of inhabitants and economic activities in these coastal regions are located near the sea.

The average population density for the EU coastal regions is 100 inhabitants per km^2. There are large differences in coastal population density among the 22 concerned EU countries. Densities over 200 inhabitant per km^2 are registered in the coastal regions of Belgium, Netherlands, the United Kingdom, Portugal, and Italy, and the highest density is observed in Malta, with more than 1200 inhabitants per km^2. The lowest densities are found in Estonia, Sweden, and Finland with less than 30 inhabitants per km^2.

The North Sea basin has by far the highest population density, with on average nearly 250 inhabitants per km^2; while the Baltic Sea and the outermost regions are most sparsely populated.

There are 194 cities with over 100,000 inhabitants within 50 km of the sea. In 2010, these urban centers were housing 38% of the inhabitants in EU coastal regions. Among the largest coastal cities are Athens, Barcelona, Istanbul, Lisbon, London, Nice, and Naples (Map 3.2), followed by the metropolitan areas of Copenhagen, Dublin, Izmir, Malaga, Marseille, Porto, and Valencia. St. Petersburg is the largest urban agglomeration by the Baltic Sea.

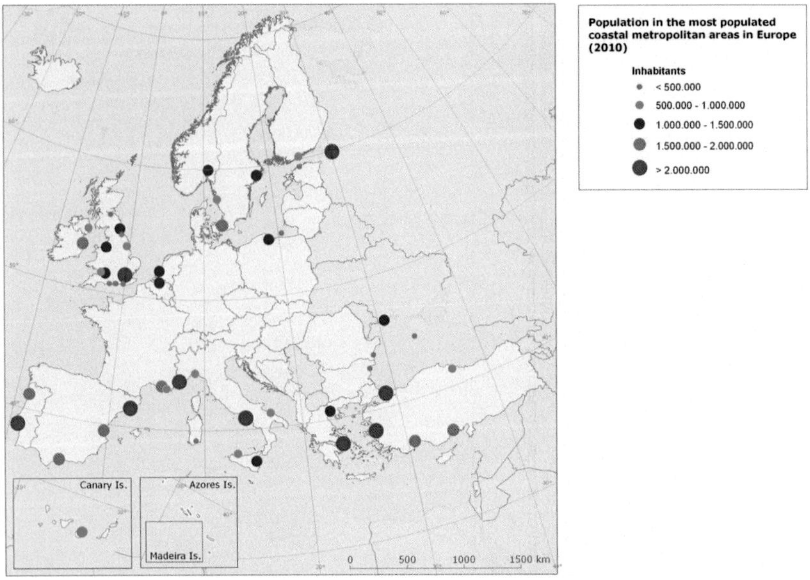

MAP 3.2 Most populated metropolitan areas in Europe, 2010. *Eurostat and National Statistical Offices, 2013.*

The northern coastline of the Mediterranean is housing up to 36% of the population in EU coastal regions, and that number would increase when coastal regions of all EEA and EFTS countries (Iceland, Liechtenstein, and Norway) are included, with some of these countries vying to be part of the European Union. In many cases, coastal regions can concentrate more than one million inhabitants even without large urban centers (Map 3.3).

The EU coastal regions bordering the North Sea have the highest urbanization rates, with 64% of the coastal population in the metropolitan areas (London, Hamburg) or densely urbanized coastal regions of Belgium, Netherlands, and Germany; as compared to only 25% of the Baltic Sea coastal population living in urban metropolitan areas.

The EU coastal outermost regions are mainly small islands of volcanic origin; this physical characteristic limits the development of urban plans to the coastlines, which concentrate the majority of the population (up to 72%).

The most homogenous concentration of population along the European coast is represented by the Spanish and French coastal provinces, which concentrate 60% and 38%, respectively, of their national population, in both the Atlantic and the Mediterranean.

European coasts have a number of population "hot spots" and "cold spots." The south of Italy and many regions of Turkey also represent two of the most populated areas, with cities like Naples, Palermo, Istanbul, Antalya, and Izmir,

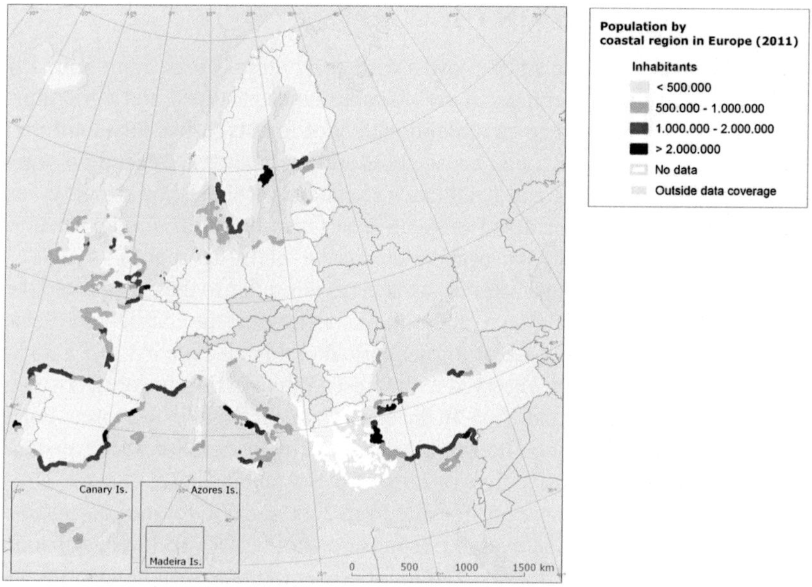

MAP 3.3 Coastal populations by European coastal regions. *Eurostat, 2013.*

among others. Some areas in Portugal, Scotland, Wales, Ireland, Germany, and Norway have a low population.

In the Baltic Sea, the coastal city capitals of Copenhagen, Riga, Stockholm, Tallinn, and Helsinki, as well as Gdansk in Poland, have population concentrations that are higher and stand in contrast to the remainder of the coasts, where population concentration is low.

Since the last population census in 2001, factors such as social and political changes, and the rapid development of the European economy from 2005 to 2008, has led to many population changes that can be observed also in coastal regions of Europe.

Several northern coastal areas have lost population due to migration to the major cities, especially to the capital cities that are located close to the sea.

On the other hand, in the Mediterranean arc from Andalusia in Spain to Provence-Alpes-Côte d'Azur in France, the coastal population has increased between 10% and 50% in some municipalities. This part of the Mediterranean contains many second residences and is a place for retirement for many people from the northern part of Europe.

Ireland and the United Kingdom, as well as the Atlantic coast of France, Belgium, the Netherlands, and Norway, are also experiencing an increase in coastal population, mainly due to the development of new infrastructures and attractive residential offers.

LAND USE CHANGES IN THE COAST

The European coasts are lined mostly with agricultural areas, represented by the changes in the percentage share of arable land, pastures, and agricultural mosaics and natural land cover, predominantly of forests (especially coniferous forests along the Scandinavian coast), nonforest seminatural vegetation represented mainly by moors and heathlands (Scotland, Iceland), or sparsely vegetated areas along the Northern European coasts near the Arctic circle. Artificial areas cover approximately 6% of the total coastal land (Figure 3.2).

The change in land coverage in coastal areas is similar to the change for all of Europe—sprawl of artificial areas is the dominant driver of the coastal-zone development, mostly at the expense of former agricultural land (Figure 3.2). Quite the opposite trend is, however, observed with forests. While in Europe forest coverage increased between 2000 and 2006, in the coastal zone it slightly decreased. Forest creation and management (although driven mainly by internal forest conversions) represents the most extensive flow of European coastal land cover change.

The total annual land use change rate from 2000 to 2006 for the coastal areas was 0.16% compared to an overall rate in Europe of 0.17%. In 19 coastal countries mapped in all CLC surveys (1990, 2000, and 2006), the change rate was 0.32% in the 1990–2000 and 0.21% in the 2000–2006 periods. This corresponds with the overall slowdown trend observed in Europe.

Urbanization: Artificial Surfaces

Most of the urbanized coastal areas are used for residential and recreation, rather than industrial or commercial, purposes (Figure 3.3). This closely corresponds to the overall situation in urban zones of Europe. Sprawl of artificial areas is the main driver of coastal zone development in Europe. It is estimated that 1347 km^2 of new urbanized areas were created from 2000 to 2006. The annual rate of urban sprawl, 0.66%, is higher than the European average (0.52%), but it decreased compared to the previous period (1990–2000).

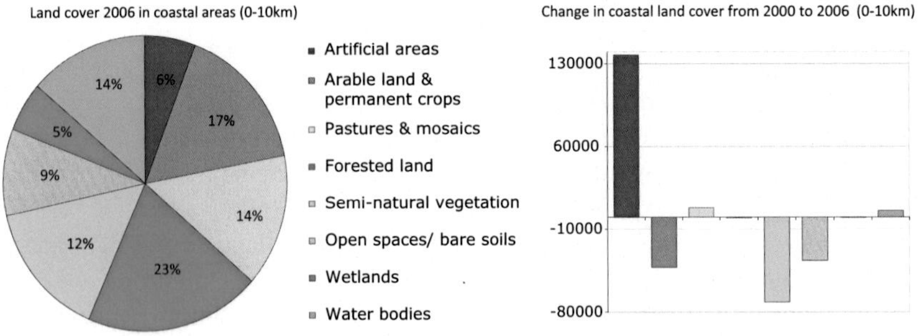

FIGURE 3.2 Land cover 2006 (left) and net change in land cover from 2000 to 2006 (right). *EEA (2013).*

The land take (Figure 3.4) is driven mostly by residential sprawl (42% of total sprawl). Other main drivers of the sprawl are developments of construction sites (21%), sport and leisure areas (14%), and commercial/industrial sites (12%). Land take proceeded mainly at the expense of former arable land (34%) and pastures (31%). The rest of newly urbanized areas came from forests and seminatural vegetation. Internal change of artificial land was represented mostly by conversions of former construction sites into developed urban areas and to a lesser extent to industrial and commercial sites. Internal urban development also occurred at the expense of green urban areas.

The expansion of artificial land has occurred all across Europe's coasts. The most pronounced increases occur in Ireland, Iceland, Southern and Eastern Spain, parts of Italy, and Cyprus, which experienced more than 10% increase of artificial land cover within the coastal zone (Map 3.4). In the latter three, this

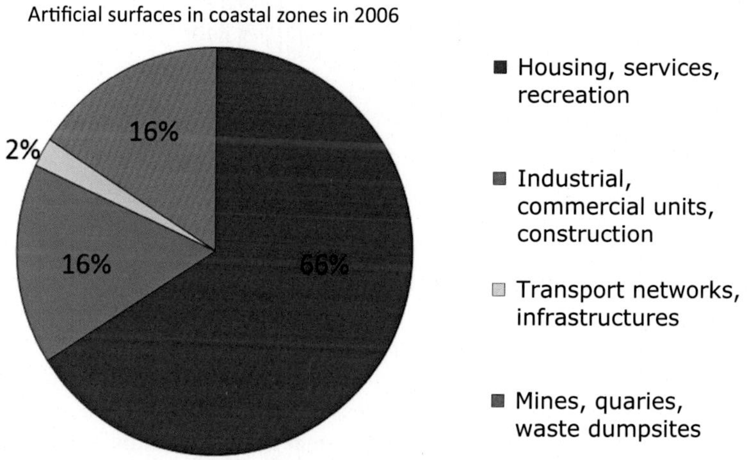

FIGURE 3.3 Artificial surfaces in coastal zones in 2006. *EEA (2013).*

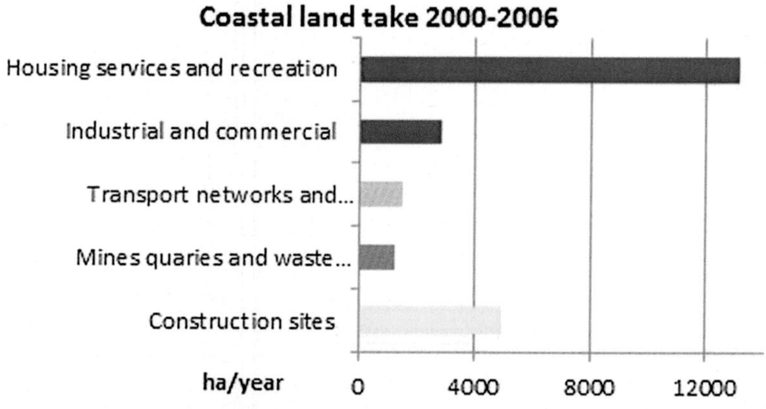

FIGURE 3.4 Coastal land take from 2000 to 2006. *From Corine Land Cover 2006, EEA.*

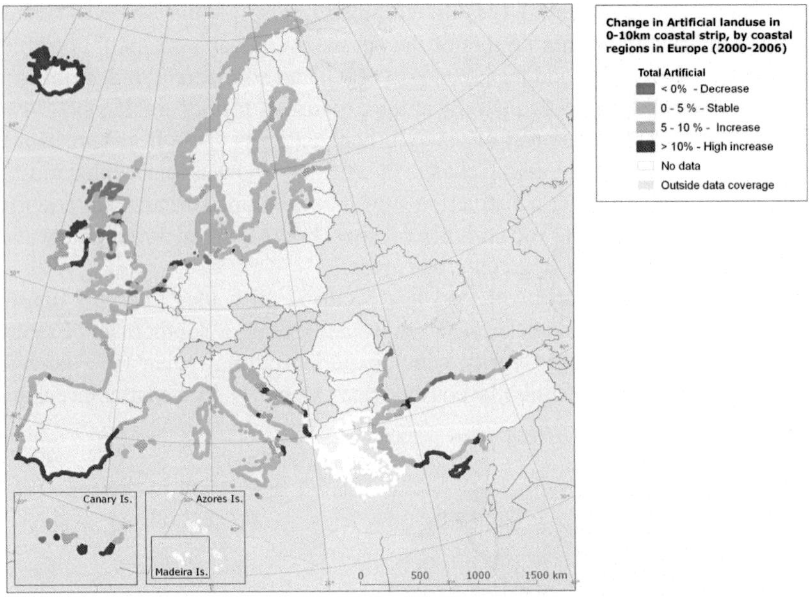

MAP 3.4 Change in artificial land use in the 0–10 km coastal zone. *Corine Land Cover 2006, EEA.*[7]

increase was likely driven by tourism, considering that 51% of bed capacity in hotels across Europe is concentrated in coastal regions[8]. Residential land expansion is predominant in Albania, Cyprus, or Ireland; while in parts of Spain, Italy, the Netherlands, Lithuania, Iceland and Croatia, industrial or commercial sprawl is dominates. In contrast, some coastal areas in the northeast of England and north of Scotland have experienced a slight decline in artificial areas.

Rural Areas: Coastal Agriculture and Forestry

As previously mentioned, almost half is used as arable land, more than 25% is made up of agricultural mosaics, and 20% is pastures (Figure 3.5). Compared to the European structure of agricultural land, in the coastal zone there is a higher share of small-scale agriculture and pastures.

Despite that the overall intensity of agricultural land cover exchange in European coastal zones slowed down significantly, compared to the previous period, 372 km^2 of arable land and permanent crops and 442 km^2 of pastures

7. Note: aggregation of land use change data per coastal region (e.g., NUTS3) can result in attribution of a change that occurred in a limited part of the region (e.g., a coastal city) to the entire region. For example, the mapping of change in Iceland is related to changes in limited areas and does not mean that artificial land use has uniformly increased across the entire coastal zone of this country.
8. http://ec.europa.eu/dgs/maritimeaffairs_fisheries/consultations/tourism/background_en.pdf.

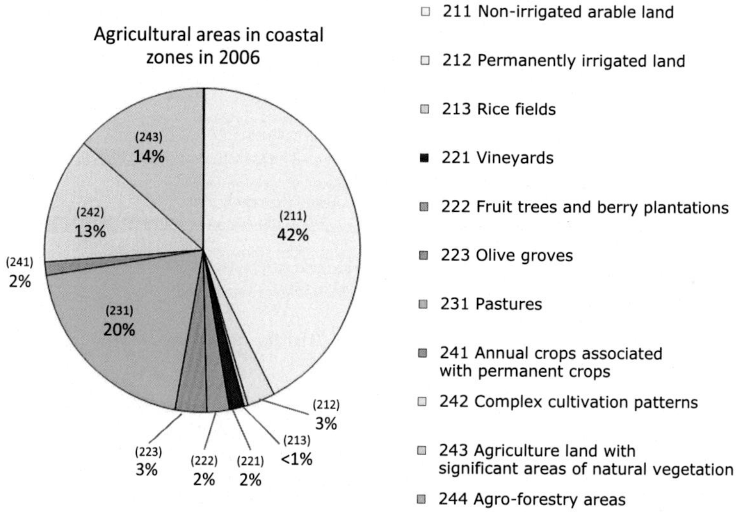

FIGURE 3.5 Agricultural areas in coastal zones in 2006. *EEA (2013).*

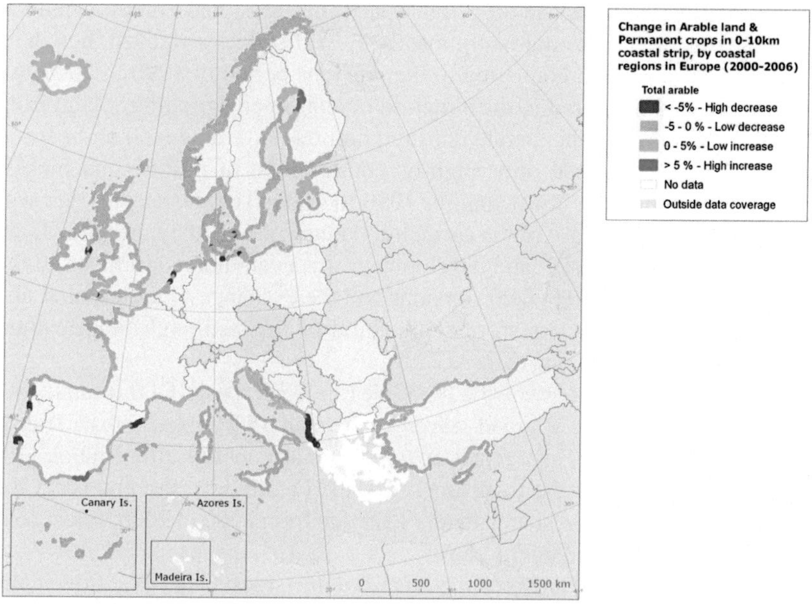

MAP 3.5 Change in arable land and permanent crops in the 0–10 km coastal zone. *Corine Land Cover 2006, EEA.*

were lost from 2000 to 2006. This applies to majority of the coastal countries. Only certain parts of Europe (Map 3.5) experienced an increase in arable land and permanent crops (Bosnia and Herzegovina, Slovenia, Croatia, and parts of Germany, Iceland, Norway, Estonia, and Finland).

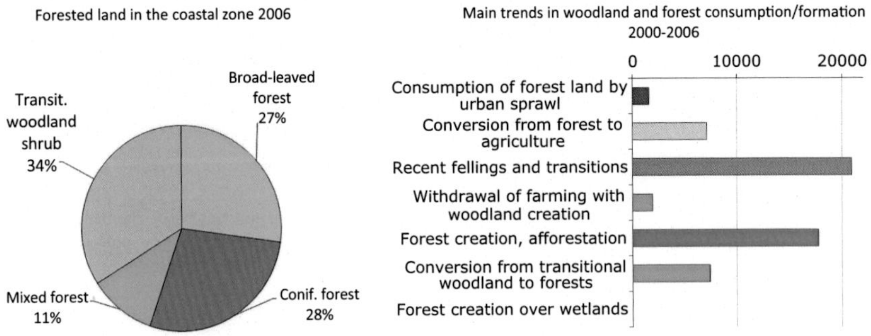

FIGURE 3.6 Forested land in the coastal zone 2006 (left) and main trends in woodland and forest consumption/formation 2000–2006 (right). *EEA (2013).*

Apart from arable land, pastures were lost mainly in Iceland, most of the United Kingdom (Wales and England), the majority of France, Italy, Ireland, Spain, Belgium, the Netherlands, Estonia, Romania, and Bulgaria. On the other hand, new pastures and agricultural mosaics were created mainly in Northern Europe.

The loss of agricultural land cover is still characterized by a consumption of all main agricultural classes (arable land, pastures, and complex cultivation patterns or agri-natural land), mostly by artificial sprawl, even though its rate slightly decreased compared to the previous period of 1990 to 2000. An interesting shift occurred in the structure of consumed agricultural land from 2000 to 2006. While the annual consumption rate of arable land and permanent cropland decreased only slightly, consumption of pasture and mosaic land was significantly lower than in 1990 to 2000. The decreasing rate was observed not only in case of the agricultural land consumed by the sprawl; the slowdown of agricultural land development was even more significant in the case of agricultural internal conversions and the exchange of agricultural and natural land (both conversion of seminatural or forested land to agriculture and withdrawal of farming).

Forests and woodlands cover 22% of the European coasts. Of this total, 40% is coniferous forest, 31% broad-leaved forest, and 15% mixed forest. Transitional woodland shrub, as a result of forest management or afforestation processes, covers 14% of the total forested land (Figure 3.6). In contrast to the increase observed in forest area in all of Europe from 2000 to 2006, forest coverage slightly decreased on the coasts.

Forest management and creation remained the largest land cover flow in the European coastal zone (Figure 3.6). Most prominent drivers of forest change were felling of tree stands and forest creation from transitional woodland, which increased twice compared to the previous period.

Beside this internal forest land rotation, the coastal forests were also lost and consumed by artificial land take. The urban sprawl converted former woodlands mainly into residential areas, construction sites, or sport and leisure facilities. Most of the new forests, besides conversion from transitional woodland, were created on burnt areas, wetlands, pastures, and agricultural mosaics.

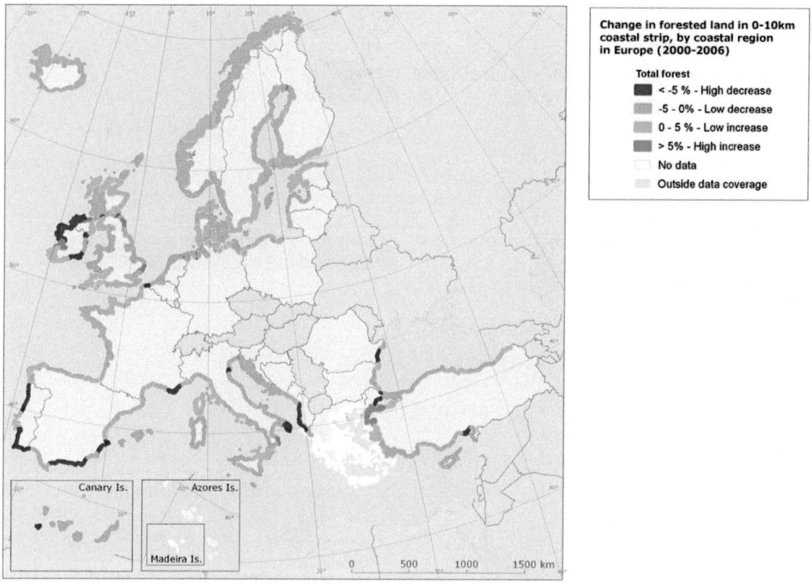

MAP 3.6 Change in forested land in the 0–10 km coastal zone. *Corine Land Cover 2006, EEA.*

Most of Europe's coast has suffered a downward trend with regard to forest areas (Map 3.6). The majority of Southern European countries such as Spain, France, and Italy have followed this trend, as well as parts of Belgium, the Netherlands, Germany, Norway, Sweden, and Finland. A forest cover increase was observed in some parts of Norway, Sweden, Finland, the United Kingdom, Ireland, Estonia, Denmark, and Southern Italy.

Natural and Seminatural Areas

Natural and seminatural vegetation, excluding forests, cover 41% of the Europe's coastal zone. Of this, 12% is covered by natural grasslands, moors, heathland, and sclerophyllous vegetation; 10% is open space with little or no vegetation (beaches, rocks, and burnt areas); 5% is wetland; and 14% is represented by inland and marine coastal waters. The latter contain lagoons, estuaries, and some fraction of coastal sea.

All natural or seminatural land-cover types experienced a decrease from 2000 to 2006 due to afforestation or sprawl of new construction sites and residential and industrial areas (Figure 3.7) at the expense of natural grassland and sclerophyllous vegetation and open spaces (burnt areas). Other important flows in coastal natural land cover exchange are seminatural rotation, extension of water courses (mostly at the expense of beaches, dunes, and sand plains), forest and shrub fires, and coastal erosion.

Growth of seminatural areas was observed only in a few countries like Ireland, Cyprus, and Latvia (Map 3.7). Most of the countries show mixed

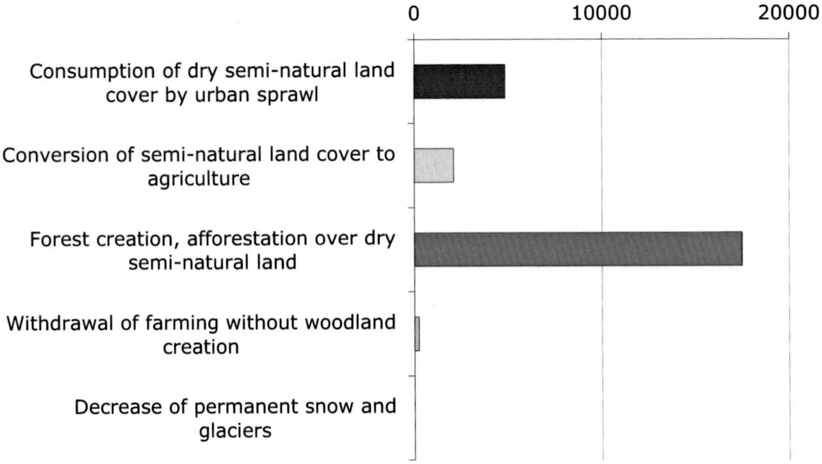

FIGURE 3.7 Main trends in dry seminatural land consumption/formation from 2000 to 2006. *EEA (2013).*

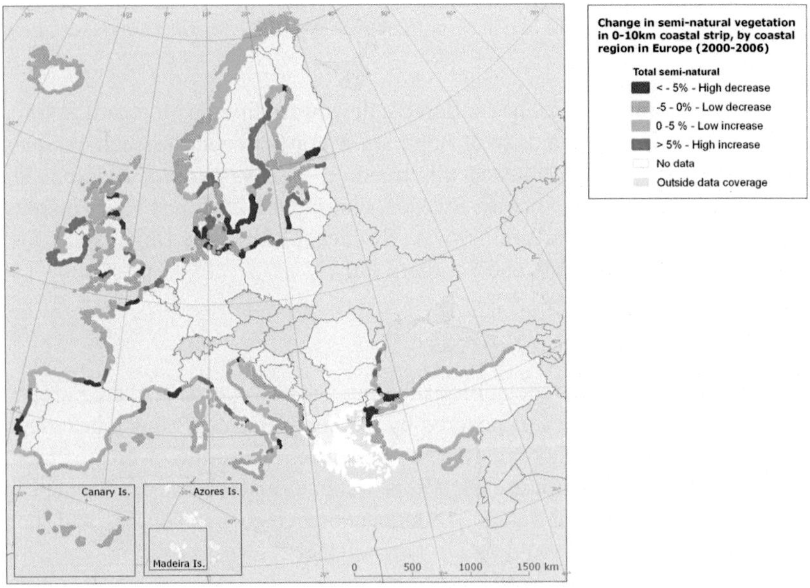

MAP 3.7 Change in seminatural vegetation in the 0–10 km coastal zone. *Corine Land Cover 2006, EEA.*

trends, with growth and reduction in different parts of the coast (e.g., Norway, Sweden, Finland, Denmark, Portugal, Belgium, Netherlands, Spain, Poland, Croatia, and Italy). However, where the growth occurred, it was less than the previous period of 1990–2000. The majority of France, Lithuania, Germany,

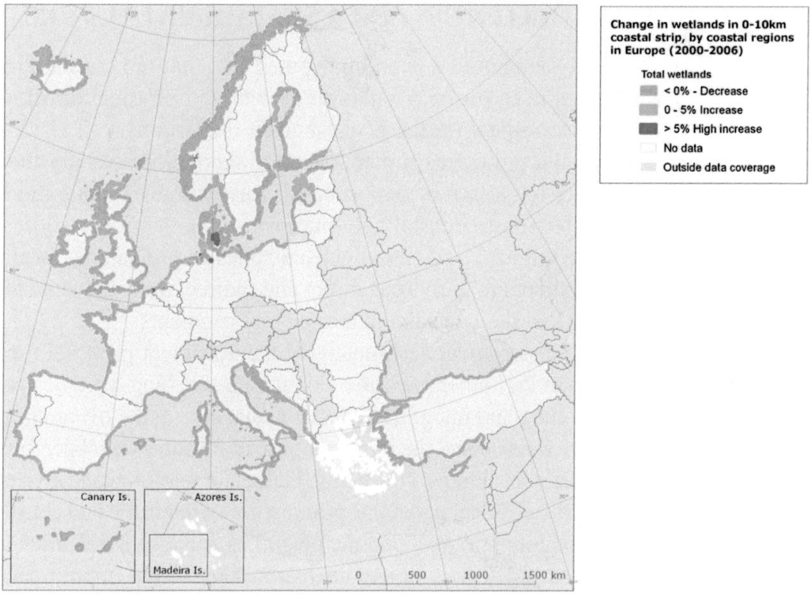

MAP 3.8 Change in wetlands in the 0–10 km coastal zone. *Corine Land Cover 2006, EEA.*

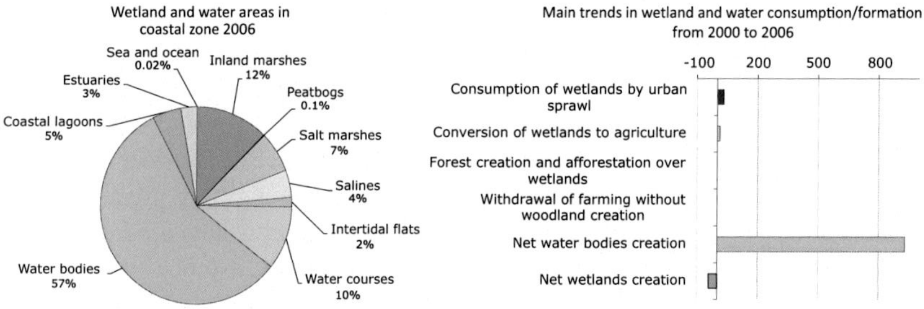

FIGURE 3.8 Wetland and water areas in coastal zone 2006 (left) and main trends in wetland and water consumption/formation from 2000 to 2006 (right). *EEA (2013).*

the United Kingdom, Turkey, Malta, Estonia, and Albania experienced a negative trend with regard to seminatural areas.

The majority of countries experienced a decrease in coastal wetland area; only in a few areas in Northern Scotland, Estonia, and Denmark did wetland coverage increase (Map 3.8). This is a result of a significant amount of European wetlands being afforested, converted to water bodies or agricultural lands (Figure 3.8).

In the coastal zone there are approximately 12,000 km^2 of intertidal flats; almost 4400 km^2 of coastal lagoons; 3700 km^2 of beaches, dunes, and sand plains; and 3000 km^2 of salt marshes.

CONCLUSIONS: CHALLENGES FOR A SUSTAINABLE FUTURE

From the data and analysis above, it is an undisputed fact that the coastal area faces a lot of challenges in terms of population growth and physical development, which has adverse impact on the ecosystem in the coastal area. Therefore, sectorial policies and policy instruments need to recognize not only their impacts on other policies but also new opportunities for synergies among them.

Existing policies referring to coastal and marine environment only address part of the diverse and complex character of coastal zones and related territorial issues, especially when there are numerous direct and indirect links between the different policies that affect the coastal and marine environments.

Following the impact assessment accompanying the document proposal for a Directive of the European Parliament and the Council establishing a framework for maritime spatial planning and integrated coastal management, the overarching problem of competition for space and the inefficient use of resources has been broken down into more specific problems, conflicting claims on maritime and coastal space, leading not only to inefficient and non-optimum use of maritime and coastal space, and suboptimal exploitation of economic potentials, but also to insufficient adaptation to climate risks and degradation of the marine and coastal environment.

In any case, planning of coastal land use and in some cases coastal waters, at the level of regions and member states, and maritime spatial planning, will continue to play a crucial role, considering the percentage of the European population living by the sea. The future of coastal regions will be heavily affected by economic developments, but increasingly also by climate change, in particular due to vulnerability to sea-level rise and extreme weather events.

This calls for a better understanding of the coastal areas and the land–sea interactions, not only on the land side but on the sea side. Historical economic interests rise over the ecosystem approach every time a policy process starts. This is a reason for ecosystem-based management approaches to be integrated into all new policies related to coastal areas, including clear processes for coordination and cooperating during the policy implementation.

The challenges mentioned need to be oriented to support the establishment of a permanent and relevant integrated coastal area management capacity that allows decision-makers and other stakeholders to commonly address the ecosystem-based management approach at the national, regional, and local level in each country, also attending to and encouraging transboundary cooperation with European and other neighboring countries.

REFERENCES

Duarte, C., Cebrián, J., 1996. The fate of marine autothropic production. Limnol. Oceanogr. 41, 1758–1766.

Duarte, C. (Ed.), Culbertson, J., Dennison, W., Fulweiler, R., Hughes, T., Kinney, E., Marbá, N., Nixon, S., Peacock, E., Smith, S., Valiela, I., 2009. Global Loss of Coastal Habitats: Rates, Causes and Consequences 1st Edition – Bilbao: Fundación BBVA.

European Environment Agency, 2010a. 10 messages for Coastal Ecosystems – Coastal Ecosystems. (EEA Report).

European Environment Agency, 2010b. EU Biodiversity Baseline. (Technical Report No 12/2010).

European Environment Agency, 2013. Balancing the Future of Europe's Coast – Knowledge Base for Integrated Management. (EEA Report No 12/2013).

McClain, M., Boyer, E., Dent, C., Gergel, S., Grimm, N., Groffman, P., Hart, S., 2003. Biogeochemical hotspots and hot moments at the interface of terrestrial and aquatic ecosystems. Ecosystems 6, 301–312.

Making the Link

By Bethany Jorgensen

Coastal Population and Land Use Changes in Europe: Challenges for a Sustainable Future ⟺ **Human Frontiers: An Act of Smuggling across Social Borders**

Alejandro Iglesias-Campos, Andrus Meiner, Kieran Bowen and Joseph Onwona Ansong

Mélanie Jouitteau

In their respective chapters, both Iglesias-Campos et al. and Jouitteau address the challenge of working across boundaries. The similarity between their approaches and discussions diverges from there.

Iglesias-Campos et al. share with us their high-level, technical, multinational perspective, outlining some of the regional challenges facing coastal zones and communities in Europe, along with their solution for more effectively planning coastal land and water use. They recognize the need to better understand coastal areas on both sides of the shoreline: on land and in the sea. Their primary recommendation calls for ecosystem-based management to be integrated into new coastal zone policies, along with clear implementation plans to ensure cooperation and coordination between communities, decision-makers, and other stakeholders. They emphasize the need for ecosystem-based management across geopolitical borders, spanning from the local to the international level.

From Iglesias-Campos et al.'s panoramic regional view, we zoom in to the level of the individual again in Jouitteau's chapter, where she uses the concept of geopolitical borders to traverse along the social boundaries we create to organize and control each other and ourselves in our societies. One such border has been constructed between scientists and nonscientists, which limits knowledge sharing and collaboration between researchers and communities, decision-makers, and stakeholders. She argues that scientists are able to act as "smugglers," transporting their knowledge to reach those on the other side of this border.

As coastal zones are examples of ecotones—transitional areas between two biomes—the social boundaries explored by Jouitteau present us with liminal borders between social categories, providing fresh perspective on the dynamics involved in crossing boundaries through collaborative work.

Chapter 4

Human Frontiers: An Act of Smuggling Across Social Borders

Mélanie Jouitteau

CNRS, UMR 5478, Université de Pau et des Pays de l'Adour, Pau, France; Université Bordeaux III, Pessac, France[1]

Chapter Outline

FRONTIERS: CONCRETE AND REAL

This first section aims to establish that if a person knows what a geopolitical border is, he or she is well-equipped to understand the categorizations power produces among humans. I will argue that the body of knowledge about the former makes correct predictions about the latter, in reviewing the paradigm of geopolitical borders and showing how these make sense in the field of categorizations among humans.

Permanent Renegotiations of Boundaries

The boundaries between human groups, some of which are spatial, are permanently subject to renegotiations. Before December 2013, women of the Great

1. My thanks go to my reviewers, including Milan Rezac as well as Pierre-Guillaume Prigent and Glen Falc'hon, for discussion about the position of pro-feminist men.

Mosque of Paris used to pray in a dedicated area within a large room with men, but the mosque authorities then decided to reduce male/female proximity by creating a dedicated space for women in the basement. Women took action to reappropriate their space with a public petition, a Facebook group, and a legal complaint.[2] This fight did not question the sexual partition of space, but it negotiated the course of the man/woman spatial boundary line.

Such negotiations are also easily identifiable with nonspatial boundaries. The dividing line between locals and immigrants in a given society varies according to changes in laws for access to employment, health care, and civil rights—each a milestone on the border between these two categories. A given immigrant can become a local insofar as his or her rights arise at full coincidence with those of other locals. A given local person can also become an immigrant: Cohen (1972) describes how, as a child of Greek immigrants integrated in Marseille, he discovered he was becoming racialized as a Jew during the Second World War. His parents had crossed the border from local immigrant to local, yet he experienced it the other way around, in discovering for himself a novel criterion of categorization. There are human categorizations that are sometimes considered irrevocable, ontologically based on criteria designated as immutable, like sex or race, for which Nature is invoked.

These separations, however, also have fluid and negotiable boundaries: border zone individuals may fall to one side or the other of the race or gender border during their lives, sometimes to their own surprise. In the United States, positive discrimination implies racial classification is founded in law. Inspection of classification criteria shows that, far from encoding an immanent Natural Order, the law uses a patchwork of conflicting and tinkered criteria that mix disparate notions of genetic descent, cultural pattern, language, or self-identification by an individual or a community. In the early twentieth century, the *one-drop rule* required that a particular person with a drop of so-called "black blood" be considered black. The drop of blood is interpreted not only as the presence of a black person among ancestors, but a logical qualification of this person as well, and so forth up the lineage tree. Additionally, these rules do change, and a person considered black in the early twentieth century by the one-drop rule may no longer be considered as such. In the twenty-first century, it is the presumed race of the mother that is transferred to the birth certificate of a child (Cloos, 2010, p. 91). It is then usually the case that after a few generations, the official race of an individual matches only that of one of his great-grandmothers. In the contemporary United States business world, the Minority Business Certification (nmsdc.org) verifies that a company is owned by a person from a "minority." This certificate concerns U.S. citizens who can prove 25% origins of any of the "black racial groups of Africa," backgrounds "from Pakistan or India," backgrounds from "Spanish-speaking regions of South America," or a person officially recognized by his or her own community in the case of Native Americans. Ancestry and self-reported cultural identification are criteria that can correlate with or contradict each other. We can scientifically observe and quantify

2. Source: their press release of December 26, 2013.

these contradictions. Death certificates in the United States are also written in terms of racial classification, which is determined either by a close relative or by a doctor or funeral director's visual estimation. This obviously leads to contradictory results (Krieger et al., 2003, cited in Cloos, 2010, p. 91), each case raising the issue of the particular outlines of the border of racial categorization: who owns a passport to which categories and what are the criteria of eligibility?

Passports and Identity Documents

A passport certifies the status of an individual, included or excluded, in relation to a group boundary. Social boundaries also use passport systems. Some highly valued degrees serve as class passports, if their production means integration to a higher class (although possibly at its margins). The property inheritance system from ascendants is a class passport based on birthright. Many official documents attest to the membership or nonmembership of an individual to a race; the certificate of nonmembership of the Jewish race required in the French state under the Vichy government is one such example (Estèbe, 1996, p. 58). Such passports also exist on the man/woman frontier. Female mutilations (clitoridectomy and infibulation, foot reduction, neck elongation, lip stretching, plastic surgery of the labia minora, mammoplasty, etc.) are typically culturally justified by a desire for sexual differentiation; a person classified as woman has to comply in order to be a "real woman." These irrevocable physical marks literally embody passports for the *Woman* category. Such passports also exist in administrative and judicial form. As detailed in Bohuon (2012), since 1968, international sports athletes in the woman category are in mandatory possession of a "Certificate of Femininity Valid for Life" (see also "sex passport" or "femininity passport," Bohuon, 2012: footnote2, p. 65). Obtaining this certificate depends upon examination of various anatomical, hormonal, or chromosomal criteria, whose definitions vary over time, which results in cases of passport withdrawals. Athlete Ewa Klobukowska had a woman's passport ("for life"), which was withdrawn and later reinstituted. She passed the test in 1964 and was twice an Olympic medallist in Tokyo. She next failed the Barr test in 1967; six doctors unanimously decided she was not physiologically a woman, and her femininity certificate was withdrawn along with her gold and bronze medals. In 1970, her case was recognized as (probably) a case of mosaicism, which means the presence of an extra Y chromosome, offering no sporting advantage over humans defined as women by possession of two X chromosomes. By decision of the International Association of Athletics Federations, her competition ban was lifted and Ewa was reassigned as *Woman* (Bohuon, 2012, p. 74). This passport directly affected her ability to work.

Work Visas

The equivalent of work visas exist in language, which serves as a marker of different social groups. Candea (2013) points out that in French, the names of poorly paid occupations have different grammatical masculine and feminine forms (shoemaker, *cordonnière, cordonnier,* or unemployed *chômeur, chômeuse*), according to the supposed biological sex of the person. Grammar adapts easily to men occupying roles traditionally assigned to women (nurse *infirmière/infirmier,* pediatric nurse, *puéricultrice/puériculteur*—with the exceptions of midwife *sage-femme*, and nanny *nounou, nourrice*). In the twentieth century, some women gained access to male jobs (president, director, researcher, engineer), but their legitimacy is still under negotiation. Language, indicative of the negotiations and battlefield for women, resists precisely where it could legitimize women's accession of these occupations. Only for low-status titles are the morphologically feminine forms of both the title and its determiner authorized (President of the Gymnastics Association, school principal, job seeker). However, the corresponding honorific titles become allowed with morphologically masculine forms (President of the Republic, Director of the Cabinet, scientific researcher):

*Madame le président de (*l'association de gymnastique /√ de la république)*

*Madame le directeur de (*l'école /√ cabinet ministériel)*

*Elle est chercheur (*d'emploi /√ au CNRS)*

It even becomes possible, when referring to the function, to refer via a masculine pronoun (*il*), which is otherwise perfectly ungrammatical in French. These exceptions to grammar rules bear the strong sociological subtext that a woman is performing a male's job. Language tags the individual as outside her assigned gender class, occupying an illegitimate space.[3]

*...**le premier ministre** [masc, sing.] [of Britain—at the time, Mrs. Thatcher], il est vrai, n'avait pas caché, en s'installant au 10 Downing Street, que la période qui s'ouvrait serait plus marquée par le sang, la sueur et les larmes que par la facilité. **Il** [masc, sing.] peut, en outre, faire valoir que nombre de ces difficultés ont été héritées des travaillistes...*

The Prime Minister, it is true, did not conceal, when she moved into 10 Downing Street, that the period then to begin would be marked more by blood, sweat, and tears than by facility. He could, moreover, rightly maintain that some of these problems belonged to the Labour legacy.

Le Monde, leader column, p. 1, 21-5-81, cited in Cornish (1986):252

Systematic exogenous labeling of a person inside a group in which she is tolerated for work reasons basically amounts to a work visa: it allows limited traffic in an otherwise forbidden area. The period of validity for the visa, and its revocability, depends on the authority regulating flows between segregated groups. Women who refuse to be assigned such a work visa develop resistance strategies.

3. See the English contrast: *John is (a waiter/#a prince) and Mary is too.* For a state of the art of feminizations in French, see Baider et al. (2007).

On January 15, 2014 in the French National Assembly, Deputy Julien Aubert addressed Deputy Sandrine Mazetier with *Madam*, followed by the masculine form for President, which she corrected with the feminized form. The male deputy next addressed her five times in his speech with the masculine form, reserving the same treatment for the female minister. Deputy Mazetier responded by addressing him with a grammatically unprecedented creation: *Mr the (feminine form) Deputy*. The next day, Deputy Aubert publicly stated that this term "violated [his] identity" and was "aggressive" and "offensive." His response is consistent with the hypothesis that this is basically the imposition of a work visa.

Immigration

Some migrants are thought to be perfectly integrated into the target group, even in cases where the border separates saliently differentiated groups. In Northern Albania, Kosovo, and Montenegro, a person categorized as *Woman* can live almost entirely a man's life (Hérault, 2009). These so-called "sworn virgins" can take a wife and have children with her regardless of the genetic material used. Sworn virgins perform traditional masculinity, providing a clear example of the migration of a group across a sex/gender border. This immigration is integrative, and there is no return option.

Other types of migration take place in the context of workplace, and raise the question of the possibility of return. In the 1970s in France, young middle- or upper-class Maoist activists were encouraged to join the working class in the factories in order to, among other things, organize class warfare (Rolin, 1996). Many people, called *les établis*, made this choice and remained workers after the economic crisis. Immigration across a class line, like geographic immigration, can be thought of under the angles of family reunification or the hope of return. Migrations across sex borders can also be observed in the workplace: women occupying places characterized as "for men," and men working as nurses or mid*wives*. Some immigrations can be of shorter duration, or clandestine, repeated or not. The ability to make undetected day trips to the other side of a border is called "passing," literally taking this spatial image we are investigating. This term applies to racialized persons passing for white, transgender, and transsexual people, social classes, or to a combination.

Tourism

Border crossing can be purely exploratory and aimed solely at personal enrichment. Tourism is overwhelmingly committed by individuals from dominant groups. Christian Seidel became the transvestite Christiane in an exploratory manner over a period of two years. He tried to broaden his understanding of the world, perceiving it from a different perspective: as a woman. As Christiane, he partied with supporters of Bayern Munich, walked in the streets of San Remo, interacted with close friends and in his heterosexual relationship

(Seidel, 2014; ZDF, 2013). The practice of "disability for a day," allowing one to experience social spaces with a wheelchair, or without sight, has the same exploratory character.

Border Crossing Signs

The boundaries of human categories are marked by specific, widely recognized signs: costumes, headdresses, and (levels of) languages. Posting these signposts sends a clear signal. These signposts are not fundamentally attached to a given border, and evolve depending on location and time. This is the case of wearing pants in Europe, which long stood as a border sign between men and women, but no longer does. Like a signpost saying "Italy 200 m" has nothing fundamentally Italian, signs and markers of social categorizations are purely conventionally attached to the group whose delimitation they signal. Pants have nothing intrinsically male or female, they mark bipedalism at most. Like the Italian signpost, it is only the border system that gives them meaning. These signs are not an expression of essence of a particular group, but they can deeply impact the body. They affect voice, posture, gait, (un)authorized gestures, how one occupies space, the amount of food that one swallows, and even the pace of ingestion. Some signs seem to so deeply pertain to a particular category that they can be tempting to essentialize, like voice pitch, a common marker of male/female borders. Such literally embodied differences have categorical internal meaning, but no more than an arbitrary frontier signpost. In the same way, a river can mark a geopolitical border. In neither the river nor in the salmon in the river is there anything fundamentally linked to the way humans from both sides decide to live their civil lives. The presence of the river is opportunistically used by humans, who consequently invest it with categorizing meaning. Physical and cultural attributes become attached to either side of categorical boundaries, with a fairly wide variation, although some constants emerge: hierarchically subordinate groups are generally associated with a greater capacity and taste for difficult, dangerous, and poorly paid work (e.g., supposed flexibility of children's fingers in carpet weaving, supposed physical endurance of slaves/colonized people/peasants, and supposed tendency of women for abnegation, etc.). The dominant groups are often associated with elements morally justifying their privileged position (e.g., intelligence, merit won generations ago, responsibility, etc.). The principal function of these signposts is to justify the demarcation of the border and the related hierarchy. They are no more intrinsically linked to a given group than other signposts.

Prohibitions and Passing Regulations, Customs Duties

Crossing from one group to another is prohibited by definition. This does not mean that crossing is impossible, but that it is potentially dangerous or expensive. A group of smugglers will set their price. Prices may rise in places over which passage is easy to block, like mountain passes or maritime straits. Three parameters are at play:

1. Tolerance of offences—the threshold beyond which a repressive system is activated.
2. Type of punitive system set in place when the tolerance threshold is exceeded.
3. Financial and human cost of the passage itself—via a public tax border system or through the private sector.

Consider the case of a human assigned to the *Man* category who crosses various social spaces with conspicuous attributes of femininity (dress, high heels, makeup, etc.). He will test the extent to which his presence is recognized and tolerated (1), and, beyond that, the type of repression triggered if any (2). All economic actors whose profits are directly related to the existence of the border and the difficulty of crossing it levy the equivalent of customs duties. The financial and human cost of the passage in (3) depends on many factors, but entire industries are built on this passage in particular: permanent depilation, facial remodeling surgery, hormone treatments, mandatory psychological evaluation in the event of a sex change operation, etc. On racialization borders, we find skin bleaching industries (cosmetics and chemical), hair straightening institutes, various surgeries designed to shape the body to a Caucasian referent type, or the administrative cost of a name change. On the border of social classes, school systems act as customs: they are given the task of regulating, and limiting, the flow of individuals and their descendants from one social class to another.

The Invisibles: Stateless Persons and Unclassifiables

A categorization criterion separating humans into two distinct groups without intersections presupposes that every human can be categorized according to this criterion. Such a system is weakened by any individual it visibly fails to categorize.

In the toughest cases, the very existence of the deviant individual is entirely forbidden. There is an international ban on states producing stateless individuals. National systems can tolerate a minimal set of "accidentally" stateless people, but if they were too numerous, it would endanger the entire system. In these rigid cases, the deviant individual has to comply to survive, and endure physical transformation if necessary.

In preindustrial societies, cultures commonly assume at least three official genders. In the Americas, the sexual physiology of the *berdaches* does not have to match man/woman-gendered activities. Roscoe (1991, p. 5) listed them in more than 130 North American tribes, in societies as diverse as nomadic Alaskan tribes and Florida city states. In contrast, in societies where only two fundamental sexes and genders are tolerated, the man/woman partition is thought of as natural and effortless. This fundamental belief is maintained at a very high cost in order to dissimulate humans most visibly failing the bipartition: intersexes and transsexuals. Reassignment to the male/female gender duality is overwhelmingly the norm for intersexes in all countries in which these operations are clinically feasible (Lahood, 2012). In France, transsexuals have access to identity documents in accordance with their chosen gender if they are subjected to enforced surgical assignment (evidence of a sex change operation), itself subjected to a psychological testing of supposed adequacy with the requested gender. In more flexible cases, presence of deviant individuals is tolerated if they are invisible, or "passing." In societies "tolerating" homosexuality, homosexuals still have to be discreet and not challenge the heterosexual norm. They are "tolerated" as long as they respect the normative dimension of the border. The outline of this border remains under the monopolistic control of an easily identifiable group (Lifshitz, 2012). The threshold can come with a quota system, beyond which "enough is enough" (enough homosexuals, women, immigrants, impoverished people). Once invisibility fades away, so-called tolerant societies can respond with great violence. Exogenous individuals may occasionally reveal their presence, to the extent that they propose a personal formula to avoid challenging the established order (humor, countersignals, participation in the repression of their peers, etc.). It is important to note that the deviant individual remains entirely in charge of the invisibilization process. Individuals fleeing their category have to take charge of completely switching their overt categorical symbols (costume, language, food habits, cultural reflexes, etc.) in order to avoid social isolation. People with disabilities quite commonly report such impossible assimilation strategies where they are tolerated to the exact extent to which they take their self-effacement (Vigand, 2011).[4]

This first part of the chapter has established that borders between social groups can be approached as fundamentally similar to geopolitical boundaries. However, be they geopolitical or otherwise, borders are not uniform and do not all produce the same effects. I now propose three parameters that distinguish the border types.

BORDER PARAMETERS OF DIFFERENTIATION

Borders differ in their corresponding evasion loopholes: clandestinity, viable escape strategies, and extent of no man's lands.

Visible Sorting Criteria and Clandestinity

Boundaries organize around a discrimination criterion. Visible criteria (morphological features of the body, skin color, shape of the face, age) do not produce the same effects as concealable/invisible criteria (shape of the genitals, sexual orientation, culture, or to a lesser extent, social origin or language). For individuals,

4. People with handicaps face one additional challenge, for to be recognized as such, and potentially helped, disability has to remain conspicuously visible. This paradox puts people in a double-bind situation, whose cost is still dependent on them. The same paradox arises for racialized persons who want to claim the benefits of positive discrimination.

the invisibility of a criterion opens strategies of clandestinization, allowing them to move in an exogenous space as long as the difference is not revealed. French colonization of Martinique and Brittany gave rise to fundamentally different colonial experiences, because the Breton people, while combining stigmas of language and social background, could access clandestinity by mastering French without identifiable accent, and dress like the settlers. People may have chosen to reveal their origins or hide them, change their minds several times, reveal to some but not others, etc. This strategy is rendered far more difficult by a non-concealable criterion like skin color. The danger of a strategy of clandestinization is the imperative of nonexistence, and it poses the question of secret recognition signs. In the case of sexual orientation categorization, individuals must also meet to live their sexual orientation. Therefore they collectively develop ways of finding each other in a crowd, while organizing their own invisibility at the same time.

Totalizing Paradigms and Interstitial Spaces

Border systems universally make totalitarian claims to categorize all available space. The claim is widely performative but also rarely true. From the point of view of a given individual, they vary according to the (un)availability of gaps and margins, to the opportunities available to escape the paradigm (even at the cost of marginalization). The borders between countries exhaustively partition the finite land space of the planet. This situation is new for humans, who have evolved for most of the known history of humanity in spaces with many relatively accessible margins for off-category individuals (deserts, steppes, forests, islands, mountains, wetlands, etc.). Withdrawing from social areas in geographical space had a cost, but was possible. In the last century, these interstitial spaces underwent a drastic reduction and the impact of geopolitical boundaries and of socially totalizing categorizations grew accordingly.

Deviant individuals undergo a symbolic decategorization out of humanity. The paradigm is that of the *Other*, of the unknown, of liminality, madness, or of subhuman monsters. A bearded woman is not considered as bordering the *Man* category, a move which would benefit her on many levels, but as bordering humanity. Intersexes testify of inhumane treatment of violent sex reassignment (Gosselin et al., 2008). In an analysis in terms of border crossings, the autobiographical story of Preciado (2008), who self-administered testosterone for a year, could be understood as a clandestine immigration process across the woman/man border, but she stressed that taking testosterone did not include her in the *Man* category: "I do not take testosterone to turn myself into a man, but to betray what society wanted to make me, to add a molecular prosthesis to my low-tech transgender identity made of dildos, text, moving images."[5] She precisely describes a

5. "Je ne prends pas la testostérone pour me transformer en homme, mais pour trahir ce que la société a voulu faire de moi, pour ajouter une prothèse moléculaire à mon identité transgenre low-tech faite de godes, de textes, d'images en movement."

passage from the *Woman* category to a new one labeled *Other*. Some transgender or transsexual people who escape the Woman category also oppose the "female to male" path, instead favoring the path from "female" to unknown," a term that best expresses the category they feel they embody (Binard, 2006, p. 398). The semantic field of robotics and digital technology is also invested. Haraway (1991) proposed the concept of cyborg, the new being overtaking the woman subject. Marginality figures often combine with spatiotemporal peripheral figures: the night, the bush, the wasteland. These concepts have in common an implied space beyond the usual, where human visions of self find a symbolic area to grow away from predetermined categories, a place where one's choices open. Logically, these exploratory positions are overwhelmingly represented in science fiction (see Ursula K. Le Guin). When a margin space offers refuge for a set of humans, their liminal visions crystallize in marooned identities and create still more borders within the margins. The power system can work around these margins provided they stay small and reasonably far apart. Some margins have higher potential to integrate individuals, and potentially short-circuit the entire system. Monique Wittig (1980) noted that the *wife* or *human reproductive* understanding of the *Woman* category excludes lesbians, and joyfully concluded that lesbians are not women, with the logical implication that lesbians were therefore free from compliance to womanhood. By doing so, Wittig opens a peripheral space with the potential to entirely bleed the *Woman* category, which by parity of argument suddenly excludes bisexuals, pansexuals, heterosexuals without children, or anyone departing the slightest from the prototypical feminine image. Ultimately, the category reduces to an abstraction—*THE Woman*—revealing its essence. The category does not exist in itself, it is merely the result of an active and arbitrary act of categorization of individuals, perpetually recreated by a global hierarchy system among humans (Butler, 1990). For those who society categorizes as women, the category manifests itself by its concrete effects (subordination, limitation of individuals, economic exploitation, etc.; see Gunnarsson, 2011).

New interstitial spaces emerge with social global changes. The Internet, for example, offers new liveability solutions for category refugees. When lacking interstices, these small margins, uncomfortable as they may be, are the only option remaining for individuals who cannot hide in plain sight; they settle on the dotted line of the border, that is, inside the borders' intersection space.

The Borders' Intersection Space

Radical geographical theory attempts to articulate the relations between spatiality and the relations of power and domination (Harvey, 2008; Soja, 2009; Gervais Lambony and Dufaux, 2009). It is in this context I note that spatialization of borders reduces their intersection areas. Some borders are inherently spatial, like geopolitical borders or the categorization between urban centers and peripheries. Some borders are only partially spatialized. Throughout 40 years of apartheid in South Africa, some areas were segregated and some shared. The man/woman border is clearly spatialized in public toilets, but heterosexuality induces sharing other spaces. The constant is that the more spatialized borders are, the harder it is to live inside their intersection space. Border spatialization implies the absence of an intersection of categorized groups but not the reverse: two nonintersecting groups may coexist in the same space provided the tight separation line is time. In an office space, management and maintenance teams can work completely separate hours, without time intersection space. Such categories can even be profoundly antagonistic. During the Rwandan genocide in mid-April to mid-May 1994, Hutu people of the Nyamata Hills "worked" every day, precisely from 9:00 AM to 4:30 PM, murdering Tutsi people. On this schedule, the surviving Tutsis hid in another space, marshes, where they took cover under papyrus or sank into the mud. Outside the Hutu "working hours," some Tutsis slept in their homes or in the village school, heading back to the marshes at 5:00 AM (Hatzfeld, 2000, pp. 81,177). The same space thus was constantly crossed by individuals from two strongly opposed groups, but during blocks of time that carefully prevented their intersection.

The temporal dimension may also create an intersection along a spatial line. At any given time, when a categorical border is drawn in a space, a person can belong to both categories at the same time. Once we add the temporal dimension to space, in accordance with the actual conditions of human life, we see that a person can go from one side of a border to the other, without "doing" anything. Passing creates, within a human, an intersection between the categories. The intersection of spatially separated borders is to be found in the story of humans who cross these borders.

Humans have both individual and collective histories. From the human point of view, belonging to two spatial categories lacking spatial intersection is thus difficult to avoid. Some humans even have several passports from different countries which officially state they belong to several nonintersecting geopolitical spaces. The importance of an exogenous space can also be purely symbolic. A human living on an island likely has a cultural symbolic system organized around sea/continent concepts. Moreover, the most watertight boundaries are precisely those that create a strong symbolic boundary elsewhere. The limit of the hypothesis (3) could come from real cases where the border is never crossable, even in thought.[6]

Multiple Intersections

With our typology of borders in hand, it is now important to emphasize that it is not empirically correct to analyze a human in terms of a single border system. A social being is always in tension between multiple borders, crossed by some, bypassed by others, categorizing him or her as dominant in one criterion and dominated in another. The result of this weaving is very different from that of addition or stacking of unique borders. How borders form between various category boundaries

---

6. Such examples could be looked at in the domain of classes and castes (cf. Delphy, 2005).

changes their social content, their impact on individuals, and therefore the possible strategies of resistance. This brings us to the study of intersectionality, the dimension that is a major academic research field, originating from Anglo-Saxon women's studies (McCall, 2005; Phoenix and Pattynama, 2006 and references therein), and irradiating now well beyond (including Dorlin, 2006, 2009; Anthias, 2013; among many others). I give here only a few examples extending those mentioned above, emphasizing the different effects of intersectionality.

Borders are designed against each other and are mutually dependent. In the case of geopolitical boundaries, with rare exceptions such as Australia, passports or national identity documents indicate the male/female. A woman who crosses gender borders automatically loses her citizenship rights. Similarly, a woman switching social class must precisely revise her gender behavior, because these are coded differently for each social class. The intersections of borders also change their alignment. In Northern European areas with significant low-income economic immigration, a racialized person will likely also suffer social stigma, be categorized as poor, and will suffer discriminations carried out on the basis of both race and social class. In emigration areas such as rural Brittany, where economic relations do not bring in new people, but rather subtract them, economic immigration is very limited. Rural lack of variation in humans' morphology means that someone nonracialized in urban space may become racialized a few kilometers away. Crossing the spatial border of labor pools, an individual can also cross the border of race categorization, which inevitably will redefine the experience of the four areas cut across by these two borders. Similarly, Bohuon (2012) points out that the criteria of what constitutes a "real woman" for sport competition specifically draws from Western culture's concept of femininity. Historically, the suspicions of non-womanhood foisted on athletes from Eastern European countries during the Cold War, next extending to all non-Westerners (Philippines, Brazil, Niger, etc.). The effects of categorization may also differ in their content. Outside of Breton industrial employment areas, racialized groups consist mostly of children adopted by local middle-class families and doctors and nurses working in hospitals. The attached social stigma is potentially different than in high-immigration areas, which in turn creates changes in the classist content associated with the racialization. In Huelva, Spain, production of winter strawberries is performed by an imported Moroccan workforce, under the "contracts in origins" negotiated between the town halls of production locations and the Moroccan state. Because Spanish producers have a gendered representation of Moroccan women as submissive, they restrict these contracts to women, and because the municipalities have a gendered representation of parenthood, they first target women with young children in order to ensure they will return to Morocco. Because producers and institutional actors have an installed colonial representation of Morocco, they impose conditions of work and, nearly, detention on workers, while enjoying the moral benefit of the white liberating symbolic position against the supposed more intense sexism of Moroccan men (Zeneidi, 2013). Inside the cramped boundaries of the workplace, trapped in a scissor-pinch at the intersection of racialization and gender assignment, Moroccan workers attempt individual liberation moves. These strategies cannot be understood without understanding precisely how the colonial and gendered dimensions of their oppression articulate and modify for them the understanding of *Woman* and *Moroccan immigrant*. The social content of the categorization borders is transformed by the interaction between the two. Social border intersections logically also affect their resistance strategies. Living in Barcelona as a white undocumented migrant profoundly changes the experience of hiding, opening the possibility of invisibility (Ressler and Begg, 2014). Some humans categorized as women in Arab societies are organized in feminist movements and meet resistance, like people in any feminist movement. The novelty is that the neocolonial context of Arab societies creates a context for the accusation that Arab feminists are colonial in essence, which profoundly modifies Arab and Islamic feminist strategies of resistance (Ali, 2012). Regardless of the real or fantasized nature of Western collaborations, this accusation is impossible to refute because only illegitimate beings have to prove their legitimacy. Such prerequisite, impossible proof marks all struggles at the intersection of multiple borders. In Northern Europe, the prerequisite imposed on Marxist feminist movements is evidence of nonbetrayal of the class struggle (Delphy, 2002, p. 174). The same presents itself in national liberation struggles. By the end of 2013 in Brittany, following massive layoffs, a social movement demonstrated in Quimper and marshalled Breton nationalist slogans claiming the right to live and work in the country. They called themselves "red bonnets" in reference to a seventeenth-century Breton struggle to abolish serfdom that was bloodily repressed by the French army. The French press and French social unionists noted the presence of Breton employers at the event and presented a class collaboration charge. Media and union structures tasked with representing the discourse of demonstrators actively made it unintelligible. French national class collaboration for resistance to the German occupation during World War II is highly praised by the same groups, showing they can otherwise easily accommodate class collaboration. Demonstrating workers in Quimper are portrayed by their own social camp like big business puppets, and in the same strategic move, veiled women are successively portrayed as puppets of the patriarchy/western world.

From the perspective of power, intersectionality in others favors division between people upon whom power is imposed: "dividing and conquering." From the perspective of an individual combining different dimensions of domination, intersectionality profoundly modifies the apprehension of domination and possible strategies of resistance.

SMUGGLERS

Never are power systems as powerful as when they are invisible or unspeakable. A geopolitical border is not invisible, and we each have a body of knowledge organized around it. Now that we have seen these borders are not fundamentally different from other categorization lines between humans, and we have a typology of borders articulated in an intersectional framework, we should be able to test this framework in a case study: the complex phenomenon of smuggling. Who are the smugglers of social relationships? Who develops a personal interest in crossing borders? What is the motivation for this activity? What is exchanged and for what price? Can smuggling be a source of political change?

Smuggling arises from supply and demand between groups or individuals on both sides of a border. Smugglers inhabit the dominant classes and develop traffic directed toward a neighboring dominated category. Popularization of science, popular education theater, progressive intellectuals, or pro-feminist men's movements are examples of smuggling where knowledge and practices are passed on from one group to another. The smugglers come mainly from near the border for two reasons: first, populations living near a border of a given class are closer to their target; they are not the privileged lot in their own group. From their point of view, smuggling activities do not put them at risk of a repression that is considered important. A number of pro-feminist men demonstrate peripheral masculinities that keep them distant from male power centers (Thiers-Vidal, 2010). Second, proximity facilitates smuggling. Near-border inhabitants have the necessary intimate knowledge of the terrain to initiate meaningful exchanges with the other side (panel signs, no man's lands, land-mine words). The French state is heavily centralized, and a human from the capital will struggle to generate any trade interest if using the words "province," "region," or "periphery," terms associated by an interlocutor with denial of identity and symbolic takeover. The lower-middle class is adjacent to the peasant or working class, and logically provides large numbers of political activists directed toward these classes. In countries where universities have moderate registration fees and low wages, university executives, scientists, and the so-called intellectual class populations are significantly more oriented toward popular classes. In contrast, the North American system organizes the debts of its students, who will repay them by purchasing very high academic salaries, which in turn ensures distance between academics and average-wage nonacademics, consequently reducing the conditions of smuggling possibilities, and subsequently the diffusion of their scholarship through the population. The proximity necessary for smuggling can also come from friendships or love, solidarity bonds, common resistance experience at another border, or family ties (when the youngest explains the Internet to grandpa). Proximity is necessary but not sufficient: social workers or bailiffs are not inherent smugglers. Various smuggling techniques allow smugglers to loosen the links between two terms of trade: in coastal marine settings, recipients recover the goods on the beach or over the water. The content of books, films, and songs can be analyzed in these terms, which poses scientists and artists as good potential smugglers. The smuggler does not always know the nature of what is exchanged: she can believe she sells forbidden goods when really people just want a glimpse of life on the other side of the border. There are different trading currencies. Chomsky (1967, 2002), for example, poses that there is a moral responsibility for intellectuals to use their analytic tools to uncover the actions of the state and reveal it to the people. In these terms, the smuggler is paid in moral value, which is itself subject to cyclical inflation or deflation. As every worker conscious of her interests, a smuggler can recycle when she finds better sources of income, or greater job/status security. Reorientation of the French leftist intellectuals from 1968 to 1981 toward right-wing and anti-revolutionary values is concomitant with the Socialist Party integrating them into power-enrichment systems. They abandoned radical political struggle and its moral value payments for income raises and security.

Does this mean betrayal is inherent to smuggling? Smuggling is doubly linked to the concept of borders in contradictory ways, both against and for the border. On the one hand, smuggling opposes the border. It creates exchanges precisely where they are banned. Having personal, political, and/or emotional reasons against the establishment of the border facilitates smuggling. On the other hand, the border is the very thing that makes smuggling possible. Smugglers develop a personal interest in its preservation. Having personal, political, and/or emotional reasons for the preservation of the border also facilitates smuggling. This paradox embodies perfectly the revolutionary paradox: one wants a border to disappear, but it implies wanting one's own disappearance (or individual transformation). It concretely means wanting to lose the conditions of possibility of the installed exchange, and lose the benefits drawn from them. Smuggling reaps the benefits of an unequal relationship. A person who can cross the border exchanges precisely with people who cannot. From the smuggler's point of view, growing in this relationship means finding moral benefits in always having the place of offer in the exchange, an insurance to interact with people he or she is not equal to. Potential disappearance of the border would make equality possible and radically redefine the terms of the trade, plausibly by removing its necessity. What, then, of individuals who specialize in smuggling? Thiers-Vidal (2010, 2013) has superbly pointed out these contradictions in his analysis of pro-feminist men's groups. Scientists-as-smugglers face the same problem. In progressive academic representation, science decrypts the world and its results are used by democratic human governments for the good of all. However, the most rewarding works for scientists are either those written in a dominant language and published in books with prohibitive prices inaccessible to citizens, or those that sell well to large industries whose work is most detached from common good. A given society might wonder what alternative political reward systems could be put in place to ensure a massive smuggling of scientific knowledge toward the population.

REFERENCES

Ali, Zahra, 2012. Féminismes islamiques. La Fabrique.

Anthias, Floya, 2013. Hierarchies of social location, class and intersectionality: towards a translocational frame. Int. Sociol. 28 (1), 121–138.

Baider, Fabienne, Khaznadar, Edwige, Moreau, Thérèse, 2007. Parité linguistique. Nouv. Quest. Fém. 26, (3).

Begg, Zany, Ressler, Oliver, 2014. The Right of Passage, Video, Général Bordure, Exhibition. Le Quartier, Centre d'art contemporain de Quimper. [10/2013–01/2014].

Binard, Florence, 2006. Lesbianisme, féminisme, genderqueerness, le Débat entre féministes essentialistes et féministes 'post-modernes'. Guyonne (dir.), Travestissement féminin et liberté(s), L'Harmattan, Leduc.

Bohuon, Anaïs, 2012. Le test de féminité dans les compétitions sportives. une histoire classée X?, éditions IXE.

Butler, Judith, 1990. Gender Trouble: Feminism and the Subversion of Identity. Routledge, New York.

Candea, Maria, 2013. Cachons ce féminin que nous ne saurions voir au pouvoir : de la résistance des FrançaisEs à la féminisation des titres glorieux. Le Nouvel Observateur (25.12.2013).

Chomsky, Noam, 1967. The responsibility of intellectuals. New York, Rev. Books 8, 3.

Chomsky, Noam, 2002. Responsabilités des intellectuels, Démocraties et marché, Nouvel ordre mondial, Droits de l'homme, (traduction de conférences données en 1996). Contref-feux, Agone.

Cloos, Patrick, 2010. La racialisation comme constitution de la différence : Une ethnographie documentaire de la santé publique aux États-Unis, thèse. Université de Montréal.

Cohen, Albert, 1972. Ô vous, frères humains (récit autobiographique). Gallimard.

Cornish, Francis, 1986. Anaphoric pronouns: under linguistic control or signalling particular discourse representations? A contribution to the debate between Peter Bosch, and Liliane Tasmowski and Paul Verluyten. J. Semantics 5, 233–260.

Delphy, Christine, 2002. L'ennemi principal, Economie politique du patriarcat, Syllepse. première édition 1998.

Delphy, Christine, 2005. 'Race, caste et genre en France', Guerre impériale, guerre sociale, Jacques Bidet (dir.). PUF, Paris.

Dorlin, Elsa, 2006. La matrice de la race. Généalogie sexuelle et coloniale de la Nation française. La Découverte, Paris.

Dorlin, Elsa, 2009. Sexe, race, classe. Pour une épistémologie de la domination. PUF, Paris.

Estèbe, Jean, 1996. Les Juifs à Toulouse et en Midi-Toulousain au temps de Vichy. Presses Universitaires du Mirail.

Gervais-Lambony, Philippe, Dufaux, Frédéric, 2009. Justice… spatiale !. Ann. Géogr. 1 (665–666), 3–15.

Gosselin, Lucie, Guillot, Vincent, Kraus, Cynthia, Perrin, Céline, Rey, Séverine, 2008. À qui appartiennent nos corps ? Féminismes et luttes intersexes. Nouv. Quest. Fém. 27, (1).

Guillaumin, Colette, 1992. Sexe, Race et Pratique du pouvoir. L'idée de Nature. Côté-femmes, Paris.

Gunnarsson, Lena, 2011. A defence of the category 'women'. Feminist Theory 12 (1), 23–37.

Haraway, Donna, 1991. Simians, Cyborgs and Women: The Reinvention of Nature. Routledge, and London: Free Association Books, New York.

Harvey, David, 2008. La géographie de la domination. Les prairies ordinaires, Paris.

Hatzfeld, Jean, 2000. Dans le nu de la vie, récits des marais rwandais. Points Seuil.

Hérault, Laurence, 2009. Les « vierges jurées » : une masculinité singulière et ses observateurs. Sextant 27, 273–284.

Lahood, Grant (réalisation) 2012. Intersexion, Ponsonby Productions.

Lifshitz, Sébastien (réalisation) 2012. Les invisibles, Zadig films, visa 129 876.

McCall, Leslie, 2005. The complexity of Intersectionality. Signs 30, 3 The University of Chicago Press, 1771–1800.

Minority business certification http://www.nmsdc.org/nmsdc/app/template/contentMgmt ContentPage.vm/contentid/1959#.Uut-wrROhfg (accessed 31.01.14).

Phoenix, Ann, Pattynama, Pamela, 2006. Intersectionality. European J. Women's Stud. 13 (3), 187–192.

Preciado, Beatriz, 2008. Testo junkie, Sexe, drogue et biopolitique. Grasset.

Rolin, Jean, 1996. L'Organisation. éditions Gallimard.

Roscoe, Will, 1991. The Zuni Man<>Woman. University of New Mexico Press.

Seidel, Christian, 2014. Die Frau in mir. Heyne Verlag, München.

Soja, Edward.W., 2009. La ville et la justice spatiale. Justice spatiale/Spatial Justice, Nanterre. http/www.jssp.org.

Thiers-Vidal, Léo, 2010. De « l'ennemi Principal » aux principaux ennemis, Position vécue, subjectivité et conscience masculines de domination. L'Harmattan.

Thiers-Vidal, Léo, 2013. Rupture anarchiste et trahison pro-féministe, écrits et échanges. Bambule.

Vigand, Philippe, 2011. Légume vert. éditions Anne Carrière.

Wittig, Monique, 1980. La pensée straight. Quest. Fém. 7, 45–53.

ZDF, 2013. Christian & Christiane. Arte productions.

Zeneidi, Djemila, 2013. Femmes/Fraises, Import/Export, Souffrance et théorie. PUF.

Part II

Developing Solutions: Challenges for Communities in the Context of Global Change

Chapter 5

Sustainable Mariculture at High Latitudes

Paul Tett, Kenny Black, Ruth Brennan, Elizabeth Cook, Keith Davidson
Scottish Association for Marine Science, Scottish Marine Institute, Argyll, Oban, UK

Chapter Outline

INTRODUCTION

In the twenty-first century there will be more people than ever on Earth. If economic growth continues, many of them will be more prosperous than their grandparents. One of the challenges of the new century is that of satisfying the resulting demand for protein. Some of this will be met by marine farming of fin-fish, shellfish, and seaweed (Duarte et al., 2009).

Nevertheless, mariculture can overload the sea's capacity to dispose of farm wastes or to provide food for shellfish. The challenge addressed in this chapter concerns the setting of safe limits to the biomass that can be farmed sustainably in coastal waters at high latitudes. This focus allows us to deal with the comparatively few types of farmed fish and shellfish that are capable of flourishing in cold stormy waters, and allows us to generalize from our experience with mariculture in Scotland, north of 55°. Similar coastal environments are emerging in Arctic coastal zones as ice-cover decreases.

MARICULTURE

Salmon and sea-trout, collectively called salmonids, are silvery carnivorous fish that spend most of their adult lives at sea, returning to rivers to breed. They have proven adaptable to rearing in large floating cages, where they are fed on pelleted food made (at least in part) from small wild fish and hence rich in organic nitrogen and phosphorus. Uneaten food and fish feces can accumulate on the sea-bed, smothering the animals that live here; in addition, the fish excrete ammonia and phosphate, which might stimulate excess growth of phytoplankton (microscopic single-celled algae).

Mussels are black-shelled bivalves that grow in clumps attached to rocks by byssus threads; they feed by pumping sea-water over their gills and filtering out phytoplankton and particulate organic matter. Mussels can be grown attached to ropes, or in suspended bags of netting, or on the seabed. Harvested, they provide an efficient means of converting phytoplankton into food, but excess cultivation may leave too little phytoplankton for the rest of the ecosystem, and the mussels give rise to a rain of (mainly inorganic) pseudo-feces onto the seabed.

REGIONS OF RESTRICTED EXCHANGE

A region of restricted exchange (RRE) is a semi-enclosed body of seawater in which exchange with the open sea is restricted by, and governed mainly by, conditions at the RRE mouth (Tett et al., 2003). Fjords, called sea-lochs in Scotland, are examples of RREs; they are glacially over-deepened river valleys that, post-glaciation, have been flooded by the sea. Fjords characterize the coasts of Scotland and Norway, which were covered in ice until about 10,000 years ago, and are opening on Arctic coasts as the ice begins to retreat here. They are sheltered and they provide good environments for salmonid and mussel cultivation. Despite their restricted entrances, circulations driven by river discharges provide continuous refreshment of their upper layers, which are cool and well-oxygenated—ideal conditions for salmonids— and which are, seasonally, rich in phytoplankton—good conditions for mussels.

In 2012, about 140,000 tons of farmed fish (mainly salmon) were harvested from Scottish sea-lochs. The UK's devolved administration in Scotland, the Scottish Government, plans to increase this to about 210,000 tons in 2020. There are also plans to double mussel-farming. An obvious question is, do the lochs have the capacity to house this increase?

CARRYING CAPACITIES

Estimating the carrying capacity of a site, water body, or region for mariculture would seem at first sight to require only ecological study. However, successful sea-farming requires collective action to fund, build, and maintain the activity, and to provide a demand for its products. Such actions are the result of social decisions, whether or not these involve expert evidence and whether or not they are made democratically or transparently.

These issues have been addressed by McKindsey et al. (2006), who set out "a hierarchical approach to determining the carrying capacity of an area for bivalve culture," a framework that can be adapted for mariculture in general. The (adapted) levels are:

1. **The physical carrying capacity** is the area that is "geographically available and physically adequate" for a given type of mariculture.
2. **The production carrying capacity** is the "optimized level of production of the target species," which might be set by the availability of food for filter-feeding shellfish or the site's ability to disperse wastes from fin-fish.
3. **Ecological carrying capacity** is the level of production that does not impact undesirably on the surrounding ecosystem(s).
4. **Social carrying capacity** takes account of the first three levels in this list, plus other stakeholder interests.

Figure 5.1 illustrates production and ecological capacities as well as introducing *economic capacity*, which might be seen as a part of the social capacity. Harvest is shown as increasing with stocking, up to a maximum, the production capacity. This maximum might be set, in the case of cultivated filter-feeding bivalves, by the local availability of planktonic food. Above this level, shellfish grow less well because they compete with each other for food. In the case of fin-fish, the level might be set by the extent to which local water movements can remove fish waste and supply oxygen, or by the need to prevent the build-up of infectious diseases. *Economic capacity* might be lower than this, because the greatest profit occurs when the market value of the harvest maximally exceeds the costs of buying and feeding the stock, plus overheads. These components of income and expenditure—and thus the stock level for maximum profit—are sensitive to many external factors.

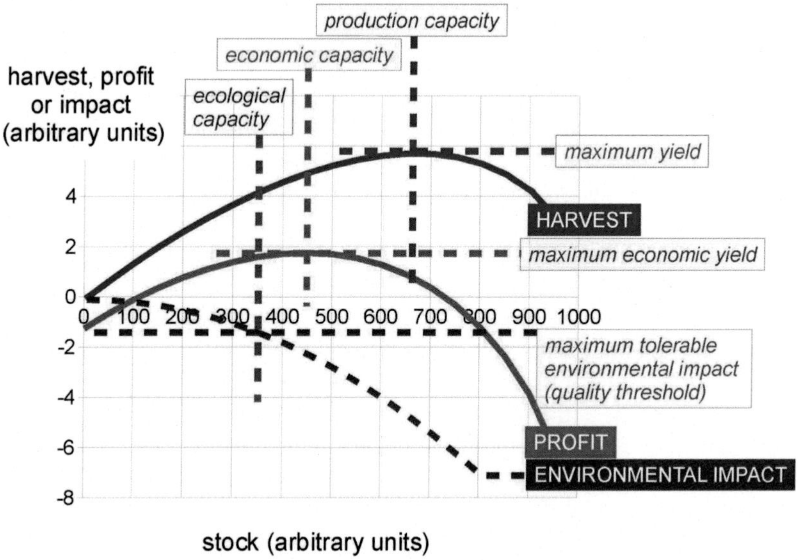

FIGURE 5.1 Several types of carrying capacity. *Ecological capacity* is set by a (possibly socially determined) threshold for maximum tolerable environmental impact. *Production capacity* is the maximum harvest obtainable at a given site. The diagram also introduces the idea of *economic capacity*, set by the maximum profit that can be extracted from a farm.

Finally, *ecological capacity* is determined by (1) the impacting activities of the farm, (2) the properties of the local ecosystem, and (3) the thresholds that are set. Item (1) is manageable by the farm. Item (3) is a complicated matter. In a permissive planning environment, the thresholds might be set by farm management, simply to ensure that environmental impact does not have knock-back effects on farm structure (e.g., through fouling), farmed organisms (e.g., through deoxygenation or enhanced blooms of harmful microalgae), or farm workers (e.g., through generation of toxic gases). In such a planning environment, wise managers—those with a view to sustainability or their insurance premiums—might wish to acquire much local scientific evidence.

Alternatively, farmers and society might rely on public regulators to set standards. This could be done "objectively"; for example, an approach based on ecosystem health (Tett et al., 2013) might tolerate any disturbance that did not impact ecosystem resilience. However, many current limits are set by social processes; for example, nutrient thresholds result from political decisions in international forums such as OSPAR (Painting et al., 2005).

SCALES

An *ecosystem* is "composed of the physical-chemical-biological processes active within a space-time unit of any magnitude" (Lindeman, 1942). However, processes tend to have dominant space-time scales, and by specifying these it is possible to identify three nested scales for mariculture (Figure 5.2).

In the case of fin-fish in floating net cages, the zone A scale is that of the seabed directly impacted by falling farm waste, and of the water immediately around the farm, where oxygen might be depleted and ammonia increased. Conceptually, zone B provides the boundary conditions for zone A—e.g., the background levels of ammonia and oxygen—and is also the scale on which nutrients may get converted into phytoplankters, because it is defined as having a water residence time on the order of days. Thus effects of sites can be cumulative.

In many cases, the zone B scale can be equated with a water body, as exemplified in the diagram, where the water bodies are, schematically, Scottish sea-lochs; residence times for upper waters are of the order of days, and boundary conditions are those at the fjord's mouth (those of the zone C scale).

Zone C is the scale of a regional sea, where water residence times are weeks or months. It is processes on this scale that set the ecological carrying capacity for a regional maricultural industry comprising many farms and sites.

SOCIALLY DETERMINED CARRYING CAPACITY

There are many examples of models that can be used to estimate ecological carrying capacity on particular scales (see review by McKindsey et al., 2006

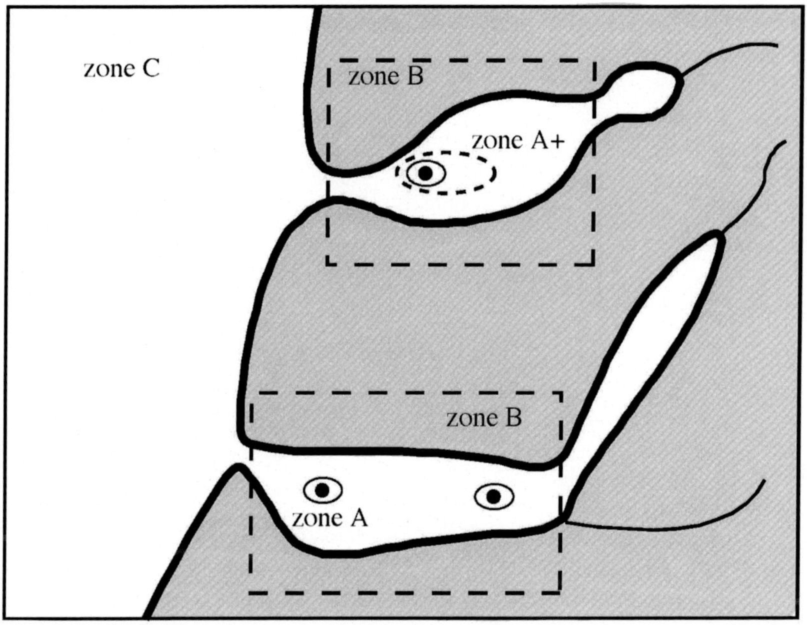

FIGURE 5.2 Scales for aquaculture (Tett, 2008). Zone A is the farm-site scale; it includes the part of the seabed that receives organic waste sinking from a farm and the part of the water column in which pollutants remain for a few hours. In tidally active waters, this water column zone is shown as A+. Zone B is the water body scale, relevant to most pelagic processes including eutrophication, where water remains for several days. Zone C is the regional scale, with residence times of weeks or months.

for shellfish farming, and Cromey et al., 2002 and Tett et al., 2011a concerning salmonids). However, *socially determined carrying capacity* is less well understood and modeled. Its complexity might best be understood in terms of *ecological, economic,* and *social licenses* (Figure 5.3). The diagram shows a local social-ecological system, centering on a maricultural activity, embedded within a larger system. In order to operate successfully and sustainably, the farm requires three sorts of license or permission:

1. **An ecological license** to ensure that the farm is not causing, and will not cause, ecological damage that degrades the system's resilience and natural functions.
2. **An economic license** to ensure that the farm can continue to operate profitably.
3. **A social license** to ensure that local people find the farm's operation to be acceptable, taking account of ecological impact, provision of employment, interests of other stakeholders, etc.

In Figure 5.3, the larger-scale system interacts with the local-scale system through ecological, economic, and social processes. Outside ecosystems provide the boundary conditions for the local ecosystem, and perhaps also the source of

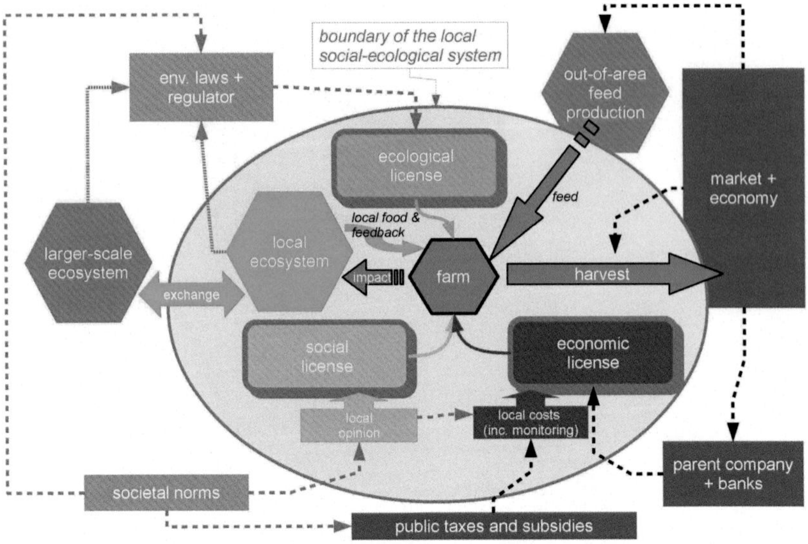

FIGURE 5.3 Issues related to social carrying capacity considered in terms of ecological, economic, and social licenses, for a local system embedded in a larger-scale social-ecological system.

fish feed. The larger-scale economy—shown here in terms of markets, owners, providers of capital, and taxes and subsidies—influences financial sustainability and, more generally, efficiency of utilization of the finite resources available to a society. Larger-scale society is shown impacting through both social norms and through formal governance. The latter may be multi-tiered, as in the case of EU Directives transposed into member state laws and implemented by regulating agencies.

REGULATION OF MARICULTURE IN SCOTLAND

The United Kingdom, with its parliament in London, fully devolves environmental and planning matters to the Scottish Government (SG), responsible to a parliament in Edinburgh since 1999. Any new development, such as a salmon farm, needs planning permission, granted by elected local authorities (which we will call county councils), who require an Environmental Impact Assessment (EIA). The Scottish Environment Protection Agency (SEPA) must issue a Controlled Activities Regulations (CAR) license to confirm that discharges from the farm will have no adverse environmental impact. The SG must grant a marine license to confirm that the farm's structure will not interfere with shipping. It will consult with the conservation agency, Scottish Natural Heritage, about impact on nature protection areas, and in future (Cook et al., 2014) to prevent import of alien species. A final complication is that the UK sea-bed is owned by the Crown Estate, an institution founded in the eighteenth century to nationalize royal land-holdings. A new farm must rent a site from this body.

Mussel farmers must also rent their sites from the Crown Estate, and get planning permission, but are otherwise less regulated. They do not require SEPA CAR licenses or SG marine licenses (so long as they avoid seaways). Small developments may not need an EIA. Whereas fin-fish regulation is characteristically restrictive, seeing farms as potentially polluting, governance is more supportive of shellfisheries. SEPA is charged with maintaining high water quality in shellfish growing areas recognized under an EU Directive. Shellfish can act as vectors of micro-algal toxins, harmless to invertebrates but poisonous to people. The UK Food Standards Agency (FSA) provides regular, publicly funded monitoring of these waters for harmful algae and their toxins, and forbids harvesting when there is a risk to human health.

DISCUSSION

Scottish regulation can be mapped to social and ecological licenses and has proven successful in ameliorating the sea-bed impact of salmon farms. Nevertheless, it suffers from several deficiencies. With one exception, it focuses on sites (zone A) rather than water-bodies (zone B). It is top-down and rule-bound. By this we mean that both SEPA and local authorities appear increasingly to work to a set of predefined criteria so as to minimize regulatory costs and the risk of legal challenges. Such procedures often frustrate citizens, and the criteria can become outdated. Finally, the EIA process, and getting permission, is slow, leading to calls for less regulation and consequent risks of exceeding the carrying capacity of sea-lochs.

The exception to the focus on sites is provided by the Locational Guidelines, SG advice to county councils and SEPA that categorizes sea-lochs according to their suitability for salmonid-farming. It does this on the basis of ecological carrying capacities estimated from seabed impact and nutrient enrichment (Gillibrand et al., 2002). However, the guidelines do not consider whole-ecosystem interactions, or synergies such as the use of cultivated mussels to remove phytoplankton generated by fish-farm waste nutrients.

Figure 5.3 suggests that the factors setting carrying capacities for mariculture are both complex and dynamic. Even if ecosystems themselves remain unchanged, social and economic change might influence both ecological quality thresholds and profitability. Whereas farm managers will have their eyes on profitability, and public regulators on environmental quality, local society (and the idea of social license) embodies a wider range of interests and thus, collectively, a more holistic view.

Ostrom (2009) argues that "common-pool" resources, such as environmental capacity for waste assimilation and food provision, are best governed locally within a larger-scale framework. What we propose is based on Ostrom's argument and the Systems Approach Framework developed during the EU research project, SPICOSA (Hopkins et al., 2012; Tett et al., 2011b). First, decisions about capacity ought, we think, to be reached locally, in local government-supported stakeholder forums dealing with whole water bodies (e.g., entire lochs). The forums should adopt the ecosystem approach (Atkins et al., 2011) and should seek synergies, such as those between mussel and salmon cultivation. Second, decisions should be flexible, to reflect changing circumstances, but should not impact on long-term sustainability; thus, forums should periodically review conclusions about capacity in light of experience. Third, scientists, regulators, and managers should use best-available ecological and economic models to provide the evidential basis on which stakeholders can reach decisions about capacity.

Scottish institutions of local government and environment protection are capable of supporting such an approach, which would, however, require changes in both policy (see Shipman and Stojanovic, 2007) and law. The UK is moving to devolve the Crown Estate in Scotland, at least to the SG and perhaps to county councils in the main maricultural regions. Ownership of the coastal zone at this level could provide a sound basis for local management.

Norway has devolved control of fish-farming to county councils. Sandersen and Kvalvik (2014) conclude that this has been good for integrated coastal zone management, but may have reduced capacity for larger-scale ecosystem-based management. This, it seems to us, is an argument for matching multilevel governance to spatial scale. For example, the SG could (in our model) retain control of regional-scale (zone C) nutrient loadings, allocating or selling quotas to water-body forums.

Finally, the requirement for the three licenses, acquired in ways appropriate to local coastal zone geography, cultures, and governance, could provide a framework for the sustainable development of mariculture in the high-latitude RREs that will be exposed as glaciers retreat.

REFERENCES

Atkins, J.P., Burdon, D., Elliott, M., Gregory, A.J., 2011. Management of the marine environment: integrating ecosystem services and societal benefits with the DPSIR framework in a systems approach. Mar. Pollut. Bull. 62, 215–226.

Cook, E.J., Payne, R.D., Macleod, A., 2014. Marine biosecurity planning: identification of best practice: a review. Scott. Nat. Heritage Comm. Rep. 748, 45.

Cromey, C.J., Nickell, T.D., et al., 2002. Validation of a fish farm waste resuspension model by use of a particulate tracer discharged from a point source in a coastal environment. Estuaries 25, 916–929.

Duarte, C.M., Holmer, M., et al., 2009. Will the oceans help feed humanity? BioScience 59, 967–976.

Gillibrand, P.A., Gubbins, M.J., Greathead, C., Davies, I.M., 2002. Scottish Executive Locational Guidelines for Fish Farming: Predicted Levels of Nutrient Enhancement and Benthic Impact. Report, 52 pp. Fisheries Research Services, Aberdeen.

Hopkins, T.S., Bailly, D., et al., 2012. A systems approach framework for the transition to sustainable development: potential value based on coastal experiments. Ecol. Soc. 17 (3) Art. 39.

Lindeman, R.A., 1942. The trophic-dynamic aspect of ecology. Ecology 23, 399–417.

McKindsey, C.W., Thetmeyer, H., Landry, T., Silvert, W., 2006. Review of recent carrying capacity models for bivalve culture and recommendations for research and management. Aquaculture 261, 451–462.

Ostrom, E., 2009. Beyond markets and states: polycentric governance of complex economic systems. In: Grandin, K. (Ed.), The Nobel Prizes 2009. Nobel Foundation, Stockholm.

Painting, S.J., Devlin, M.J., et al., 2005. Assessing the suitability of OSPAR EcoQOs for eutrophication vs ICES criteria for England and Wales. Mar. Pollut. Bull. 50, 1569–1584.

Sandersen, H.T., Kvalvik, I., 2014. Sustainable governance of Norwegian aquaculture and the administrative reform: dilemmas and challenges. Coastal Manage 42, 447–463.

Shipman, B., Stojanovic, T., 2007. Facts, fictions, and failures of integrated coastal zone management in Europe. Coastal Manage 35, 375–398.

Tett, P., 2008. Fishfarm wastes in the ecosystem. In: Holmer, M., Black, K., et al. (Eds.), Aquaculture in the Ecosystem. Springer, Netherlands, pp. 1–46.

Tett, P., Gilpin, L., et al., 2003. Eutrophication and some European waters of restricted exchange. Cont. Shelf Res. 23, 1635–1671.

Tett, P., Gowen, R., et al., 2013. A framework for understanding marine ecosystem health. Mar. Ecol. Prog. Ser. 494, 1–27.

Tett, P., Portilla, E., Gillibrand, P.A., Inall, M., 2011a. Carrying and assimilative capacities: the ACExR-LESV model for sea-loch aquaculture. Aquacult. Res. 42, 51–67.

Tett, P., Sandberg, A., Mette, A. (Eds.), 2011b. Sustaining Coastal Zone Systems. Dunedin Academic Press, Edinburgh.

Making the Link

By Omer Chouinard

Sustainable Mariculture at High Latitudes	⟺	**Coastal Governance Solutions Development in Latvia: Collaboration Communication and Indicator Systems**
Paul Tett, Kenny Black, Ruth Brennan, Elizabeth Cook and Keith Davidson		*Raimonds Ernsteins, Janis Kaulins, Ilga Zilniece and Anita Lontone*

Paul Tett et al. demonstrate that Scottish regulations can be mapped to social and ecological licenses and have proven successful in ameliorating the sea-bed impact of salmon farms. Tett et al. provide examples of models used to estimate ecological carrying capacity on particular scales.

However, they underline that, of the models discussed, socially determined carrying capacity is less well understood and modeled. Its complexity might best be explained in terms of ecological, economic, and social licenses. Additionally, the authors mention that it has several deficiencies. It is top-down and rule-bound. By this they mean both the Scottish Environment Protection Agency and local authorities appear increasingly to work to a set of predefined criteria so as to minimize regulatory costs and the risk of legal challenges. Such procedures often frustrate citizens, but with adjustments to better account for local particularities, the three-license system may offer a framework for sustainable mariculture.

The contribution of Raimonds Ernsteins et al. focuses on the application of the environmental collaboration communication model in three coastal municipalities in Latvia in order to assess and advise on the situational development of coastal climate risk communication specifications. The authors emphasize that coastal communication has improved during recent years, but difficulties with information, perceptions, and gaps with related behavior should be studied further.

Innovative communication approaches have to be designed by involving all complementary communication instruments and all target groups, especially mediators. They insist that besides general and regular coastal governance and practice management issues, more and more such interactions as floods, erosion, and natural and social impacts on local coastal municipalities are becoming serious concerns. From their point of view, the most significant flood risk factors, particularly in the context of climate change and sea level rise, are wind surges caused by strong winds or storms as well as rapid, lasting rainfalls.

Both chapters provide examples of possible models for two different yet critical components of coastal zone management, with Tett et al. working to ensure the sustainability of fish farming through a tripartite licensing system, and Ernsteins et al. striving to improve coastal risk communication to increase resilience in coastal communities.

Chapter 6

Coastal Governance Solutions Development in Latvia: Collaboration Communication and Indicator Systems

Raimonds Ernsteins, Janis Kaulins, Ilga Zilniece, Anita Lontone
UNESCO Chair in Sustainable Coastal Development, University of Latvia, Riga, Latvia

Chapter Outline

INTRODUCTION AND BACKGROUND

Coastal governance processes are still in development in Latvia at the national and local levels. At the local level, 16 coastal municipal administrative territories stretch along the Latvian coast of the Baltic Sea for 500 km of mainly sandy beach and forested coastal environment. Recent national coastal governance developments, concluding with the National Coastal Spatial Development Guidelines of 2011, required more active and advanced approaches. These were initiated by various stakeholders, especially local/national NGOs, associations of coastal municipalities, and obviously researchers/academics from different research fields. However, coastal sustainable development governance (SDG) has still not been recognized as everyday practice. Our coastal research evaluations have identified two main problem areas in governance of coastal socio-ecological systems: (1) **lack of coastal integrity understanding** and (2) **lack of collaboration**, both within and between stakeholder groups (Ernšteins, 2008; Ernšteins et al., 2012a, 2014).

These problems must be taken into account when strengthening existing governance instruments as well as developing new ones, especially those aimed at improving coastal knowledge, skills, attitudes, motivation, and behavior for more pro-environmental activities and co-labor developments (Daniels and Walker, 2001; Michelsen and Godemann, 2007; Ernšteins, 1999). The need is to gain stakeholders understanding toward enhancement of **coastal communication** in the widest possible meaning of this term, and to learn more adequately from existing experiences of known environmental communication models and approaches (Cox, 2010; Corbett, 2006; Daniels and Walker, 2001; Day and Monroe, 2000)

Coastal communication instruments have been improved during recent years and can be used as effective tools at both the national and local levels, but gaps between information, perceptions, and behavior certainly do exist and should be studied further (Ernšteins, 2006a, 2010; Cox, 2010). An innovative communication process has to be designed by involving **all complementary communication instruments and all target groups**. This is no longer to be seen as a simple matter of improving coastal information dissemination, better facilitation of stakeholder participation, or more interactive work in formal or informal education, etc., but as maximizing **complex, integrative, and complementary approaches** used in proactively applying diverse and wide communication instruments concerning information, education, participation, and pro-environmental behavior (Ernšteins, 1999, 2006b, 2010).

In addition to general and regular coastal governance and practice management issues, increasing problems are caused by floods and erosion in coastal municipalities. The most significant flood risk factors are those arising from climate change and sea-level rise. Household/stakeholder views of these changing risks have to be taken into account. All of these and related coastal issues have been studied in Latvia via the development of collaborative communication approaches and instruments, taking as an example the environmental collaboration communication model (Ernšteins, 1999, 2006b) accepted also by the Environmental Ministry as a guideline document (MEPRD, 2001) and later in 2004 and 2014 included in national environmental policy plans.

This research into coastal governance and communication included the development and application of the coastal-climate-risk-communication systems and coastal sustainability indicator systems (ISs). The latter is one of the most frequently missing aspects of coastal communication. **Case study research (CSR)** integrative methodologies were developed for particular research-and-development (R&D) projects between coastal municipalities and the university (Ernšteins, 2006a). These projects were collaborative as far as possible in particular local circumstances. The methodologies included traditional questionnaire surveys (inhabitants/households) and structured in-depth interviews with municipal leaders/planners and main stakeholders, as well as focus group interviews and analyses of documents. Coastal communication and governance proposals were afterward elaborated and delivered to municipal leaders and stakeholders, and tested where it was possible at least with roundtable discussions and/or focus group interviews.

COLLABORATION COMMUNICATION MODEL FOR COASTAL DEVELOPMENT

The CSR studies in Latvian coastal municipalities were planned around the environmental collaboration communication model. We evaluated its voluntary and successful realization in some Latvian coastal municipalities during the last decade (Ernšteins, 2008; Ernšteins et al., 2010), in order to develop it further and also to adjust for coastal-climate-risk communication between all stakeholders (Figure 6.1).

The developed model, with its four complementary components—**information, education, participation, and environmentally friendly behavior**—is shown in Figure 6.2. This shows the links between coastal communication tasks as a cycle (Ernšteins, 2008). Within the particular cycle oriented to an environmental/coastal issue case, this ensures the development of practical case-based coastal awareness components. Within a multicycle integration it leads to the development of motivated self-experience and personal practice, and so enhances general coastal sustainability awareness.

The results of these exercises have been measured as changes in knowledge and practical skills, understanding and ability to solve problems, attitudes (especially concerning self-regulation), motivation and readiness for concrete action, and particularly self-experience, for case-related target groups as well as individuals.

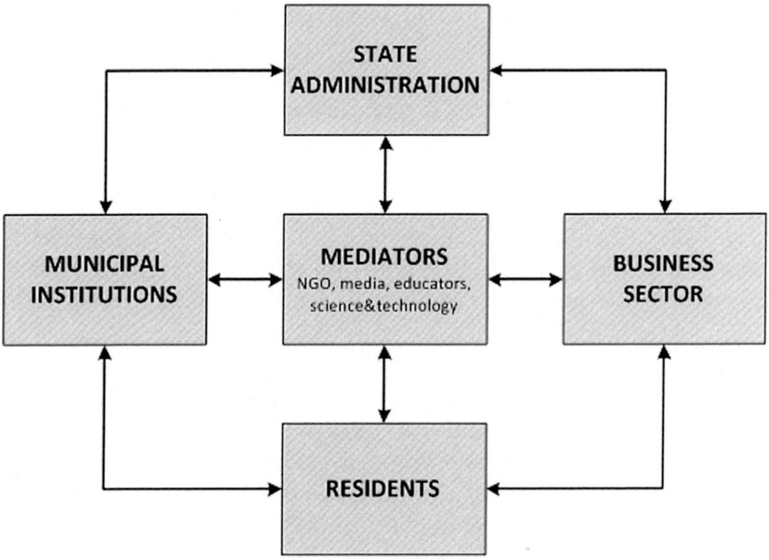

FIGURE 6.1 Key target groups for coastal sustainable governance (Ernšteins, 1999).

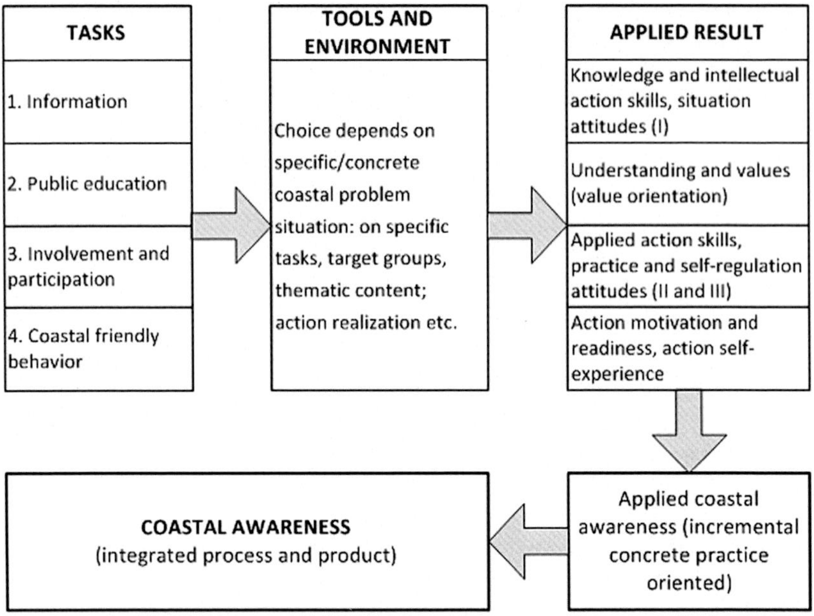

FIGURE 6.2 Incremental coastal communication process: four-step cyclical model. *Adapted from Ernšteins (1999, 2006b).*

FOUR-PART COASTAL COMMUNICATION: CLIMATE CHANGE AND RISK APPLICATIONS

CSR using this model was done at three coastal municipalities in Latvia (Saulkrasti, Salacgrīva, and Ventspils), within a framework of coastal governance processes and stakeholder collaboration. The studies differed in the main topic of coastal communication. Communication was evaluated in terms of its educational, informative, participatory, and environmentally friendly action aspects, and the type and extent to which all target groups, in each area of potential coastal threats, were involved.

The CSR in Saulkrasti municipality was focused on **rural territories' coastal risk communication** (University of Latvia, 2010). The first element in coastal risk communication involves the identification of existing natural and technological risk. Next, the possible risk-prone geographical areas have to be analyzed, and the public has to be informed of the potential impacts of the natural risks. This information step in risk communication is closely linked with the education step, since an area's residents must receive adequate education/training about natural environmental risks, along with instructions in case of emergencies. To this end, a mechanism for municipal risk education and coordination has to be established, including education through the news media. The next steps in ensuring effective coastal risk communication are participation and action/behavior, directed toward coastal risk prevention, response to emergency situations, and emergency recovery operations. All of this needs to be incorporated into planning and other policy documents, and the information and plans need to be made systemically, and made available to the public.

To successfully organize, supervise, and complete the coastal risk-communication process, the following were necessary preconditions:

- An understanding by municipal leadership and the general public of the importance of identifying, analyzing, and communicating environmental risks.
- A change in public behavior toward more active participation in environmental risk identification and communication processes.
- Coordinated and effective communication and collaboration among all involved actors and stakeholders.

The CSR in Salacgriva municipality was directed toward **climate change adaptation and communication issues** (University of Latvia, 2011; Ernšteins et al., 2012b). At the beginning it was ascertained that Latvian coastal municipalities do not use their internal potential for communication to achieve balance among economic, social, and environmental developments. Although many municipalities have set sustainable development as their objective, individual developments often take place in isolation (Dolan, 2006; Ernšteins et al., 2012a). Municipalities that have successfully implemented sustainable management practices do not use the opportunity to promote their experience, which would foster positive image-building. Municipalities underestimate the need for, and potential benefits of, promoting the participation in the decision-making process of all interest groups (Ernšteins et al., 2014).

To ensure climate change adaptation (CCA) in coastal municipalities, certain communication procedures have to be established based on several approaches. The organizational communication approach takes place within a municipality and is aimed at achieving conceptual agreement on integrating the CCA approach into development planning, everyday work, and communication and cooperation practice. The interest group communication approach describes a process in which a municipality promotes the development of a communication and collaboration network for the local interest groups so as to strengthen the municipal internal collaboration potential for the promotion of CCA. The self-experience communication approach is based on local interest groups and interested individuals who are involved in the CCA development. It encourages them to compile and disseminate accounts of their experiences of environmentally friend-behavior. The complementary communication approaches that were explored in Salacgriva and recommended for CCA communication development included:

- Climate communication as science communication development.
- CCA communication integration into all municipal sector plannings.
- Application of environmental marketing communication in CCA.
- Self-organization and participation development of mediator groups.

- CCA instruments communication (including information, education, participation, and development of pro-environmental behavior) as a CCA sectorial application.

The CSR in the Ventspils municipality was based on the communication of **coastal town and industrial risks** (Kalniņa et al., 2013). Ventspils is a coastal town with several high-risk companies that are located within its port and involved in cargo shipment operations. The port is connected with the railway network that runs through densely populated areas of the town. The maritime climate is part of everyday life for coastal residents, and natural coastal risks are not sufficiently assessed. Insufficient public information on environmental risks and action planning for households, were the main reasons for the elaboration of environmental risk communication. Among the reasons for Ventspils residents' inactivity are the overreliance on local government activities and lack of awareness of their own responsibilities. Good collaboration has been established between the main target groups and the local government, but the cooperation potential with the mediation sector is underused. The target groups differ in their opinions in relation to the responsibilities of environmental risk management and communication.

As a result of the cooperation project, environmental risk communication should be integrated into special risk management as well as into the general environmental and development planning process and the documentation of the local government. All appropriate communication means and channels and their potential should be fully applied to provide open access to environmental risk-related public information, combined with diverse public education and involvement, which will provide an environmental-friendly, risk-reducing, and adequate emergency communication action plan. To support the implementation of communication of coastal risks, three main policy areas were identified:

- Collaboration between all involved target groups, ensuring environmental risk communication, as shown in Figure 6.1.
- Application of all four collaboration communication components, as shown in Figure 6.2.
- Use of the full range of the governance instruments for environmental risk communication.

INDICATOR SYSTEMS FOR COASTAL COMMUNICATION AND GOVERNANCE

Coastal ISs need to be developed as instruments for coastal communication and governance. The theoretical basis for IS design and building for coastal sustainable development governance (IS SDG) has been insufficiently developed, as well as insufficiently linked with the practice of integrated governance for coastal areas/municipalities, in Latvia. Consequently, we carried out research into SDG indicators (Kauliņš et al., 2011) and models of IS and their development in the municipal development planning process (Ernšteins et al., 2009; Sano and Medina, 2012). The relationship of indicator selection with indicator integration among sustainable development dimensions and governance levels was established, and was formulated as **horizontal and vertical integration** of IS. The results of the research were tested by implementing IS SDG in the municipal governance of Saulkrasti (Ernšteins et al., 2011). Recommendations were produced for the implementation of IS at different levels of coastal sustainability governance (Kaulins, 2015).

Models described previously for IS formation were mainly based upon functional analysis of the target system (Bossel, 1999) or a problem analysis of the territory (Rydin et al., 2003). They were seldom directly linked with the governance process of the territory and thus with the content of particular development planning documents that are important in the making of legal governance decisions (Ghosh et al., 2006; Moreno-Pires and Fidélis, 2012). Our model for the place of indicators for SDG in municipal integrated planning is shown in Figure 6.3.

In relation to sustainability dimensions, there are different types of issues to evaluate. Indicators can be divided into four groups:

1. **Subsectoral** indicators, which describe a certain course of action, related to a certain sector within one sustainability dimension.
2. **Sectoral** (or one-dimensional) indicators, which describe a course of action or its group, related to the whole sustainability dimension.

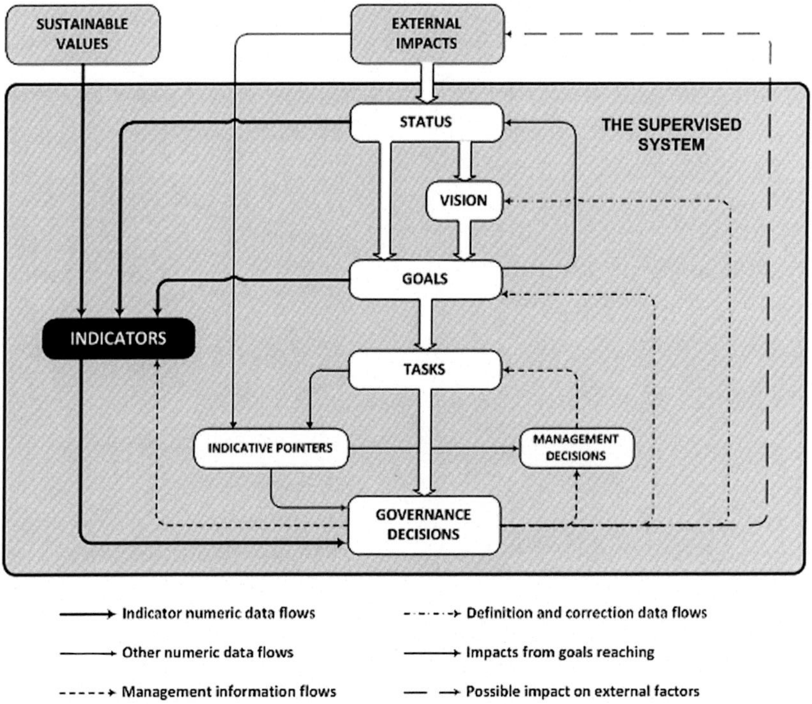

FIGURE 6.3 The place of indicators for sustainable development governance in the municipal integrated development planning model (Kaulins, 2015).

3. **Integrative** indicators, which describe courses of action related to integrative problem areas, i.e., including at least two sustainability dimensions.
4. **Integral** also **strategic** indicators, which describe the main and the general indicators of the governed system that characterize the given governance system as a whole and/or in comparison with similar systems.

Indicator selection is associated with choices of variables. Such choices can be formalized algorithmically, as shown in Figure 6.4.

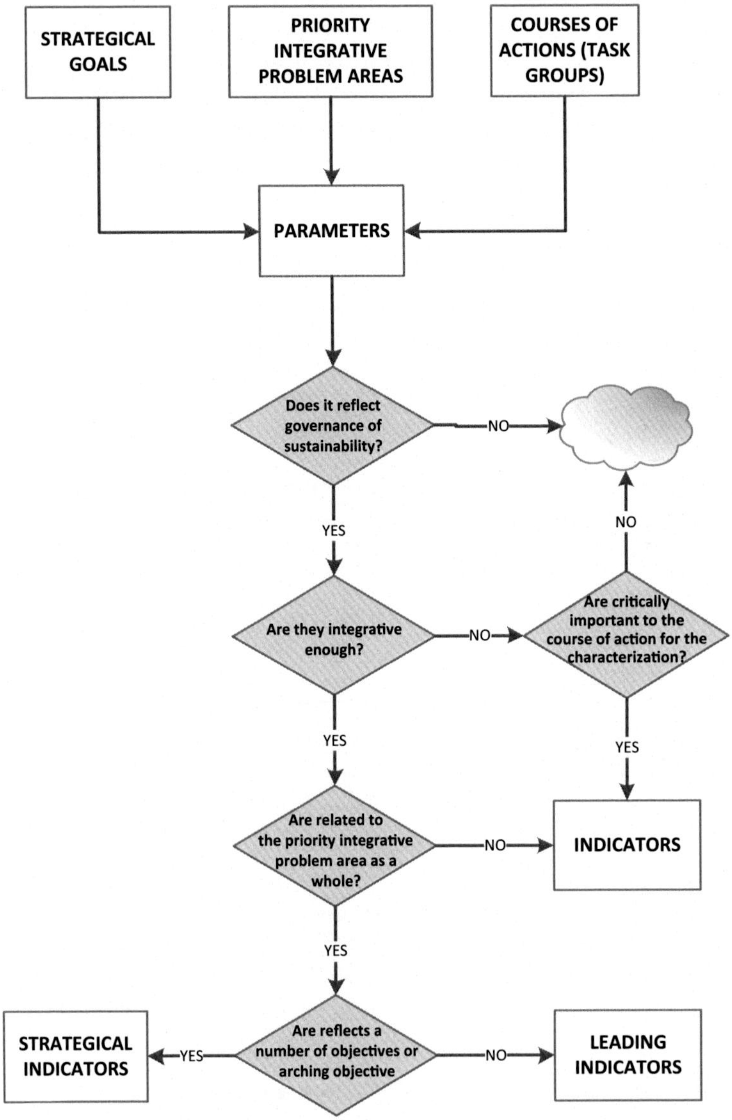

FIGURE 6.4 Selection of indicators for the development planning process (Kauliņš et al., 2011).

As a result of this research (Kauliņš et al., 2011, 2015; Ernšteins et al., 2011), a functioning system of indicators for the governance of sustainable development was designed, produced, and implemented as the direct supervision instrument for the governance and strategy of municipalities. This took place for the first time in Latvia in 2013. The IS has been incorporated into the main local development planning document: the municipal sustainable development strategy of the Saulkrasti coastal municipality (Saulkrasti, 2013a). This document recently became a mandatory requirement by the national government, although the legal requirement does not specify an IS. This voluntarily designed and formally approved IS now provides the entire communication basis of the municipality's strategies supervision and assessment mechanism. This IS implementation system comprises the following (Saulkrasti, 2013b):

1. A structured list of indicators, grouped by sustainability dimensions and integrative problem areas (67 indicators in all); see Table 6.1.
2. Training of the municipality's personnel in use of the IS, as data sources and responsibility for calculation were determined for each indicator.
3. Approval of the IS (after public consultations) by the resolution of the local administration, as a part of the mandatory sustainable development strategy of the municipality.
4. Production of a users' manual for the IS and indicators for sustainable governance monitoring, in which instruction about result processing and reporting for the needs of decision-makers and other public target groups was also provided.

During 2014–2015, the first full-scale assessment of the governance of sustainable development is taking place. The results will be included in the annual public report of activities of the municipality, and publicized separately as a municipal sustainability report. Following this implementation of an IS SDG, we were able to assess it from several thematic and territorial viewpoints. This also resulted in IS thematic elaboration. The viewpoints were as follows:

1. For CCA strategy for the Salacgrīva coastal municipality,
2. In a guideline document for the governance of a healthy and environmentally friendly food life-cycle for the Salacgrīva coastal municipality.
3. For coastal risk communication planning for the Saulkrasti coastal municipality.

The IS proved to be interesting for various main stakeholders as well as municipal administrations. Once they have obtained official approval they are to be integrated within the municipal sustainable development governance system. A draft version of such an integrated IS system has been designed and prepared for Salacgriva and is now been considered by the municipal council.

TABLE 6.1 Saulkrasti Coastal Municipal Indicator System: Content List of Indicators Grouped by Sustainability Dimensions and Thematic Groups (Saulkrasti)

Natural Environment (D)	Economic Environment (E)
D1. Status of the "green web" (3a) D2. Real and potential loads on environment by public utilities (5) D3. Air quality and impact on climate change (4) D4. Surface water quality, coastal zone (2) D5. Land use for development (2) D6. Natural risks (2) D7. Common indications (1)	E1. Economically active people (1) E2. Budget of municipality (2) E3. Entrepreneurship and employers (3) E4. Traffic routes (4) E5. Local port development (3) E6. Tourism characteristics (6)
Social Environment (S)	**Governance and Communication (P)**
S1. Health care characteristics (2) S2. Support for the cultural environment (2) S3. Social care and social security (3) S4. Education system (3) S5. Household life quality (4)	P1. Municipal governance (3) P2. Municipal information and communication (3) P3. Activities in the nongovernmental sector (2)
Integral (Strategic) Indicators (I)	
I1. Number of inhabitants (1) I2. Complex indexes (2)	

[a]*Number of separate indicators developed for the each thematic group mentioned here.*

COASTAL OBSERVATORY NETWORK: COASTAL COMMUNICATION SYSTEM DEVELOPMENT

The seacoast of Latvia forms part of the territory of common interest of the Baltic Sea countries, but has its own peculiarities. Our suggestion is to establish an international communication system for the observation of coastal sustainability, subdivided into national components—observatories of coastal sustainability. The system could be established based upon a jointly designed and agreed-upon list of coastal indicators in the Baltic region countries, adapted to the conditions of specific territories, considering the peculiarities of the coast and the information acquisition at each governance level in each Baltic region state. The system's operations should make use of applied and academic research and be usable for information preparation for decision-making systems aiming at HELCOM and other regional agencies, institutions, and networks, and at the national and local governance levels. This is a proposal elaborated upon by coastal researchers of universities in Sweden, Finland, Estonia, and Latvia in partnership with various types/levels of coastal authorities. It will, ideally, become a consistent and reliable source for public communication about and for the sustainability of the coastal socioecological systems of the Central Baltic region.

ACKNOWLEDGMENTS

R&D data were collected and elaborated upon, and the paper prepared, within the framework of the Latvian National Research Program Project on Environmental Diversity and Sustainable Governance (SUSTINNO, 2014–2017). We acknowledge the part played in the realization of the previous CSR projects mentioned, by university contributors—Liga Zvirbule, Valdis Antons, Diana Sulga, Sintija Kursinska, and Daiga Stelmahere—as well as master's-degree students Solvita Muceniece, Selina Abelniece, Anete Sturma, Madara Kalnina, Liga Vasarina, Olga Kocmarjova, and Mara Lubuze.

REFERENCES

Bossel, H. (Ed.), 1999. Indicators for Sustainable Development: Theory, Method, Applications. A Report to the Balaton Group. IISD, p. 138.

Corbett, J., 2006. Communicating Nature: How We Create and Understand Environmental Messages. Island Press, Washington.

Cox, R., 2010. Environmental Communication and the Public Sphere. Thousand Oaks, SAGE Publ., p. 455.

Daniels, S.E., Walker, G.B., 2001. Working through Environmental Conflict: The Collaborative Learning Approach. Praeger, Westport.

Day, B.A., Monroe, M.C., 2000. Environmental Education & Communication for a Sustainable World: Handbook for International Practitioners. http://www.envcomm.org/Publication.htm.

Dolan, H., 2006. Climate Change, Coastal Communities & Health//Presentation for the Coastal Community Health Summit "Coastal Community Health: A Social Crisis. 20 pp.

Ernšteins, R., 1999. Environmental science education development. In: Latvian. (Videszinātniskās Izglītības Attīstība). Monography, Publisher "Vide", Rīga, p. 342.

Ernšteins, R., 2006a. Partnerships between municipalities and universities as means to promote regional and local sustainable development in Latvia. In: Adomssent, M. (Ed.), Higher Education for Sustainability. Germany VAS Publishing, Frankfurt, pp. 245–250.

Ernšteins, R., 2006b. Local agenda 21 process facilitation: environmental communication and self-experience development in Latvia. In: Filho, L.W., Ubelis, A., Berzina, D. (Eds.), Sustainable Development in the Baltic and Beyond, Frankfurt, Peter Lang Europaischer Verlag der Wissenschaften, pp. 305–318.

Ernšteins, R., 2008. Sustainable coastal development in Latvia: collaboration communication and governance imperative. Book. In: Ernšteins, R., Jurmalietis, R. (Eds.), Sustainable Coastal Development: Colaboration Governance. University of Latvia Academic Publishing, Rīga, pp. 159–178.

Ernšteins, R., 2010. Sustainable coastal development and management: collaboration communication and governance. Human resources. Lithuania, Klaipeda University, Klaipeda J. Soc. Sci. 3, 247–252.

Ernšteins, R., Kauliņš, J., Kudreņickis, I., 2009. Sustainable coastal development indicator system studies in Latvia. 3rd International Conference on Sustainable Development and Planning, Sustainable Development and Planning IV, WIT Transactions to Ecology and the Environment, vol. 120, WIT Press, pp. 653–664.

Ernšteins, R., Kuršinska, S., Štāls, A., Zīlniece, I., Rudzīte – Griķe, M., Lagzdiņa, Ē., 2010. Environmental management integration into municipal development process: collaboration communication imperative. In: Proceedings,Conference. University of Applied Sciences, Valmiera, pp. 53–63.

Ernšteins, R., Kauliņš, J., Līce, E., Štāls, A., 2011. Integrated coastal management for local municipalities in Latvia: sustainability governance and indicator system. WIT Trans. Built Environ. 149, 29–40.

Ernšteins, R., Antons, V., Stals, A., Lubuze, M., Šulga, D., Kursinska, S., Lice, E., 2012a. Towards complementary municipal and social resilience understanding: stakeholder training on coastal sustainability governance and communication. In: 12th International Multidisciplinary Scientific Geoconference SGEM 2012', Proceedings, Bulgaria, Albena, pp. 1007–1014.

Ernšteins, R., Lontone, A., Zvirbule, L., Antons, V., Zīlniece, I., Kauliņš, J., Vasariņa, L., 2012b. Climate change adaptation integration into coastal municipal development: governance environment and communication preconditions. In: 12th International Multidisciplinary Scientific Geoconference SGEM 2012, Proceedings, Bulgaria, Albena, pp. 1077–1084.

Ernšteins, R., Lontone-Ieviņa, A., Kauliņš, J., Strazdiņš, J., Šteinberga, Z., Kudreņickis, I., Zīlniece, I., Ķepals, A., 2014. Municipal climate change adaptation governance in Latvia: cross-sectoral and multi-instrumental understanding. Reg. Form. Dev. Stud. 3 (14), 40–52.

Ghosh, S., Vale, R., Vale, B., 2006. Indications from sustainability indicators. J. Urban Des. 11 (No. 2), 263–275.

Kalniņa, M., Zīlniece, I., Ernšteins, R., 2013. Environmental risk communication in coastal municipality: Ventspils case (in Latvian). In: Proceedings, 15th International Research Conference. Liepaja University, Liepaja, Latvia, pp. 427–435.

Kaulins, J., 2015. Indicators for sustainable development governance. Summary of Doctoral thesis in Environmental science (sub-branch - Environmental Governance), University of Latvia, Riga, 36 p.

Kauliņš, J., Ernšteins, R., Kudreņickis, I., 2011. Sustainable development indicators for integrated coastal management: definition area and spatial properties. WIT Trans. Ecol. Environ. 144, 299–311.

Michelsen, G., Godemann, J. (Eds.), 2007. Handbuch Nachhaltigkeits kommunikation: Grundlagen und Praxis. Oekom, Muenchen, p. 940.

Ministry of Environmental Protection and Regional Development (MEPRD), 2001. National Environmental Communication and Education Strategy. Republic of Latvia, Riga.

Moreno-Pires, S., Fidélis, T., 2012. A proposal to explore the role of sustainability indicators in local governance contexts: the case of Palmela, Portugal. Ecol. Indic. 23, 608–615.

Rydin, Y., Holman, N., Wolff, E., 2003. Local sustainability indicators. Local Environ. 8 (No. 6), 581–589.

Sano, M., Medina, R., 2012. A systems approach to identify sets of indicators: applications to coastal management. Ecol. Indic. 23, 588–596.

Saulkrasti municipality: strategy of sustainable development for 25-year perspective, 2013a. In: Latvian. KBLC Ltd., 2013, pp. 90. (Saulkrastu novada ilgtspējīgas attīstības stratēģija 25 gadu perspektīvā. SIA KBLC).

Saulkrasti municipality: strategy of sustainable development for 25-year perspective, 2013b. In: Latvian. Manual for Using of Sustainable Development Governance Indicator System, KBLC Ltd., pp.160. (Saulkrastu novada ilgtspējīgas attīstības stratēģija 25 gadu perspektīvā. Attīstības un ilgtspējības monitoringa indikatoru sistēmas lietošanas rokasgrāmata.).

University of Latvia, 2010. Coastal communication policy planning for coastal municipalities: Saulkrasti municipality (in Latvian). In: Ernsteins, R. (Ed.), University-municipality Research-development Project, Riga, p. 298.

University of Latvia, 2011. Climate change adaptation policy planning for coastal municipalities: Salasgriva municipality (in Latvian). In: Ernsteins, R. (Ed.), University-municipality Research-development Project, Riga, p. 248.

Making the Link

By Paul Tett

Coastal Governance Solutions Development in Latvia: Collaboration Communication and Indicator Systems	⟺	Geoengineering Coastlines? From Accidental to Intentional
Raimonds Ernsteins, Janis Kaulins, Ilga Zilniece and Anita Lontone		*Martin D. Smith, A. Brad Murray, Sathya Gopalakrishnan, Andrew G. Keeler, Craig E. Landry, Dylan McNamara and Laura J. Moore*

Humans are attracted to the coastal zone, but satisfaction from life here has to be balanced against special risks. Historically these have often been from other people, such as the pirates and slave-traders raiding Mediterranean coasts during lawless periods. But always, if usually in the background of coastal dwellers' minds, have been threats from natural phenomena. These include shock events, such as storm surges and tsunamis, and long-term pressures, such as rising sea levels associated with natural and anthropogenic planetary warming.

These two chapters, and others (e.g., Chouinard et al. in Chapter 2), are concerned with the resilience of the coastal socioecological system to such events and processes. The chapter that you have just read (Ernsteins et al.) considered how to improve communication and planning in coastal governance. The next chapter (Smith et al.) deals with physical aspects of coastlines in relation to these threats. Left alone, coasts adjust to the erosion and deposition of material, and so can respond to sea-level rise by moving inland. Humans interfere locally with this natural response, by coastal protection or beach nourishment. However, protection at one point can cause increased erosion elsewhere. Thus there is a need for governance on a scale appropriate to coastal dynamics, including the large scale implied by the word "geoengineering."

Geoengineering is a word that has had a bad rap, because it has been used to argue that humanity need not worry about burning fossil fuels if the thermal consequences of the resulting increase in greenhouse gases can be nullified—for example, by injecting sunlight-scattering particles into the stratosphere. Smith et al. argue that we already geoengineer coastlines: we should simply think about how to do this in a joined-up fashion.

Taken together, the two chapters should also prompt reflection on scales for management of people and natural processes in the coastal zone, and on the coupling of such scales: that is, on the word "Integrated" in the phrase "Integrated Coastal Zone Management."

Chapter 7

Geoengineering Coastlines? From Accidental to Intentional

Martin D. Smith[1], A. Brad Murray[1], Sathya Gopalakrishnan[2], Andrew G. Keeler[3], Craig E. Landry[4], Dylan McNamara[5], Laura J. Moore[6]

[1]Nicholas School of the Environment, Duke University, Durham, NC, USA; [2]Department of Agricultural, Environmental and Development Economics, The Ohio State University, Columbus, OH, USA; [3]University of North Carolina Coastal Studies Institute, Wanchese, NC, USA; [4]Department of Agricultural & Applied Economics, University of Georgia, Athens, GA, USA; [5]Department of Physics & Physical Oceanography, University of North Carolina, Wilmington, NC, USA; [6]Department of Geological Sciences, University of North Carolina, Chapel Hill, NC, USA

Chapter Outline

INTRODUCTION

Are humans geoengineering whole coastlines? This may sound like a ridiculous question at first blush because human interventions along the coast are often local, and decisions about them are decentralized. Yet studies of coastal geomorphology increasingly demonstrate that local coastline features can have nonlocal effects that propagate over large spatial scales. The economics of coastal management shows that local economic and political conditions influence decisions to stabilize the coast, and these decisions, in turn, feed back on coastal economies through impacts on the physical system. These physical impacts produce spatial externalities that can extend large distances along the coast. The tight coupling of physical processes and economics combined with the large scale of coastal geomorphology implies that local coastal management decisions in one community can affect not only adjacent communities but also ones far away. When a coastline is viewed as a spatially extended, coupled human–natural system, it is clear that humans are in fact engineering whole coastlines, albeit haphazardly. Like it or not, we are practicing geoengineering without an explicit intent to achieve particular outcomes, such as patterns of coastline change and coastal development.

Much of the academic debate in coastal management is polarized between two extremes: one that presumes all communities should be saved and that we will continue to do business as usual, and the other that advocates a complete and total retreat from development at the coast as we know it (Pilkey and Young, 2008). For example, economic analyses of beach replenishment over the short term (25–50 years) invariably find that the benefits engendered by protection of property and support for coastal recreation justify replenishment costs (Bell, 1986; Parsons and Powell, 2001; Landry et al., 2003; Smith et al., 2009), whereas many coastal geologists and environmentalists have espoused support for climate adaptation strategies that embrace retreat in the face of erosion (Riggs and Ames, 2003; Pilkey and Dixon, 1996). To some extent, the two extreme views attempt to address different issues: one is focused on the near term while the other highlights the reality of the long-term future.

Decisions along developed coastlines in the next century will need to address both short- and long-term consequences. A coupled systems approach can help us to navigate a path from business-as-usual to a future in which physical forcing in the coastal zone is likely to be very different from today.

Recognition of the many spatial externalities of coastal management and the inherent coupled nature of the human–coastline system raises numerous difficult and potentially uncomfortable questions. For example: How do we want the coastline to look in the future? Where and to what extent do we want to defend coastal communities, existing patterns of development, and the associated infrastructure? Are some communities more worth "saving" or "protecting" than others? Typically, it is not within the purview of coastal managers and policymakers to ask these types of questions, and certainly not on a large spatial scale. Instead, managers, stakeholders, and policymakers are tasked with considering the desired state of the slice of coast for which they are responsible, not acknowledging (or perhaps not understanding) that decisions in their domains ultimately influence the entire coast.

From uneasy questions comes the potential for hope: our evolving understanding of coupled human–natural coastline systems allows us to examine how different future states of the coastal system arise from different sets of human decisions (under a range of climate and socioeconomic scenarios). Moving forward requires an explicit recognition that we are now geoengineering coasts accidently. Although individual communities act purposefully, we lack a coordinated effort that builds on existing knowledge of the interdependence of decentralized coastline interventions. Armed with a willingness to include coupled systems science in coastal planning decisions, we can move toward intentional rather than accidental geoengineering of our coasts.

BACKGROUND

The coastal environment is a region of focused human activity, with associated population changes, buildings and real estate improvements, and supporting infrastructure (Small and Nicholls, 2003). With this dense human settlement, the coastline has some of the highest property values and can be a significant source of tax revenue for local governments (AIR, 2008). Coastal land losses from erosion, storm damage to property and infrastructure, loss of human life, and disruption of social and economic activity have long threatened coastal economies, as the recent disaster Hurricane Sandy highlights. To manage vulnerability, humans engage in coastline alterations such as seawall construction and beach nourishment—placement of sand to widen the beach and increase dune height. U.S. federal expenditures through 2009 on beach nourishment alone amounted to $2.9 billion (NOAA, 2009). And total annual coastline protection expenditures in Europe were estimated to be $4 billion in 2001 (Nicholls et al., 2007). Economists often view the natural environment as an exogenous forcing in property markets and recreation/tourism decisions, while coastal scientists tend to view human shoreline modifications as exogenous perturbations to natural coastal dynamics. As we will describe, however, studies of coastal geomorphology and coastal economies increasingly paint a picture of these systems as strongly coupled at intermediate (year to decade) time scales. At these time scales, human manipulations alter regional patterns of coastline change, which ultimately affect future human coastline modifications. This two-way interaction between natural coastline dynamics and human agency makes human-occupied coastlines tightly coupled dynamical systems. At longer time scales, changes in both climate forcing and socioeconomic conditions will influence the coupled system (Figure 7.1). Next we introduce the geomorphic and economic dynamics, as well as the climate and societal forcing that drive the coupled dynamical system.

Geomorphic System Dynamics

Long-term shoreline change arises primarily from gradients in alongshore and cross-shore sediment transport. Alongshore sediment transport involves movement of sand in the alongshore direction by breaking waves and associated currents; annual net transport volumes can be large, up to a million m^3 passing through a given cross-shore transect. Alongshore gradients in this long-term net sediment flux produce a net gain or loss of sediment locally that tends to shift the shoreline seaward (gain) or landward (loss). Because large-scale coastline shapes (kilometers or larger) tend to create gradients in net alongshore flux (different coastline orientations experience different sets of wave conditions), these gradients then cause long-term shoreline change (Lazarus et al., 2012) that reshapes coastlines. On an approximately straight stretch of coastline oriented in an arbitrary direction, even subtle curvatures can lead to progressive shoreline change, such that on most coastline stretches, convex-seaward areas tend to erode (or erode more rapidly than the large-scale average) while

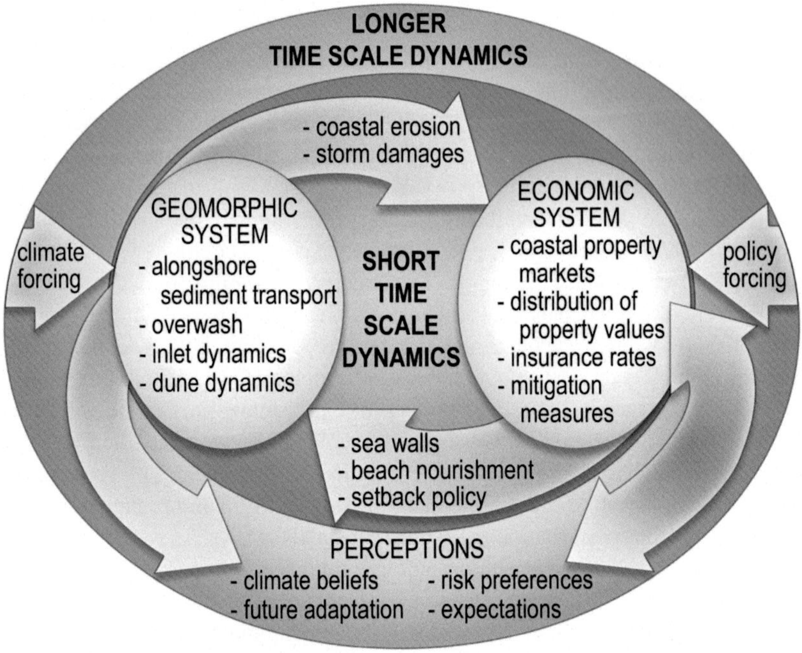

FIGURE 7.1 Coupled coastal geomorphic-economic system.

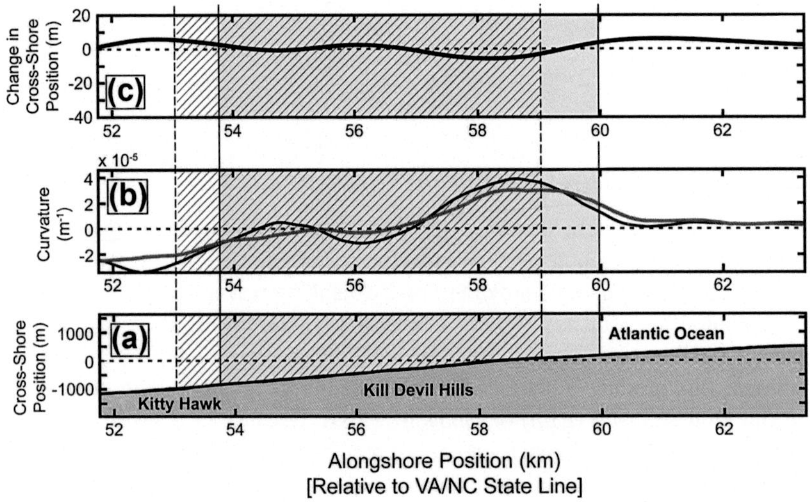

FIGURE 7.2 Example of shoreline change related to subtle coastline curvatures. The panel (c) shows the shoreline change over a 9-year span (1996–2005 on the Outer Banks of North Carolina, USA). The panel (b) shows shoreline curvatures at the beginning (black) and end (red) of the 9-year span, with positive (negative) values indicating seaward convexity (concavity). The panel (a) shows the approximately straight coastline shape. During this time span, areas seaward convex (concave) curvature correspond with shoreline erosion (accretion)—a commonly observed relationship on scales of kilometers or greater. *After Lazarus and Murray (2011).*

concave-seaward stretches tend to build seaward (or erode less rapidly); see Figure 7.2 (Lazarus et al., 2012; Ashton and Murray, 2006a,b). With more complex coastline shapes, when one portion of a shoreline protects other portions from waves approaching from some angles, shoreline changes at one location can affect shoreline changes at long distances, up to tens of kilometers away (Slott et al., 2010; Ells and Murray, 2012) (Figure 7.3).

The alongshore-transport-related component of coastline change is superimposed on long-term coastline changes related to sea-level rise (SLR), which involve movement of sediment in the cross-shore direction (Moore et al., 2010; Wolinsky and Murray, 2009). Although nearly every storm that brings large waves to a shoreline causes temporary erosion as sand is taken offshore, that erosion is usually reversed after the storm, as fair weather waves sweep the sand back toward the beach (List et al., 1997). During strong storms, however, the combination of storm surge and waves can erode away existing coastal dunes, and carry sand landward of the beach and dune. That sand can be deposited on a barrier island (a process called overwash). Sediment deposited via overwash on a barrier island increases the elevation of the island, at a long-term rate that tends to equal the rate of SLR in the natural state (Moore et al., 2010), which is beneficial because it helps to keep the island above sea level. When sand is transported past the island and into the back-barrier bay, in a process called inundation (Sallenger, 2000), however, it does not contribute to island elevation. In either case—whether sand is retained on top of the island (overwash) or is washed farther landward (inundation)—that sand is effectively removed from the beach and nearshore seabed system, causing net movement of the shoreline in the landward direction. In the long term (averaging over the characteristic return period for storms that cause overwash or inundation), landward movement of the shoreline combined with overwash deposition on the landward side of a barrier island causes island migration. Relative to overwash, inundation events tend to cause a greater rate of long-term island migration, since sand is lost from the whole barrier-island system in those events.

SLR increases the frequency of overwash and inundation events (by tending to make dune and island elevations lower relative to sea level). Therefore, increasing the rate of SLR intensifies the long-term rate of sand removal from the nearshore system, increasing the likelihood of barrier island migration. The rate of shoreline erosion that results depends approximately linearly on the rate of SLR, but it also depends on a host of other factors, including the erodibility and composition of the underlying substrate (Moore et al., 2010) as well as interactions between beach processes and vegetation growth ("ecomorphodynamic" mechanisms) that influence the maximum size of coastal dunes and the rate at which dunes recover following disturbance by storms (Durán and Moore, 2013).

Human management of coastal dunes and manipulation of coastal processes and erosion rates can further complicate the relationship between waves, currents, storms, changing sea levels, and barrier island response.

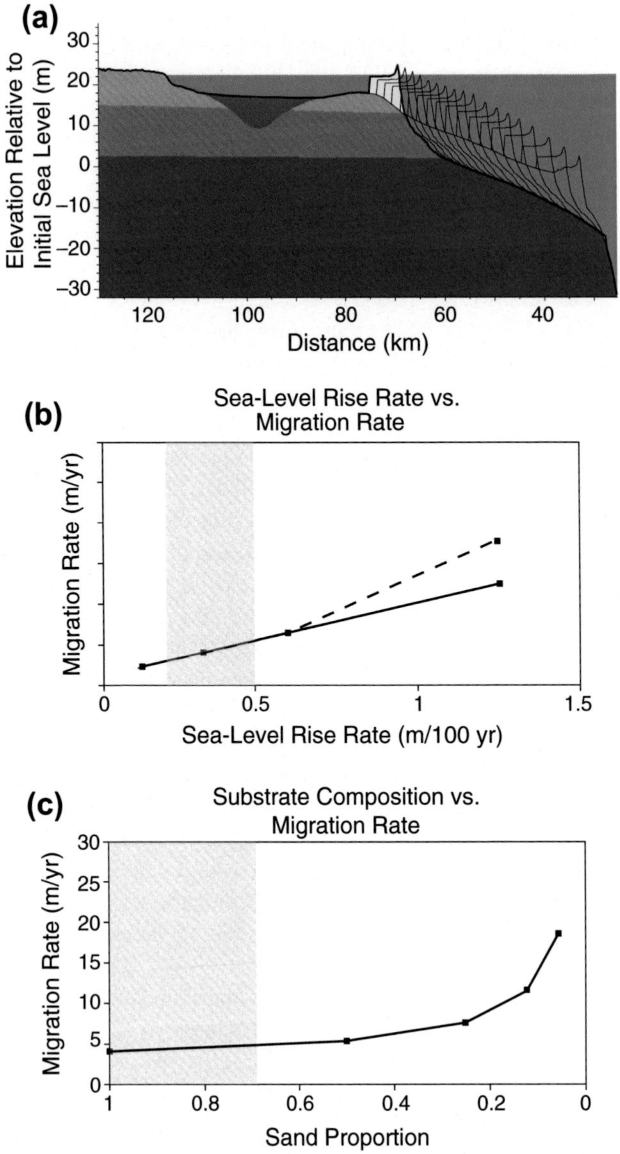

FIGURE 7.3 (a) Example GEOMBEST simulation of barrier island response to sea-level rise. (b, c) Results of sensitivity analyses, which show the effect of changes in the rate of sea level rise (for constant total sea level rise, solid line; and for constant sea level rise rate, dashed line, in (b) and substrate composition (c) on barrier island migration rate. Across the range of simulations in (B) (constant sea level rise rate and simulation time) total sea level at the end of the simulation varies. In these cases, the barrier migrates across different lengths of the underlying substrate and substrate slope effects also affect migration rate. Gray bars indicate best estimates for the range of most likely Holocene values in the North Carolina Outer Banks. *After Moore et al. (2010).*

Socioeconomic System Dynamics

The coastal environment provides an array of service flows for the coastal economy. Beach and dune sediments provide erosion and storm protection for coastal development, esthetic values for coastal residents, space for coastal recreation and leisure activities, and natural habitat for flora and fauna (Landry, 2011). Flux of coastal sediments can thus affect coastal economies and prompt intervention in coastal dynamic processes. Advocacy and support of coastal sediment management practices reflect the diversity of local economic activities, distribution of political ideologies, and aspects of local culture and knowledge (McNamara and Keeler, 2013; Phillips and Jones, 2006; Cooper and McKenna, 2008).

Residential and commercial development of coastal barriers has responded to growing demand for coastal recreation (due to increasing disposable income and influx of retirees), and development at the shore facilitates access and enjoyment of coastal amenities. In real estate markets, sale prices reflect the aggregation of expected flows of value over the life of the property, where future flows are discounted. Prices reflect a property's use value due to factors like square footage, lot size, unit type, location, and aspects of the local neighborhood (e.g., crime rate, school district). Prices also encompass flows of value from environmental amenities and disamenities, which can be quantified and related to individuals' underlying tastes (Rosen, 1974). The values of recreational amenities and storm protection provided by wide beaches, for example, are capitalized into property values. Hedonic property models can be used to estimate economic values stemming from coastal resource quality and to predict how coastal real estate markets respond to changes in the physical geomorphic system. These models have been used extensively to measure the impact of beach width (Brown and Pollakowski, 1977; Landry et al., 2003; Landry and Hindsley, 2011) and storm risks (Bin et al., 2008) on coastal property values. The distribution of property values across spatial locations ultimately depends on economic agents who decide where to locate, sorting into different properties and markets according to their income, their preference for amenities as well as their assessments and preferences about erosion and storm risk.

Beaches also support economic values of coastal recreation for visitors and tourists (Bin et al., 2005; Lew and Larson, 2008; Landry and McConnell, 2007). Studies of recreation demand indicate economic benefits of coastal erosion management accruing to visitors, many of whom do not own or rent coastal property (Parsons et al., 1999; Whitehead et al., 2008; von Haefen et al., 2004). There is also potential that non-visitors derive value from management of coastal erosion, whether due to potential future use, vicarious use of others, bequest motives, or simply the existence of coastal habitat (Silberman et al., 1992; Shivlani et al., 2003). Lastly, management of coastal sediments and the associated effects on visitation and tourism can have significant impact on business opportunities and economic activity (creating jobs and enhancing household income and tax revenues).

The linkage between coastal geomorphology and human use of the coastal environment is multifaceted and complex. The economic value of coastal property, recreation, and tourism is directly dependent upon amenity flows of beaches and dunes (storm and erosion protection and recreational, leisure, and aesthetic services) that vary over time as coastal geomorphology changes. Both current management actions and expectations about future actions are critical determinants of economic activity, including real estate values, investment patterns, and visitation levels. Take property values, for instance. With active shoreline stabilization (e.g., via beach nourishment), property values that reflect benefits from beach nourishment may also tend to simultaneously influence subsequent nourishment decisions and therefore beach width. Accounting for these dynamics of beach width and knowledge and expectations of individual home buyers can dramatically affect estimates and interpretation of coastal amenity values (Landry and Hindsley, 2011; Gopalakrishnan et al., 2011). Feedbacks in the coupled coastal-economic system are further supported by empirical analysis of beach nourishment in North Carolina (Gopalakrishnan et al., 2011), where predicted time intervals (years) between nourishment events are closer to observed intervals when beach amenity values used to parameterize a dynamic model of optimal beach management incorporate feedbacks (Figure 7.4).

In addition to the relatively slow changes associated with chronic coastline erosion, large storm events can have a punctuated, negative effect on property values, though values often recover over time (Bin and Landry, 2013). Storm surge and the large waves that coincide with storm events can cause widespread destruction of coastal property and infrastructure. The degree to which these damages impact coastal economies is strongly tied to both plan-view physical coastline characteristics, such as coastline position and associated fronting beach width, and vertical coastline features, such as height above mean sea level and dune elevation.

Climate Forcing

The Intergovernmental Panel on Climate Change (IPCC 2013, Fifth Assessment Report) estimates that global sea level will rise between 28 and 61 cm by the year 2100. These predictions explicitly exclude the effects of ice sheet flow and are generally considered lower bounds on the amount of SLR that will occur. Several factors, including local changes in land surface elevation (e.g., due to subsidence, glacioisostatic adjustment) and changing ocean currents (Ezer et al., 2013), will cause relative changes in sea level to vary from place to place. Globally, warming will also likely increase the frequency of the strongest hurricanes (Knutson et al., 2010; Bender et al., 2010), with warming of mid-latitude coastal oceans increasing landfalling storm intensity for the U.S. Atlantic coastline, in particular (as has possibly already been observed; Komar and Allan, 2008).

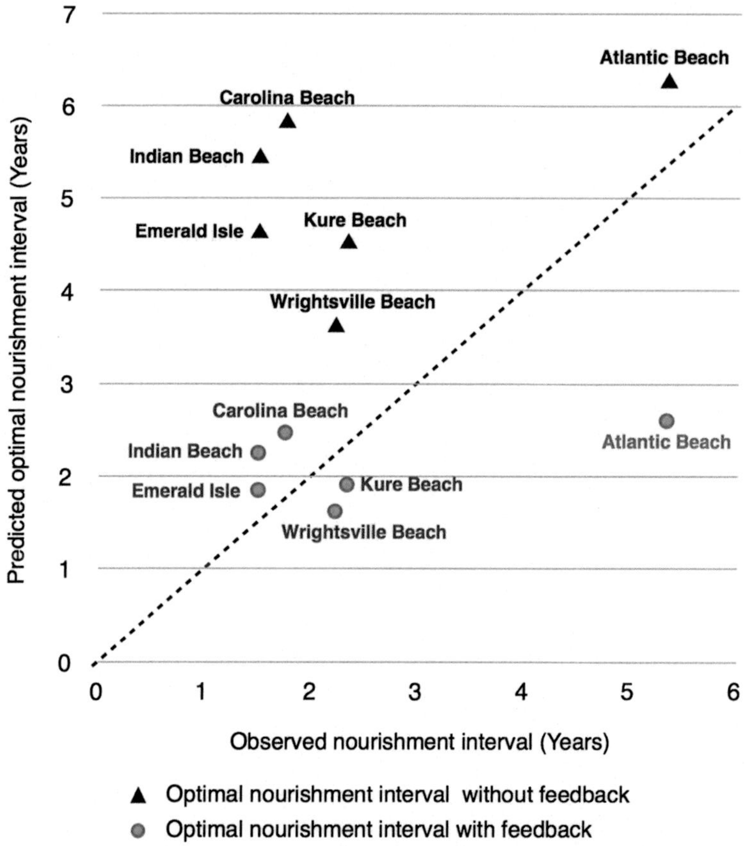

FIGURE 7.4 Observed and predicted nourishment intervals in six North Carolina beaches. Predicted optimal intervals calculated using beach values that incorporate endogenous feedback (green) are closer to the 45-degree (dashed) line where observed and predicted nourishment intervals are equal. Predicted intervals using amenity values that do not incorporate feedback (brown) are farther from the 45-degree line in five out of the six beaches studied. The discrepancy between the observed and predicted interval for Atlantic Beach in the feedback scenario reflects the fact that replenishments of Atlantic Beach are often due to dredge spoil disposal from nearby locations and not recorded as nourishment.

Socioeconomic Forcing

Not all relevant features of the socioeconomic environment are characterized by two-way couplings to the geomorphic system; some socioeconomic conditions are exogenous to, but nonetheless important drivers of, coastline change. Growth in demand for coastal real estate is the obvious example. It is estimated that over half the world population is concentrated in the coastal zone covering just 10% of the earth's land surface (NOAA, 2009), and this trend is persisting despite growing concerns about coastal hazards. As the size of the population living at the coast increases, the value of the built environment rises accordingly. And as the value of property at risk increases, the set of economically justifiable engineering interventions along the coast expands. Similarly, high real estate values and large coastal populations can put pressure on policymakers to address coastal hazards. The State of North Carolina, for example, had banned hard structures on the coast from 1985 until 2011, when it passed legislation allowing for a limited number of groynes to be built adjacent to tidal inlets. It remains to be seen whether economic pressures for property protection will expand this reversal, as more and more coastal property and infrastructure face greater risks.

Public policies exogenous to specific coastal locations can also have strong effects on economic choices in coastal property markets, and through these effects, serve as drivers of the choices communities make about altering the physical coastline. Examples in the United States include flood insurance, disaster relief, and investments in and maintenance of transportation and utility infrastructure. Climate hazard insurance covering coastal flooding is currently managed primarily by the federal government through the National Flood Insurance Program (Dixon et al., 2006; Michel-Kerjan, 2010). Higher insurance rates and greater limitations in coverage can increase incentives to invest in shoreline stabilization to reduce financial risk (Botzen et al., 2009). In other circumstances, however, less favorable insurance coverage can reduce the value of coastal investment, lowering incentives for shoreline engineering. Disaster relief has similarly ambiguous impacts on choices about defensive engineering. High expectations about disaster relief payouts can directly reduce the value of shoreline stabilization by decreasing financial risk from climate-related events. Larger disaster relief expenditures, however, tend to encourage development patterns that increase the overall value and vulnerability of the built environment and can therefore incentivize shoreline engineering.

Policies regarding investment in transportation and utilities infrastructure and support for shoreline engineering are central to the coupled system's evolution. Such investments are affected by local, state, and national policies, and influence diverse aspects of the costs and benefits of owning coastal property. Investments that tend to reduce the risk and hassle of interruptions in basic services will increase the value of coastal property and therefore tend to increase incentives to protect property through shoreline engineering. Shoreline engineering has historically been financed by both local and national entities, with the balance shifting toward local financing in recent years. Federal policy continues to affect the local cost of engineering projects, both through decisions about explicit and implicit subsidization and through regulatory policies that affect the difficulty and expense of implementing and monitoring engineering projects.

In the remainder of this chapter, we explore several dimensions in which coastal geomorphology and coastal economies are coupled. These couplings raise difficult questions about where, when, and in what manner humans should intervene in coastline change. We highlight the challenges involved in making purposeful decisions that avoid inadvertent and uncoordinated geoengineering of our coastlines.

ALONGSHORE CONNECTIONS: COMMUNITIES AFFECT EACH OTHER

Human manipulations can affect coastlines at least as much as natural processes do. Coastal engineers have long understood that human interventions to stabilize beaches can have spatial spillovers at the local level. A groyne in one location, for instance, alters alongshore sediment transport and has direct implications for adjacent locations. Similarly, a beach nourishment (beach extension/construction) project that creates a subtle seaward bump in the shoreline can increase the rates of net sediment delivery to adjacent beaches. When towns implement localized stabilization policies, not accounting for the physical and economic implications of these dynamic spatial interactions, it results in suboptimal outcomes relative to coordinated management of the coastline (Gopalakrishnan et al., 2014).

Over scales up to kilometers, such effects are well studied, and there are examples of stabilizations that have triggered striking coastline offsets in a matter of decades (Figure 7.5). There is also increasing evidence that stabilization of an eroding shoreline—either through beach nourishment or hard structures—can affect long-term rates of shoreline change over surprisingly long distances up to tens of kilometers (e.g., Slott et al., 2010; Ells and Murray, 2012).

These nonlocal effects arise for two reasons. First, plan-view coastline shape tends to evolve dynamically in the natural state, and pinning the location of the shoreline in one or more places affects nearby shoreline shapes and orientations, which in turn affects the alongshore sediment flux into more distant coastline stretches, changing their shape. This mechanism for long-range effects of localized human manipulation applies on coastlines of any shape, including approximately straight coastlines (e.g., Van den Berg et al., 2011). Whether a shoreline is stabilized through beach nourishment or through hard structures, the coastline updrift will experience lowered erosion rates (relative to the case without the stabilization; Figure 7.6); stabilization, over the long term, will create a bump in the coastline that reduces the sediment flux passing the stabilized point by changing the coastline orientation locally. The changes in shoreline orientation and sediment transport rates then propagate progressively farther updrift as the duration of stabilization increases.

The downdrift effects of nourishment and hard structures, however, differ. In the presence of stabilization by hard structures, sediment flux into the downdrift shoreline segments is reduced, which increases erosion rates (relative to the case without stabilization). Again, the effect (increased erosion) propagates progressively farther with time. When an area is stabilized with nourishment, however, the sediment flux into downdrift sections of the shoreline is not limited to the reduced flux coming in from updrift; because stabilization is accomplished by adding sand as fast as it is removed (over the long term, averaging over multiple nourishment cycles), the rate at which sediment is transported into downdrift shoreline segments is just determined by the shoreline orientations. Sediment flux will tend to increase

FIGURE 7.5 Coastline offset in Ocean City, MD and Assateague, MD (view to the south). Navigation jetties blocked alongshore sediment transport, causing the shoreline to build seaward to the north of the structures (updrift) and to erode landward south of the structures (downdrift).

in the downdrift portion of the coastline bump, so that downdrift areas also erode less rapidly than they would have without the stabilization.

The second mechanism by which local shoreline stabilization produces a long-range effect arises on coastlines that are more complexly shaped such that one part of the coastline protrudes sufficiently seaward to affect the waves reaching other parts of the coastline (i.e., producing a "wave shadow"). Protruding coastline features include rocky headlands, but also emergent features (e.g., "cuspate capes," such as found on the North Carolina coastline) on sandy shorelines that form because of feedbacks arising from patterns of alongshore sediment flux (Ashton et al., 2001). Shoreline stabilization can affect how far seaward these sandy protrusions extend, thereby causing shifts in the zone of wave shadowing. Changes to patterns of wave shadowing alter patterns of alongshore sediment flux, which in turn produce changes in patterns of shoreline erosion and accretion. Initial work examining long-range effects of localized shoreline stabilization focused on the heavily developed Carolina coastline (Figure 7.6, panel (a)).

Wave-shadowing effects propagate instantaneously over the alongshore scale of the coastline feature—up to 100 km in the cuspate-cape example (Figure 7.6, panel (a)). However, through the propagation of shoreline-orientation effects alone, localized stabilization can affect remote locations over distances of tens of kilometers over human time scales (e.g., Figure 7.6, panels (b) and (c))—whether the coastline has a complicated or simple shape. As a result, the entire coastline is a coupled system, and spatial externalities must be considered for the entire system and not just locally.

Dwindling availability of easily accessible sand resources, and the resulting increases in the cost of shoreline engineering projects, also has the potential to link coastal communities over long distances. Two communities on a complex

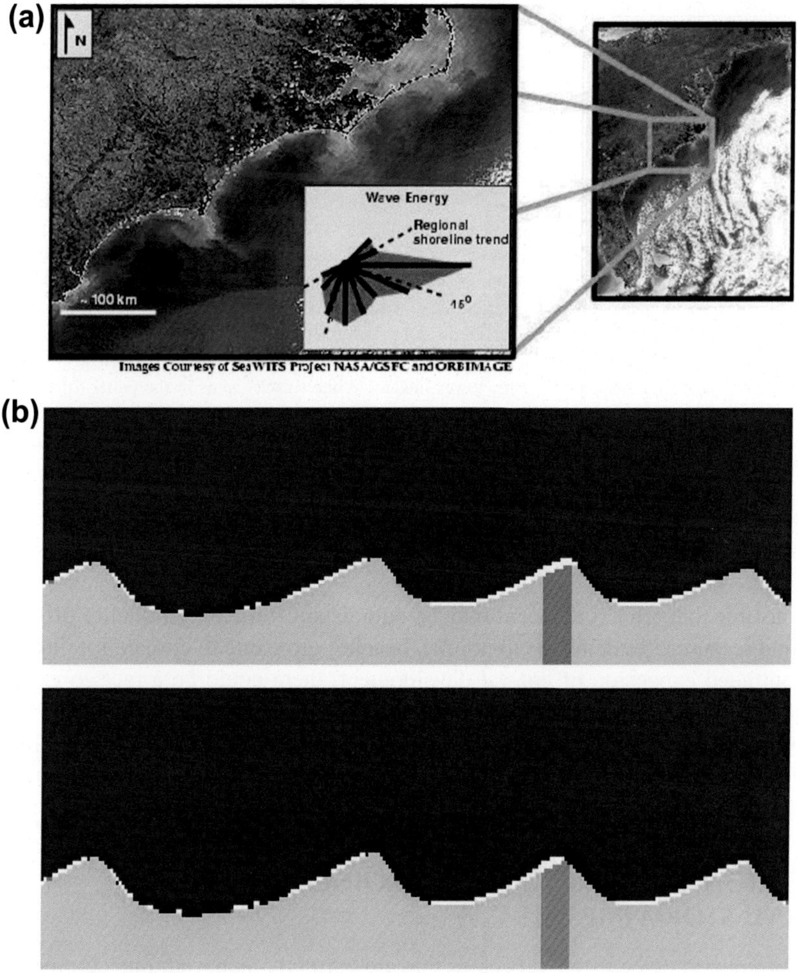

FIGURE 7.6 CONT'D

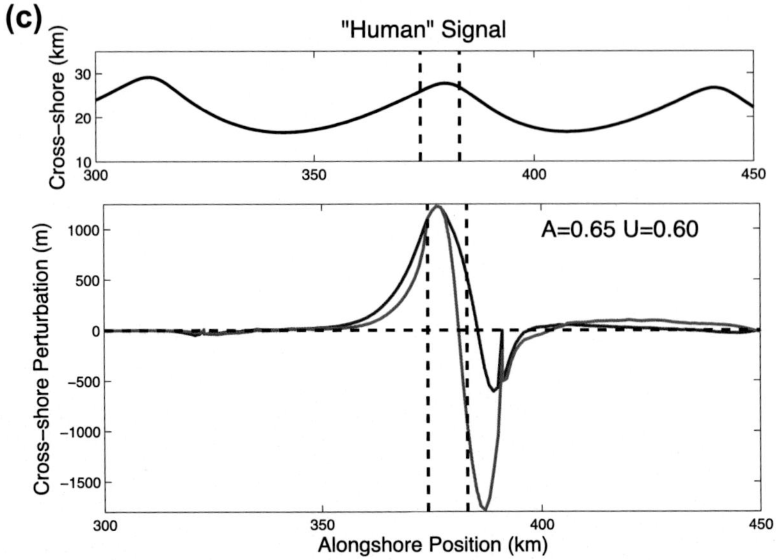

FIGURE 7.6 Example of long-range effects of local stabilization. The panel (a) shows the Carolina Capes. *(After Ashton et al., 2001.)*, which motivate the case-study examination of human influences on large-scale coastline change on cuspate-cape coastlines. The panel (b) depicts the differences between different coastline-change scenarios: one in which the coastline shape changes freely in response to a change in wave climate (a different distribution of wave-approach directions related to changing storm statistics (Slott et al., 2006; McNamara et al., 2011; Moore et al., 2013), and two in which shoreline erosion is prevented in the location of the dark stripe, either with nourishment (brown stripe, upper image) or hard structures (gray stripe, lower image). White (black) areas in the panel (b) indicate where the shoreline ended up farther seaward (landward) after a century of stabilization than in the nonhuman-influenced model run. The panel (c) shows this "human signal" graphically, highlighting that the propagating shoreline-orientation effects can cause changes in shoreline position updrift, and that shoreline-orientation effects and wave-shadowing effects combine to produce tens of meters of shoreline change, and even as many as tens of kilometers (downdrift). The blue line and red line indicate changes with nourishment and seawalls, respectively (b) and (c) after Ells and Murray (2012).

coastline that are not adjacent may be equidistant from a high-quality offshore sand resource. As demands to nourish beaches grow due to climate forcing and continued expansion of coastal development, there could be a race to exploit common-pool sand resources or a potential market for offshore sand. We may see communities trading sand resources. Indeed, some high property value locations in Florida have depleted local offshore sand resources and are engaging in bargaining with distant communities for their offshore sand (Alverez, 2013).

CHRONIC VERSUS ACUTE: ONGOING EROSION AND STORM IMPACTS

There are in essence two processes that are currently managed: one chronic and continuous, and another discrete or punctuated. Shoreline stabilization primarily represents a reaction to chronic, prolonged shoreline erosion, caused by gradients in wave-driven alongshore sediment transport and SLR. Chronic erosion, accumulating over decades, brings the shoreline into contact and conflict with coastal development (even when the development was originally built with a seaward buffer of open land). Despite some variability over time, alongshore transport is a continuous process, rather than something that occurs during punctuated storm events. Although beach erosion does occur during storms, the sand that moves offshore during a storm moves back onshore after the storm, undoing most of the storm-related erosion. Thus, even though alongshore transport is especially strong during storms, cumulative shoreline erosion accumulates over years to decades, spanning multiple storms (and calmer periods). Rebuilding beaches in the cross-shore dimension through beach nourishment is largely a response to this ongoing, approximately continuous process as well as the continuous process of SLR.

In contrast to continuous processes, storms are discrete, stochastic events that can bring large waves, storm surge, and wind—causing sediment to move in the cross-shore direction, and possibly creating fundamental change in configurations of coastal development. Storms highlight the importance of cross-shore processes and the vertical dimension of beachfront development. Increasingly, augmenting the vertical dimension of the beach with nourishment is seen as a strategy for storm damage mitigation.

However, efforts to protect coastal development in the short-run (e.g., building seawalls or artificially tall and continuous sand dunes) alter or prevent overwash during all but the most extreme storms. Construction of dense development itself can also alter or prevent overwash. When extreme storms occur, coastal development, infrastructure, and coastal protection affect the pattern and volume of sediment flux. When overwash occurs on a developed coastline, humans typically remove it quickly (to exhume roads and other infrastructure), bulldozing sand back to the beach to rebuild a protective dune line or trucking it elsewhere. All of these human influences on sand deposition by overwash prevent natural processes from increasing island elevation (Magliocca et al., 2011)—increasing long-term vulnerability from erosion and flooding.

In addition, because overwash removes sediment from the beach and nearshore system, tending to move the shoreline landward, combating storm processes that occur in the vertical dimension affects coastline dynamics in the horizontal dimension. For example, when economic development is concentrated in particular alongshore locations, the prevention of overwash and subsequent nourishment to protect economic development causes significant variation in relative alongshore coastline positions on an otherwise straight coastline (McNamara and Werner, 2008a,b; Figures 7.7 and 7.8).

By preventing some of the horizontal erosion that would otherwise have occurred, limiting overwash may contribute to a seaward bump in the coastline relative to an unprotected community. The extension of a segment of coastline seaward will, over the timescale of decades, propagate alongshore changes to shoreline dynamics. The policy-relevant result is that protecting a community and its housing stock from storm impacts leads to potentially significant spatial externalities in the alongshore direction.

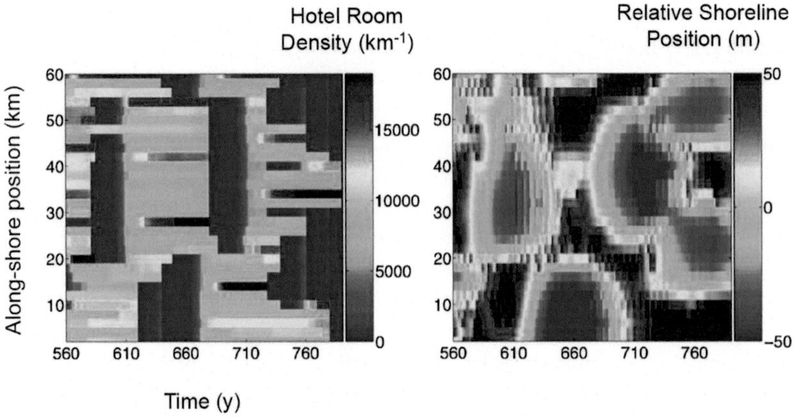

FIGURE 7.7 Interactions between barrier island evolution and resort communities. Hotel density (left panel) and relative cross-shore shoreline position (right panel) are plotted against alongshore position and time. Areas with increased density in hotels have engaged in nourishment activities that build tall dunes to prevent overwash and widen the beach to maintain coastline position against erosion. Both actions cause the shoreline in high-density hotel regions to extend further seaward relative to surrounding areas with smaller hotel densities. Over time, defending those seaward positions becomes more difficult as sand is redistributed preferentially to concave-seaward portions of the coastline, and the lack of overwash and associated elevation gain raises the potential hazards from severe storm surges. Ultimately, communities are abandoned at locations that have been defended for several decades. *From McNamara and Werner (2008a).*

TEMPORAL SCALE

In the long term, SLR hastens both chronic landward erosion and tends to increase the frequency of acute storm effects. However, with shoreline positions moving back and forth during storms and intervening calm weather, and with the cumulative shoreline change related to large-scale gradients in alongshore transport (several kilometers and up), shoreline response to SLR is hard to pick out from shoreline variance and other trends. Similarly, gradual changes in the frequency and magnitudes of storm surge, waves, and winds—whether related to increases in SLR rates or to changes in storm frequency and intensity—are difficult to identify and appreciate. Only over time scales of decades and longer do climate-change-related effects really matter.

Despite these long time scales, the likely effects of climate change tend to push time scales of coastal geomorphology closer to meaningful economic time scales. Already, sea levels are high enough, storms are frequent and intense enough, and alongshore sediment transport is sufficiently powerful to trigger human responses to defend the shoreline. SLR, changing wave climates, and intensification of tropical storms all have the potential to speed up physical coastline changes (Slott et al., 2006; Moore et al., 2013). This acceleration, in turn, could influence the kind and scale of adaptive responses by humans due to higher erosion rates in developed areas. Moreover, these elements of climate

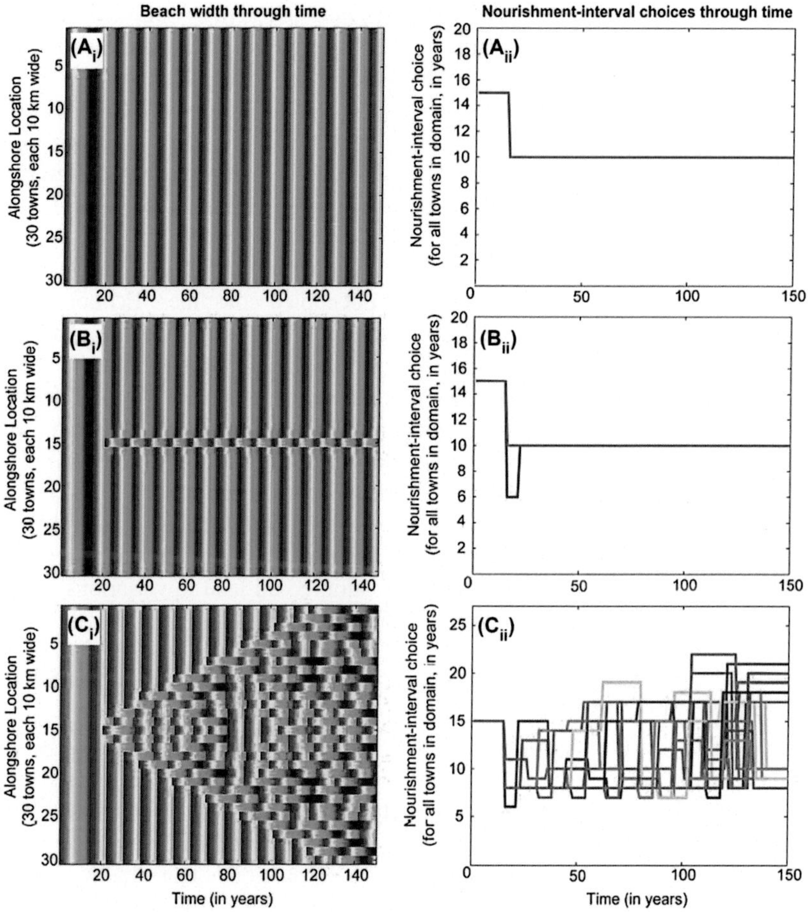

FIGURE 7.8 Patterns of beach nourishment depend on background shoreline erosion rates. Left panels (A_i, B_i, C_i) show beach widths at different alongshore locations through time, related to beach nourishment activities. Right panels (A_{ii}, B_{ii}, C_{ii}) show the nourishment intervals that result from an optimization calculation (Smith et al., 2009; Gopalakrishnan et al., 2011), based on local perceptions of beach erosion rates in recent years (Lazarus et al., 2011), at each 10-km town in the domain. In the top panels (A_i, A_{ii}), nourishment intervals are uniform alongshore. In the middle panels (B_i, B_{ii}), a relatively low shoreline erosion rate (0.5 m/year), forcing the middle town to nourish out of turn, does not affect decisions in neighboring towns. In the bottom panels (C_i, C_{ii}), under a higher erosion rate (1.5 m/year, which could represent the effects of increased rates of sea-level rise), perturbing the middle town, by altering the short-term erosion rates felt by neighboring towns, sets off a chain reaction that leads to a chaotic spatial/temporal pattern of nourishment activities. *From Lazarus et al. (2011).*

change have the potential to interact with one another. With SLR, for example, the same net impact of storm surge could come from a less intense storm. A more intense storm on top of SLR has the potential to generate an even more catastrophic storm surge.

As climate forcing intensifies and potentially accelerates coupled morpho-economic dynamics, there is potential for emergence of large-scale human/coastline phenomena on meaningful economic time scales. This is the essence of inadvertent geoengineering: human responses to climate-driven risks can produce coastlines with qualitatively different and unintended characteristics. Indeed, climate forcing that raises the background erosion rate can destabilize an entire coastline as individual communities pursue independent shoreline stabilization strategies (Lazarus et al., 2011).

Another important time scale is how human perceptions change and, in turn, motivate different behaviors along human-occupied coastlines. Economically driven manipulations to the coastline in part reflect human perceptions of long time-scale physical processes. There can also be significant differences in these perceptions over large spatial scales. For example, the state of North Carolina recently attempted to regulate projections of future rates of SLR so that they remained within historical rates, while other regions such as New York City are discussing fortification against some of the worst-case projections for SLR. As we have shown, coastline alterations can have unintended long-range effects, which means that variations in the perception of the physical processes and subsequent variations in coastline modification could lead to variations in the unintended consequences of coastline geoengineering.

JURISDICTIONAL DIVISIONS

The question of which jurisdictional level—e.g., federal, state, or local—should regulate the environment has long interested economists and political scientists and has spawned a substantial literature in environmental federalism. The general thinking in economics is that highest jurisdictional level (federal) is the right level for problems that involve pure public goods that society at large enjoys, whereas the lowest level (local) makes sense for problems where local jurisdictions capture all of the benefits and costs of regulating environmental quality.

Most real-world situations are somewhere in between, where benefits of environmental quality flow (Oates, 1999, 2001) disproportionately to the local jurisdiction, but there are significant spillovers into other jurisdictions that may warrant higher-level regulatory coordination. The problems of stabilizing coastlines and adaptating to climate change reflect classic tensions in the environmental federalism literature. For example, the spatial externality of the groyne with immediate impacts through alongshore sediment transport on a neighboring community is a typical example of a spillover that requires making rules above the local jurisdiction. Construction of a sea wall to protect private beachfront property can have almost immediate and deleterious effects on public beaches. In fact, stabilization decisions frequently take place with funding and decision-making at both the municipality and county level. Similarly, maintaining a wide beach generates amenities that directly benefit

FIGURE 7.9 Beach nourishment of Sandy Key, Florida. *Image source:* http://www.csc.noaa.gov/archived/beachnourishment/.

local homeowners—and that are capitalized into the value of the property they own—but also spill over to benefit beachgoers who travel from outside the local jurisdiction. A wide, nourished beach can accommodate many more beachgoers than a narrow, unnourished beach wedged between existing development and the sea (Figure 7.9).

While these features fit neatly into the environmental federalism taxonomy, there are unique features of coastline stabilization that highlight more vexing cross-jurisdictional externalities. First, the allocation of funds at the federal level has the potential to generate unintended spatial-dynamic externalities. In the United States, a significant portion of funds for beach nourishment come from federal spending. Projects are funded and approved on a case-by-case basis, and the fact that federal commitments through the Army Corps of Engineers have historically been made for very long time periods (typically 50 years) has created a legacy of previous decisions that impedes federal consideration of spatial-dynamic feedbacks or prioritization of future projects. A project that may pass a benefit-cost test in a vacuum (or even have the highest benefit-to-cost ratio among a set of proposed projects) could produce a net loss as a result of feedbacks produced through alongshore sediment transport and the propagation of large-scale spatial features.

Second, communities are linked to each other through exploitation of common sand resources. Particularly along complex coastlines, such as the cuspate coast of the Carolinas, communities that may be far away from each other along the shoreline may have similar proximity to economically recoverable and high-quality offshore sand resources. Because no individual community owns these resources, there may be incentives to overexploit them or race to exploit them as the race to fish unfolds in open access settings. A shift from predominantly federal to more local funding for nourishment will tend to exacerbate this race-to-nourish effect, as availability of cheap sand directly affects the local tax burden.

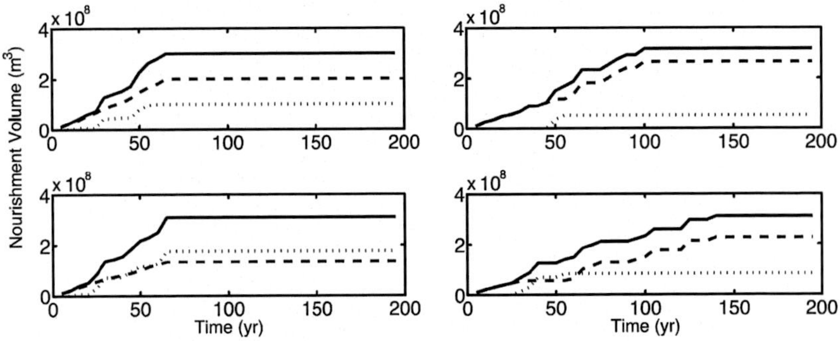

FIGURE 7.10 Exhaustion of sand resources when higher-property-value and lower-property-value communities exploit a common pool. Lines depict cumulative nourishment volumes for two towns with different property values (dashed, higher property value; dotted, lower property value; solid, total), under four different scenarios of nourishment sand prices and spatial patterns of wealth relative to spatial patterns of background erosion rates. The left panels depict scenarios in which the cost of sand for nourishment remains constant through time, while the right panels show cases in which sand costs increase as the common pool dwindles in volume. The top panels depict scenarios in which the higher-property-value towns are located in areas with higher background erosion rates, while the lower panels show the opposite spatial relationship. In these experiments, the two towns are located near each other, but on opposite sides of a cuspate cape that is being affected by a changing wave climate (which produces the contrasting background erosion rates). *From McNamara et al. (2011).*

An additional complication is the presence of vacation home owners whose property is taxed locally but who live elsewhere and are unable to vote in local elections that may affect nourishment decisions. These behaviors create another form of spatially uncoordinated behavior that could then propagate spatial features over large scales. In numerical model simulations where two communities compete for common sand resources and the cost of acquiring sand for nourishment rises with dwindling supply, the lifetime of available sand depends strongly on the interaction between patterns of property value and erosion rates (Figure 7.10).

Specifically, when high-property-value towns are located in regions of high erosion, sand resources shared by high- and low-property-value towns can dwindle rapidly. This raises the possibility that combinations of property value and background erosion rates at one location could affect the ability of towns in other locations to defend against shoreline erosion.

Complicating the above jurisdictional confusion that makes efficient management of spatial feedbacks problematic is a set of federal policies that have direct implications for shoreline stabilization. Poststorm disaster relief and federal flood insurance, as well as environmental review under the National Environmental Policy Act for consistency with federal environmental statutes, all have significant effects on the design options and long-term flow of costs and benefits of beach engineering projects. These policies are designed to meet a range of goals in protecting public goods, mitigating moral hazard, and correcting market failures. They have the effect of changing the incentives and choices of individual property owners and local/regional governments in complex and unpredictable ways.

DISCUSSION

The tight coupling of human and natural systems described in this chapter points to the need for coordinated decision-making to set rules and incentives for coastal management. Unfortunately, the processes and feedbacks that characterize the system do not fit into existing political, economic, or regulatory structures. Our understanding of the geomorphological effects of uncoordinated human actions on the coast is improving rapidly, but there is little indication that we are similarly improving our ability to use this knowledge to produce better outcomes. A shift from accidental to intentional geoengineering of the coast could enhance the well-being of coastal communities.

Although use of coupled systems science presents opportunities relative to the status quo, our understanding of these systems is nascent, and our ability to practice intentional geoengineering of whole coastlines inevitably will be limited. Harnessing our coupled systems science to make better decisions should not inspire overconfidence in predictions about coastline changes and associated effects on coastal communities. Tremendous uncertainty will remain even if coastal planners incorporate the best science available. When scientists have such uncertainty about managed dynamical systems, calls for adaptive management are ubiquitous. As humans intervene in the coastline, one naturally would wish to learn from that experience and apply it to future decisions. And one even may wish to design coastline interventions with the intent of gaining knowledge about how the coupled system works. The conceptual appeal of adaptive management is undeniable. However, its practice in a coupled coastline system with multiple interacting spatial and temporal scales is far from clear. Practical implementation requires determining what constitutes real information about the functioning of the coupled coastline system. If we shift toward coordinated management of coasts informed by coupled systems thinking, how do we evaluate the outcomes that we observe? Which changes can meaningfully be attributed to human decisions (whether directly or through feedbacks in the spatial dynamics); which changes are attributable to external forcing (physical or socioeconomic); which changes are truly random shocks to the system; and what time scale do we use to assess the impacts of our planning decisions? The scientific challenges are daunting, but the alternative to this path is to inherit a geoengineered coastline that our society did not choose.

REFERENCES

AIR, Worldwide Corporation, 2008. The Coastline at Risk: 2008 Update to the Estimated Insured Value of U.S. Coastal Properties, AIR Worldwide Corporation.

Alvarez, Lizette, August 24, 2013. Where Sand Is Gold, the Reserves Are Running Dry. The New York Times. http://www.nytimes.com/2013/08/25/us/where-sand-is-gold-the-coffers-are-runningdry-in-florida.html?_r=0.

Ashton, A., Murray, A.B., 2006a. High-angle wave instability and emergent shoreline shapes: 1. Modeling of capes, flying spits and sandwaves. J. Geophys. Res. 111 (F04011). http://dx.doi.org/10.1029/2005JF000422.

Ashton, A., Murray, A.B., Arnault, O., 2001. Formation of coastline features by large-scale instabilities induced by high-angle waves. Nature 414, 296–300.

Ashton, A.D., Murray, A.B., 2006b. High-angle wave instability and emergent shoreline shapes: 2. Wave climate analysis and comparisons to nature. J. Geophys. Res. 111 (F04012).

Bell, F.W., 1986. Economic policy issues associated with beach renourishment. Policy Stud. Rev. 6, 374–381.

Bender, M.A., Knutson, T.R., Tuleya, R.E., Sirutis, J.J., Vecchi, G.A., Garner, S.T., Held, I.M., 2010. Modeled impact of anthropogenic warming on the frequency of intense Atlantic hurricanes. Science 327 (5964), 454–458. http://dx.doi.org/10.1126/science.1180568.

Bin, O., Crawford, T.W., Kruse, J.B., Landry, C.E., 2008. Viewscapes and flood hazard: coastal housing market response to amenities and risk. Land Econ. 84 (3), 434–448. http://dx.doi.org/10.3368/le.84.3.434.

Bin, O., Landry, C.E., 2013. Changes in implicit flood risk premiums: empirical evidence from the housing market. J. Environ. Econ. Manage. 65 (3), 361–376.

Bin, O., Landry, C.E., Ellis, C.L., Vogelsong, H., 2005. Some consumer surplus estimates for North Carolina beaches. Mar Resour. Econ. 20 (2), 145–161. http://www.uri.edu/cels/enre/mre/mre.htm.

Botzen, W.J.W., Aerts, J.C.J.H., van den Bergh, J.C.J.M., 2009. Willingness of homeowners to mitigate climate risk through insurance. Ecol. Econ. 68 (8–9), 2265–2277. http://dx.doi.org/10.1016/j.ecolecon.2009.02.019.

Brown, G.M., Pollakowski, H.O., 1977. Economic Valuation of shoreline. Rev. Econ. Stat. 59, 272–278.

Cooper, J.A.G., McKenna, J., 2008. Social justice in coastal erosion management: the temporal and spatial dimensions. Geoforum 39 (1), 294–306.

Dixon, L., Clancy, N., Seabury, S.A., Overton, A., 2006. The National Flood Insurance Program's Market Penetration Rate: Estimates and Policy Implications. RAND Corporation, Santa Monica, CA.

Durán, O., Moore, L.J., 2013. Vegetation controls on the maximum size of coastal dunes. Proc. Natl. Acad. Sci. 110 (43), 17217–17222.

Ells, K., Murray, A.B., 2012. Long-term, non-local coastline responses to local shoreline stabilization. Geophys. Res. Lett 39 (19), L19401.

Ezer, T., Atkinson, L.P., Corlett, W.B., Blanco, J.L., 2013. Gulf Stream's induced sea level rise and variability along the US mid-Atlantic coast. J. Geophys. Res. Oceans 118 (2), 685–697.

Gopalakrishnan, S., McNamara, D., Smith, M.D., Murray, A.B., 2014. Decentralized Management Hinders Coastal Climate Adaptation: The Spatial Dynamics of Beach Nourishment. Duke Environmental and Energy Economics Working Paper Series No. EE, 14–04.

Gopalakrishnan, S., Smith, M.D., Slott, J.M., Murray, A.B., 2011. The value of disappearing beaches: a hedonic pricing model with endogenous beach width. J. Environ. Econ. Manage. 61 (3), 297–310. http://dx.doi.org/10.1016/j.jeem.2010.09.003.

von Haefen, R., Phaneuf, D., Parsons, G.R., 2004. Estimation and welfare analysis with large demand systems. J. Bus. Econ. Stat. 22 (2).

Knutson, T.R., McBride, J.L., Chan, J., Emanuel, K., Holland, G., Landsea, C., Held, I., Kossin, J.P., Srivastava, A.K., Sugi, M., 2010. Tropical cyclones and climate change. Nat. Geosci. 3 (3), 157–163.

Komar, P.D., Allan, J.C., 2008. Increasing hurricane-generated wave heights along the U.S. East coast and their climate controls. J. Coastal Res. 24 (2), 479–488. http://dx.doi.org/10.2112/07-0894.1.

Landry, C.E., 2011. Coastal erosion as a natural resource management problem: an economic perspective. Coastal Manage. 39 (3), 259–281.

Landry, C.E., McConnell, K.E., 2007. Hedonic onsight cost model of recreation demand. Land Econ. 83 (2), 253–267.

Landry, C.E., Hindsley, P., 2011. Valuing beach quality with hedonic property models. Land Econ. 87 (1), 92–108.

Landry, C.E., Keeler, A., Kriesel, W., 2003. An economic evaluation of beach erosion management alternatives. Mar. Resour. Econ. 18 (2), 105–127.

Lazarus, E., Ashton, A., Murray, A.B., 2012. Large-scale patterns in hurricane-driven shoreline change. In: Sharma, A.S., Dimri, V.P., Bunde, A. (Eds.), Extreme Events and Natural Hazards: The Complexity Perspective. American Geophysical Union.

Lazarus, E.D., Brad Murray, A., 2011. An integrated hypothesis for regional patterns of shoreline change along the northern North Carolina Outer Banks, USA. Mar. Geol. 281 (1), 85–90.

Lazarus, E.D., McNamara, D.E., Smith, M.D., Gopalakrishnan, S., Murray, A.B., 2011. Emergent behavior in a coupled economic and coastline model for beach nourishment. Nonlinear Processes Geophys. 18 (6), 989–999. http://dx.doi.org/10.5194/npg-18-989-2011.

Lew, D.K., Larson, D.M., 2008. Valuing a beach day with a repeated nested logit model of participation, site choice, and stochastic time value. Mar. Resour. Econ. 23 (3), 233.

List, J.H., Sallenger Jr., A.H., Hansen, M.E., Jaffe, B.E., 1997. Accelerated relative sea-level rise and rapid coastal erosion:: testing a causal relationship for the Louisiana barrier islands. Mar. Geol. 140 (3–4), 347–365.

Magliocca, N.R., McNamara, D.E., Murray, A.B., 2011. Long-term, large-scale morphodynamic effects of artificial dune construction along a barrier island coastline. J. Coastal Res. 27 (5), 918–930. http://dx.doi.org/10.2112/jcoastres-d-10-00088.1.

McNamara, D., Keeler, A., 2013. A coupled physical and economic model of coastal real estate response to climate risk. Nat. Clim. Change 3 (6), 559–562.

McNamara, D.E., Werner, B.T., 2008a. Coupled barrier island–resort model: 1. Emergent instabilities induced by strong human-landscape interactions. J. Geophys. Res. 113 (F01016). http://dx.doi.org/10.1029/2007JF000840.

McNamara, D.E., Werner, B.T., 2008b. Coupled barrier island–resort model: 2. Tests and predictions along Ocean City and Assateague Island National Seashore, Maryland. J. Geophys. Res. Earth Surf. (2003–2012) 113 (F1).

McNamara, D., Murray, A.B., Smith, M.D., 2011. Coastal sustainability depends on how economic and coastline responses to climate change affect each other. Geophys. Res. Lett. 38. http://dx.doi.org/10.1029/2011GL047207.

Michel-Kerjan, E.O., 2010. Catastrophic economics: the National Flood Insurance Program. J. Econ. Perspect. 24 (4), 165–186.

Moore, L.J., McNamara, D.E., Murray, A.B., Brenner, O., 2013. Observed changes in hurricane–driven waves explain the dynamics of modern cuspate shorelines. Geophys. Res. Lett. 40 (22), 5867–5871.

Moore, L.J., List, J.H., Williams, S.J., Stolper, D., 2010. Complexities in barrier island response to sea level rise: insights from numerical model experiments, North Carolina Outer Banks. J. Geophys. Res. 115 (27). http://dx.doi.org/10.1029/2009JF001299.

Nicholls, R.J., Wong, P.P., Burkett, V., Codignotto, J., Hay, J., McLean, R., Ragoonaden, S., Woodroffe, C.D., Abuodha, P.A.O., Arblaster, J., 2007. Coastal systems and low-lying areas. In: Canziani, O.F., Parry, M.L., Palutikof, J.P., van der Linden, P.J., Hanson, C.E. (Eds.), Climate Change 2007: Impacts, Adaptation and Vulnerability. Contribution of Working Group II to the Fourth Assessment Report of the Intergovernmental Panel on Climate Change. Cambridge University Press, Cambridge, UK, pp. 315–356.

NOAA, 2009. Beach Nourishment: A Guide for Local Government Officials – Historical Expenditures for Beach Nourishment Projects: Geographical Distribution of Projects and Sources of Funding. Coastal Services Center website: http://www.csc.noaa.gov/beachnourishment/html/human/socio/change.htm (accesssed 2013.).

Oates, W.E., 1999. An essay on fiscal federalism. J. Econ. Lit. 37 (3), 1120–1149.

Oates, W.E., 2001. A Reconsideration of Environmental Federalism, Discussion Paper 01–54, Resources for the Future. Washington, DC.

Parsons, G.R., Massey, D.M., Tomasi, T., 1999. Familiar and favorite sites in a random utility model of beach recreation. Mar. Resour. Econ. 14 (4), 299–315. http://www.uri.edu/cels/enre/mre/mre.htm.

Parsons, G.R., Powell, M., 2001. Measuring the cost of beach retreat. Coastal Manage. 29 (2), 91–103.

Phillips, M.R., Jones, A.L., 2006. Erosion and tourism infrastructure in the coastal zone: problems, consequences and management. Tourism Manage. 27 (3), 517–524.

Pilkey, O.H., Dixon, K.L., 1996. The Corps and the Shore. Island Press, Washington, DC.

Pilkey, O.H., Young, R., 2008. The Rising Sea. Island Press, Washington, DC.

Riggs, S.R., Ames, D.V., 2003. Drowning the North Carolina Coast: Sea-level Rise and Estuarine Dynamics, Raleigh, North Carolina.

Rosen, S., 1974. Hedonic prices and implicit markets - product differentiation in pure competition. J. Polit. Econ. 82 (1), 34–55.

Sallenger, A.H., 2000. Storm impact scale for barrier islands. J. Coastal Res. 16 (3), 890–895.

Shivlani, M.P., Letson, D., Theis, M., 2003. Visitor preferences for public beach amenities and beach restoration in South Florida. Coastal Manage. 31 (4), 367–385.

Silberman, J., Gerlowski, D.A., Williams, N.A., 1992. Estimating existence value for users and nonusers of New Jersey beaches. Land Econ. 225–236.

Slott, J.M., Murray, A.B., Ashton, A.D., 2010. Large-scale responses of complex-shaped coastlines to local shoreline stabilization and climate change. J. Geophys. Res. Earth Surf. 115, F03033. http://dx.doi.org/10.1029/2009jf001486.

Slott, J.M., Murray, A.B., Ashton, A.D., Crowley, T.J., 2006. Coastline responses to changing storm patterns. Geophys. Res. Lett. 33 (L18404). http://dx.doi.org/10.1029/2006GL027445.

Small, C., Nicholls, R.J., 2003. A global analysis of human settlement in coastal zones. J. Coastal Res. 19 (3), 584–599. http://dx.doi.org/10.2307/4299200.

Smith, M.D., Slott, J.M., McNamara, Dylan, Brad Murray, A., 2009. Beach nourishment as a dynamic capital accumulation problem. J. Environ. Econ. Manage. 58 (1), 58–71.

Van den Berg, N., Falqués, A., Ribas, F., 2011. Long-term evolution of nourished beaches under high angle wave conditions. J. Mar. Syst. 88 (1), 102–112.

Whitehead, J.C., Dumas, C.F., Herstine, J., Hill, J., Buerger, B., 2008. Valuing beach access and width with revealed and stated preference data. Mar. Resour. Econ. 23 (2), 119–135. http://www.uri.edu/cels/enre/mre/mre.htm.

Wolinsky, M.A., Murray, A.B., 2009. A unifying framework for shoreline migration: 2. Application to wave-dominated coasts. J. Geophys. Res. Earth Surf. 114 (F01009).

Making the Link

By Jean-Paul Vanderlinden

Geoengineering Coastlines? From Accidental to Intentional	⟺	**Remote Sensing Solutions to Monitor Biotic and Abiotic Dynamics in Coastal Ecosystems**
Martin Smith, A. Brad Murray, Sathya Gopalakrishnan, Andrew G. Keeler, Craig E. Landry, Dylan McNamara and Laura J. Moore		*Andrea Taramelli, Emiliana Valentini and Loreta Cornacchia*

Smith and colleagues raise a critical question: how local actions, in this instance coastline stabilization, can be implemented while taking into account the impacts of local action on multiple scales, be they temporal or spatial. They argue in favor of multi-cross-scale coordination that is one of the central and most challenging tenets of coastal zone management: how to make visible the locally invisible. In light of the foreseen impacts of climate change, this challenge needs to be addressed urgently as the dynamics of teleconnection will increase as the time-frames involved shorten.

So how do we turn the locally invisible into a locally visible? How do we set the conditions for successful cross-scale coordination? How does one adopt the already mentioned bird's eye view? An answer is found in the next chapter, from Taramelli and colleagues. In this chapter, the authors describe and analyze how remote sensing, combined with subpixel techniques, may allow for the monitoring, maybe someday automated, of coastal zone seascapes. The case studies that are presented show how, through the mobilization of scaled knowledge, one is empowered by being able to assess the impacts of actions "here and now" and "there and tomorrow."

Chapter 8

Remote Sensing Solutions to Monitor Biotic and Abiotic Dynamics in Coastal Ecosystems

Andrea Taramelli[1,3], Emiliana Valentini[1], Loreta Cornacchia[1,2]
[1]*ISPRA - Istituto Superiore per la Protezione e Ricerca Ambientale, Rome, Italy;*
[2]*Royal Netherlands Institute for Sea Research (NIOZ), Yerseke, The Netherlands;*
[3]*IUSS, Istituto Universitario di Studi Superiori, Pavia, Italy*

Chapter Outline

INTRODUCTION

Coastal ecosystems are hugely important for ecological and economic reasons, but are at the same time very vulnerable to degradation caused by human activities (such as dredging and habitat destruction), and threats linked to climate change (increase in sea-level, flooding risks, occurrence of extreme events), which can ultimately increase erosion in coastal areas. Remote sensing (RS) datasets are increasingly considered to satisfy reporting obligations under the conservation directives (Lengyel et al., 2008; Vanden Borre et al., 2011). This powerful tool for making spatially detailed habitat types maps (Roughgarden et al., 1991; Wickland, 1991; McDermid, 2005) takes advantages of the spatial and temporal resolution of satellite data, and it is increasingly popular as a potential tool to aid in decision making for planning and management over large areas (Melesse et al., 2007).

In this context, land- and seascapes are complex systems, and scientists have tried to apply general systems theory to them (e.g., Chorley, 1962; Tebbens et al., 2002). While the dynamics of these systems are as yet little understood, there is increasing evidence for feedback mechanisms between geomorphology and ecology at different spatial scales (microphytobenthos, salt marshes, arid and dune vegetation), resulting in alternate stable states (e.g., van de Koppel et al.,

2001; Weerman et al., 2011). Empirically, certain "laws" have been found to hold in such self-ordered systems (Scheidegger, 1994; Taramelli et al., 2014a), such as power laws (Newman, 2005; Rietkerk and van de Koppel, 2008 and references therein). Similar power law distributions have been observed in patch size distributions of microphytobenthos or microalgae-dominated microbial mats in intertidal flats (e.g., Weerman et al., 2012). Interestingly, researches in both arid (Kéfi et al., 2007; Taramelli et al., 2012) and wetland (Rietkerk et al., 2004; Weerman et al., 2012) ecosystems suggest that vegetation spatial patterns can be used as possible indicators of ecosystem degradation in response to biotic or environmental stress (cf. also Scheffer et al., 2009). Deviation from a power law can be interpreted as a sign of an upcoming shift in the ecosystem.

There is an increasing need for robust methodologies, which can integrate in situ knowledge, data acquisition, and monitoring activities. It is clear that in order to understand the complex, nonlinear dynamics of coastal ecosystems, and more specifically the impact of external forces (natural and anthropogenic) on these dynamics, good process knowledge, based on the acquisition of accurate and complete datasets of ecosystem attributes, is indispensable. Because of the highly dynamic and heterogeneous nature of coastal ecosystems, both in space and in time, and the difficulty to obtain quantitative data over larger areas, RS is ideal and cost-efficient to obtain high-resolution data.

The scientific community can now take advantage of the free availability of images with high temporal resolution and lower spatial resolution (e.g., MODIS, AVHRR) (Maiti and Bhattacharya, 2009) as well as the set of multispectral medium-resolution Landsat (MSS, TM, ETM+) with a medium temporal one. In this sense, and considering the spatial and temporal prerequisites of this study, which requires the use of satellite imagery for characterizing the past land cover succession processes through the examination of a multitemporal image series, the use of Landsat TM and ETM+ imagery seems to be the most appropriate, as this satellite family has been continuously providing images since the 1970s from three main satellites. The coming generation of earth observation (EO) operational satellite instruments, like Copericus-Eumetsat-ESA's Sentinel-1 and 2 (launched early 2014), and NASA's Landsat-8 LDCM will provide increased possibilities to the coastal monitoring due to their improved spectral, spatial, and radiometric resolution, and their operational sense. These developments will in principle enable more sophisticated procedures for a better retrieval of climate relevant parameters.

These insights will be exploited, showing different examples of the RS processing chain, in order to improve ecosystem analysis on the basis of the integration of different sensors and spatial resolutions (both new and previous missions), and to understand the scale-dependency of the observed parameters and spatial patterns, and their spatial and temporal evolution in a broader geomorphological and ecological context. The time-series analysis will provide new interpretations of coastal evolution and ecosystem degradation based on eco-geomorphological maps and spatial patterning of coastal vegetation. This will provide us with a baseline for the prediction of the response of coastal systems to global change phenomena (e.g., sea-level rise, increased erosion).

METHODS

The aim of the RS methodology is to locate and characterize important coastal parameters for different regions using multitemporal data from the multi/hyper-spectral sensors, as well as topographic and bathymetric elevation, SAR phase/amplitude, and LiDAR data. The identification of different indicators is based on spectral properties, topography/landforms (low topography), disturbed areas (agricultural, construction), vegetation distribution, and field knowledge. In addition, dune vegetation, salt marshes, microphytobenthos cover, and submerged vegetation can each be found adjacent or stand-alone in different study sites (Figure 8.1).

In coastal areas, where reflectance spectra vary considerably at scales comparable to or smaller than the instantaneous field of view (IFOV), the spectral

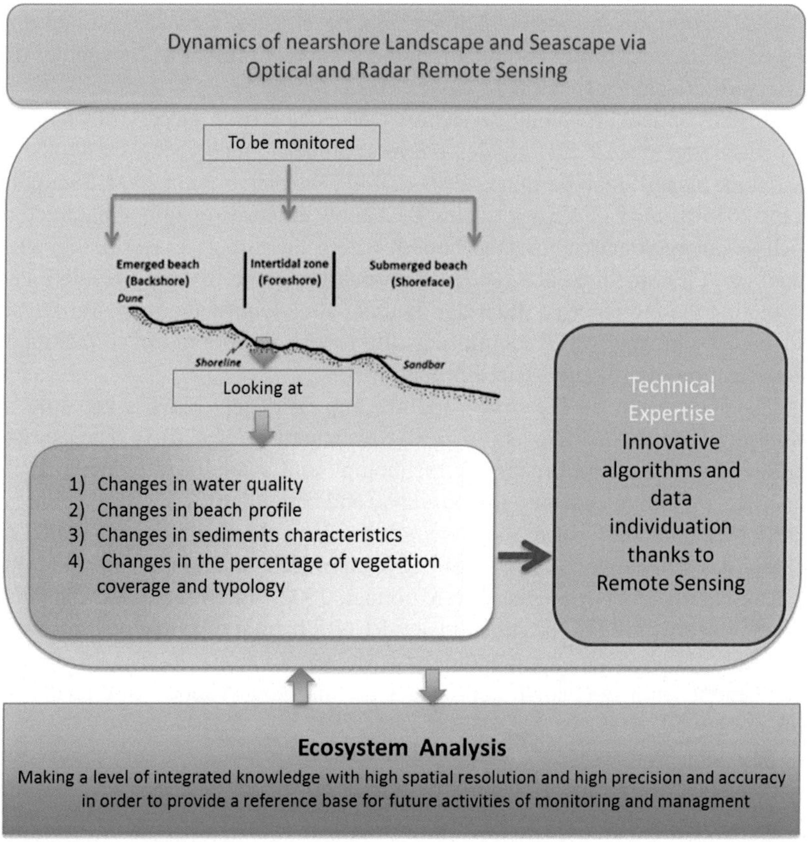

FIGURE 8.1 Study logic approach for coastal environments that can be considered as mosaics of interacting ecosystems (Forman and Gordon, 1986). This view implies relationships between landscape and seascape in more (i.e., emerged nearshore) or less (i.e., submerged nearshore/shallow waters) evident feedbacks.

reflectance of an individual pixel will generally not resemble the reflectance of a single class (dark surface, soil or vegetation), but rather a mixture of the reflectances of two or more classes within the sensor's IFOV (Small, 2004). The concept of the mixing space was developed to facilitate analysis of mixed reflectance with linear mixture models. The linear mixture problem can be treated using linear inverse theory. Linear mixture models are based on the observation that, in many situations, radiances from surfaces with different "end-member" reflectances mix linearly in proportion to their area within the IFOV. This observation has enabled the development of a systematic methodology for spectral mixing analysis (SMA) (Adams et al., 1986), which has proven successful for a variety of quantitative applications with multispectral imagery (e.g., Smith et al., 1985; Elmore et al., 2000; Taramelli and Melelli, 2009; Valentini et al., 2015). If a limited number of distinct spectral end-members can be identified, then it is possible to define a mixing space where mixed pixels can be described as linear mixtures of these end-members. Given sufficient spectral resolution, a system of linear mixing equations can be defined and solved for the best-fitting combination of end-member fractions for each of the observed reflectance spectra.

Besides optical RS techniques, active sensors have been successfully applied in coastal areas: SAR investigates the morphometric rates using the DInSAR technique named small baseline subset (SBAS; Berardino et al., 2002; Taramelli et al., 2014b), and LiDAR backscatter data allow the investigation of parameters such as emerged and submerged bottom (using the airborne laser bathymetry roughness; Taramelli et al., 2013). Preliminary surface mapping results, also correcting the influence of the water column, already provide inputs for physical process models while monitoring differences in bottom cover facilitates identification of different surface properties.

The integration of EO from satellites and field data plays a key role in assessing biodiversity (Nagendra, 2001; Nagendra et al., 2013). The combination of satellite data processing technology and spatially referenced ancillary data, including vegetation, topography and bathymetries, climate, geology, landforms and bedfororms, sedimentology (e.g., McDermid and Franklin, 1995), has proved to be a cost-effective approach to classify and quantify coastal parameters (Donoghue and Mironnet, 2002). Interpreting EO data and ground observation for Natura 2000 and EUNIS habitat type mapping requires basic knowledge from expert ecologists (McDermid et al., 2005) and can further benefit from innovative approaches in this field (Vanden Borre, 2011b; Valentini et al., 2015).

Although field spectral library collection is not a reliable and repeatable procedure, it allows the collection of properties of physical surfaces that are generally repeatable (e.g., Eastwood et al., 1997; Taramelli et al., 2013). Field work could produce errors from the sampling grid that overlooks features between adjacent survey sites compared with the systematic analysis of RS images integrated with field measurements (Green et al., 2012), but this problem can be ameliorated by using field spectral radiometer measurements and ground observations for image calibration purposes (Lucas et al., 2002). Nevertheless, when comparing proximal and RS measurements, the spatial scales involved are drastically different (Bian, 1997), and expert knowledge can be the key to relating scales, sensors, and the environmental variability. While the large scale can be accounted for by studying changes in key variables as computed by the most recent models (both morphodynamic and bio-climatic), regional and local scales can be studied by applying a dynamical downscaling of coastal evolving characteristics using RS and geostatistical GIS-based analytical procedures, integrating multisensory and multidisciplinary approaches (Figure 8.2).

CASE STUDIES

Foce Bevano, Italy

Foce Bevano is a small fluvial outlet located in the northern Adriatic, near the town of Ravenna. The Bevano is a river with a catchment area of 92.5 km^2 (Balouin et al., 2006) and a total length of 34 km. The river mouth presents ecologically important habitats, such as wetlands, pinewoods, sandy beaches, and sand dunes. This relatively minor watercourse forms a small-scale estuarine system.

Over the past 50 years, the Bevano area has undergone major morphological changes (Armaroli et al., 2013; VV. AA., 2009). The mouth underwent a rapid northwards migration due to marine processes and a low-energy fluvial regime. Also, subsidence phenomena lowered Foce Bevano by about 21 cm from 1984 to 2005 and decreased the slope toward the sea (VV. AA., 2009; Vicinanza et al., 2009).

Landsat TM and ETM+ imagery (U.S. Geological Survey, 2003) were analyzed to obtain the main spatial and temporal patterns of change in the study site between 1992 and 2000 (Taramelli et al., 2014a). The basic methodology was to use RS time-series to analyze (1) trends in coastal evolution over time, and (2) time series of spatial patterns in coastal vegetation for improved ecosystem analysis. More specifically, changes in vegetation size and density cover were used to characterize the coastal area evolution by combining multisensor spaceborne RA (Figure 8.3).

Due to the subsidence phenomenon affecting the area, a time-series of SAR (synthetic aperture radar) data (ERS-1/2 satellites) was used to produce deformation maps of coastal morphology through the SBAS algorithm (Berardino et al., 2002). Change detection analysis and Empirical orthogonal functions (EOFs) (Bjornsson and Venegas, 1997; Hannachi et al., 2007; Lorenz, 1956; Taramelli et al., 2012) were, respectively, used to measure the variation in vegetation cover over the whole study period, and to quantify spatial variability patterns and the main temporal evolution trends year by year.

The Bevano area appears strongly changed in terms of halophytic and riparian vegetation (Figure 8.3). This increase in dune vegetation has taken place where the mouth of the estuary has morphologically changed during the last two decades. In

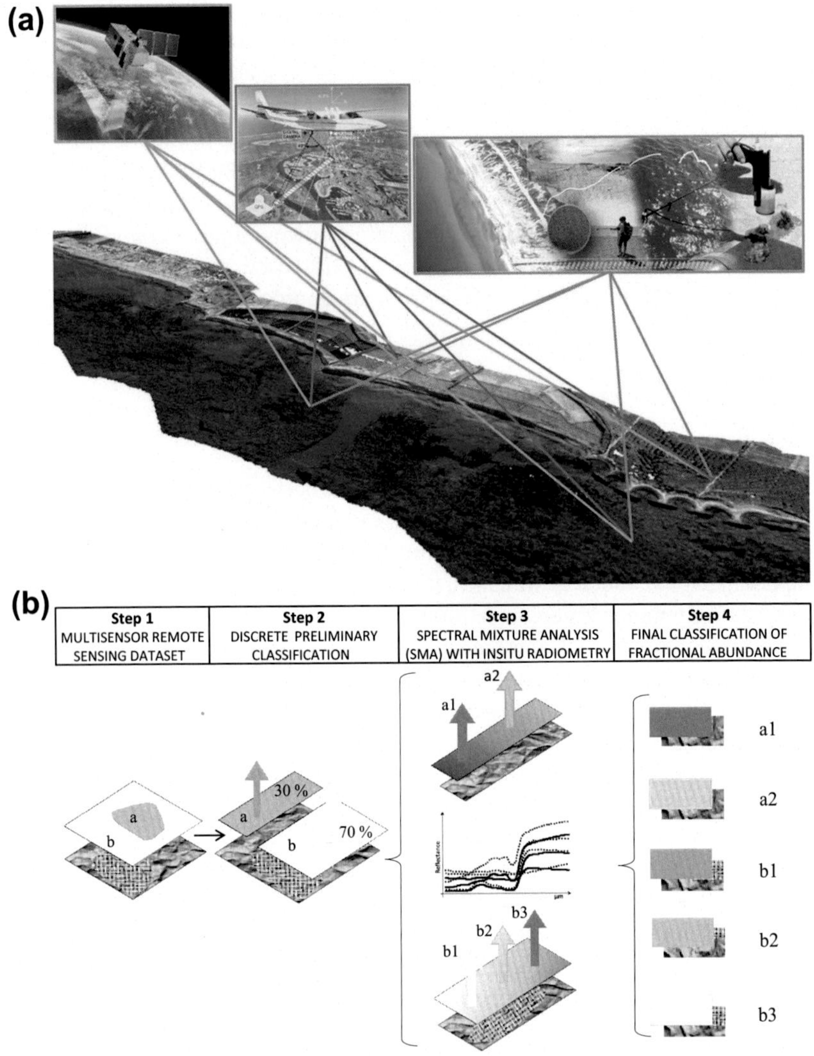

FIGURE 8.2 (a) Remote sensing data fusion processing chains include space-borne, airborne, and field sensor integration. Different scales analyzed with RA active and passive sensors (airborne, satellite, and in situ) generates multisensory and multidimensional datasets of signals. (b) An example of a processing chain that integrate three typologies of datasets. *(From Valentini E., 2013, Ph.D. thesis).* Starting from an airborne hyperspectral image and LiDAR altimetry (Step 1), a discrete classification can separate the major spectral variability (Step 2). By coupling the linear spectral mixture models (LSMAs) with single and multiple spectra from field spectral radiometry (Step 3), the number of detectable targets can be increased at the subpixel level and fractional abundance for each pixel can be classified by using thresholds of abundance (Step 4).

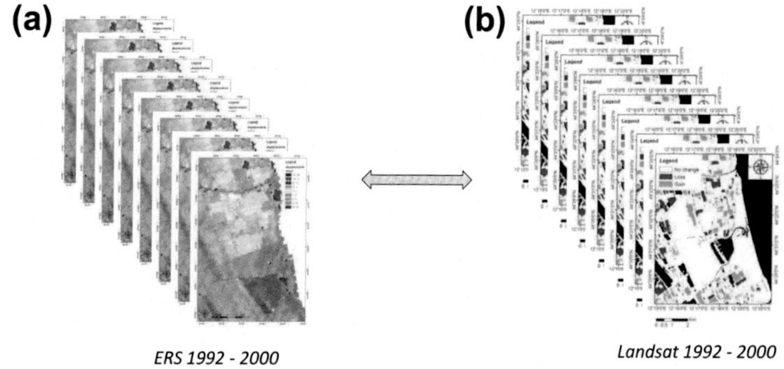

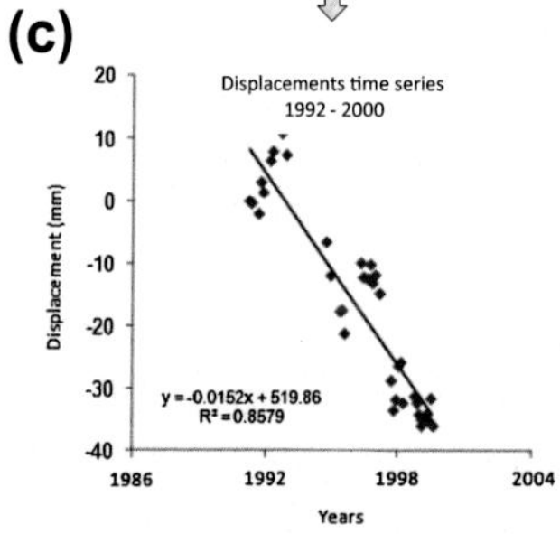

(d)

FRACTION	N. OF PIXEL	PERCENT
LOSS	504900	15,08%
NO CHANGES	2673900	79,84%
GAIN	170100	5,08%

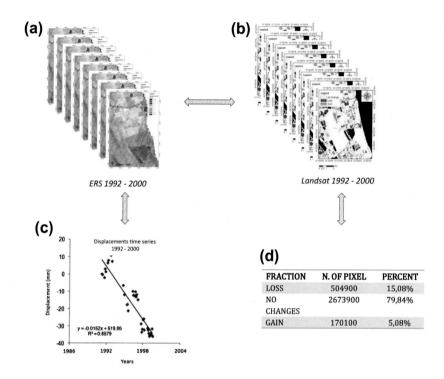

FIGURE 8.3 Example of multitemporal analysis in coastal sandy beaches (Foce Bevano, Adriatic sea) based on subsidence trends and vegetation changes (radar, multispectral). (a) Mean velocity maps (cm-year) time series. (b) Vegetation fraction change analysis using fractions maps. (c) Time series (mm/year) at the Foce Bevano test sites. (d) Gain and loss statistics within the difference image of the fraction maps and the NDVI.

particular, the loss and gain values at these sites correspond to the high degree of change. The multitemporal morphological analysis (subsidence) showed higher displacement rates where wetland hydrology is more strongly influenced by anthropic activities (Taramelli et al., 2014b). At these sites, 9% of loss in total vegetation value is used as an accurate threshold to delineate the signature of the vegetation loss within the subsidence pixels. Vegetation, is a good indicator of salinity and piezometric level of the area, showing clear gain and loss near subsiding pixels; this is in accordance with changes in both elevation and moisture/salinity that can shift vegetation from halophytic to non-halophytic, making communities unstable.

Saeftinghe Salt Marsh, Scheldt Estuary

Saeftinghe is a salt marsh and nature reserve located in the Scheldt estuary, in the southern oligohaline portion at the Dutch-Belgian border. In the 1930s, only 25% of the area was covered by salt marsh vegetation, while the rest consisted of mudflats and tidal channels (Vandenbruwaene et al., 2012). At the present time,

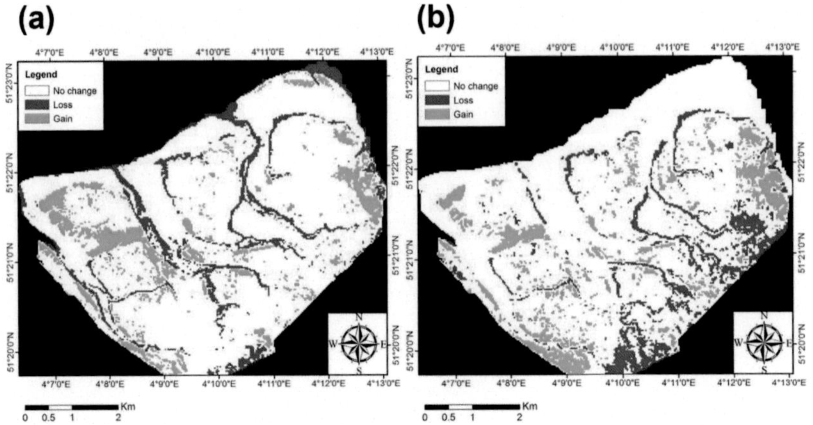

FIGURE 8.4 Change detection from NDVI (a) and SMA fractions map (b) between 1986 and 2006, for the Saeftinghe salt marsh. Gain pixels (in green) are pixels with increased vegetation cover during the study period; loss pixels (in red) are pixels with decreased vegetation cover.

the proportions are reversed: Saeftinghe is formed by 30% of mudflats and channel systems, while salt marsh vegetation covers the remaining 70% of the surface.

Landsat TM and ETM+ imagery (U.S. Geological Survey, 2003) were used to analyze the main spatial and temporal patterns of vegetation change in the study site between 1986 and 2006 (Taramelli et al., 2014a). The change detection through NDVI and SMA vegetation fraction maps revealed a marked increase in salt marsh due to the colonization by pioneer species and the development of extensive new areas of vegetation in previously bare tidal creeks (Figure 8.4, "gain"pixels in green on the map).

The general positive trend in vegetation development is also inferred from the empirical orthogonal function analysis on the SMA vegetation fraction maps: areas of new vegetation colonization are clearly identified, while the inverse change from vegetated to bare tidal flat areas is rarely detected. The reason is probably linked to the high tidal range and suspended sediment concentration in the salt marsh, allowing high sediment accretion rates and low risk of vegetation loss due to sea level rise (Wang and Temmerman, 2013). The tidal range also provides a constantly high drainage, ensuring favorable plant growth conditions.

New insights into estuarine salt marsh evolution have been emerging from the study of vegetation patterns: the size-frequency distributions of macrophyte patches extracted from medium-resolution satellite RS data of the Saeftinghe marsh appear to follow a power law with varying scaling exponent (Figure 8.5). This power law distribution is encountered at all different structural levels of herbaceous vegetation colonizing the salt marsh.

As a last step, the main trends of vegetation evolution in the Saeftinghe salt marsh were related to the patterns of climate variations, and in particular

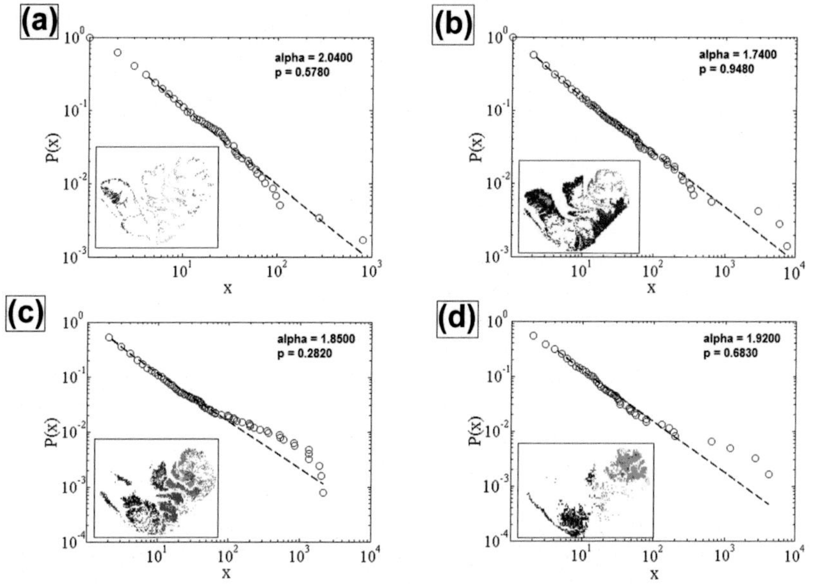

FIGURE 8.5 Size-frequency distribution of vegetation pattern areas (x) in the Saeftinghe salt marsh, extracted from fraction maps on SPOT-4 imagery. Log–log plots from (a) to (d) each correspond to a structural level of herbaceous vegetation, ranging from the dynamic pioneer zone in (a) to the stable, dense vegetation in (d). For each graph, the inserts on the bottom left corner are maps of the Saeftinghe marsh, showing the location of each vegetation zone. P-values greater than 0.1 indicate that a power law is a good fit to the data (Clauset et al., 2009).

in annual mean sea-level changes over the same time frame. Results show a similar trend of annual sea level and vegetation evolution during the time series between 1986 and 2006, showing a possible influence of water levels on changes in vegetation cover. The Saeftinghe salt marsh can be regarded as an example of a dynamic system responding adaptively to sea level changes, with potential benefits to society by providing continued coastal protection.

CONCLUSIONS

In a rapidly changing environment, traditional knowledge of physical processes (both biotic and abiotic) is insufficient to adequately deal with threats to, and opportunities for, societies in coastal areas. These challenges, such as floods, droughts, salinization, changes to environments as a result of urbanization, and changes to ecosystems as a result of climate change, are inherently probabilistic with often-unknown uncertainty bounds. To move further in research this may involve the development of new computational techniques and models that integrate physical, environmental, and societal systems. The objective of this chapter was to demonstrate the value of EO-based information products for the emerging sector of ecosystem services valuation and to enhance the science and practical application of ecosystem services assessment within coastal area vulnerability analyses.

GLOSSARY

AVHRR Advanced very high resolution radiometer (AVHRR) instruments are a type of spaceborne sensor that measures the reflectance of the earth in five wide spectral bands. Most AVHRR instruments are carried by the National Oceanic and Atmospheric Administration (NOAA) family of polar orbiting platforms (POES).

Copernicus Copernicus, previously known as GMES (Global Monitoring for Environment and Security), is the European program for the establishment of a European capacity for earth observation.

DInSAR The DInSAR technique compare images that are acquired in different times (temporal baseline), and measures the changes of the surface. These measures are shown by a series of colored bands, the so-called fringes or interferogram. The electromagnetic waves used are characterized by an alternation of crests spaced about 5 cm apart; this distance is called the wavelength.

End-member Spectral end-members are defined as spectrally "pure" features (e.g., vegetation, soil, etc.). Pure spectral end-members are usually defined under idealized in situ or laboratory conditions where reflectance spectra are acquired using a portable spectroradiometer focused only on a single surface (e.g., a single leaf from a maple tree). When in situ measurements are not possible, spectral end-members can also be derived from "pure" features in the imagery

EOF In statistics and signal processing, the method of empirical orthogonal function (EOF) analysis is a decomposition of a signal or dataset in terms of orthogonal basis functions that are determined from the data. It is the same as performing a principal components analysis on the data, except that the EOF method finds both time series and spatial patterns. The term is also interchangeable with the geographically weighted PCAs in geophysics

ESA The European Space Agency (ESA) is an intergovernmental organization dedicated to the exploration of space, with 20 member states.

EUMETSAT EUMETSAT is an intergovernmental organization founded in 1986. Its purpose is to supply weather and climate-related satellite data, images, and products—24 h a day, 365 days a year—to the National Meteorological Services of the member and cooperating states in Europe, and other users worldwide.

IFOV The IFOV is a measure of the spatial resolution of an RS imaging system. Defined as the angle subtended by a single detector element on the axis of the optical system, the IFOV and the distance from the target determine the spatial resolution. A low-altitude imaging instrument will have a higher spatial resolution than a higher-altitude instrument with the same IFOV.

InSAR SAR interferometry (InSAR) compares ("interferes") two images acquired from slightly different positions (spatial baseline), producing three-dimensional images of the earth's surface, measuring the topography.

Landsat ETM+ The Landsat Enhanced Thematic Mapper Plus (ETM+) sensor onboard the Landsat 7 consists of eight spectral bands with a spatial resolution of 30 m for bands 1 to 7.

Landsat MSS The Multispectral Scanner (MSS) is one of the original Landsat imaging sensors and has flown on Landsat satellites 1 through 5.

Landsat TM The Landsat Thematic Mapper (TM) is a multispectral scanning radiometer that was carried on board Landsats 4 and 5.

Landsat 8 LDCM The Landsat 8 Data Continuity Mission (LDCM) operates in the visible, near-infrared, short-wave infrared, and thermal infrared spectrums providing moderate-resolution imagery, from 15 to 100 m.

LIDAR The light detection and ranging (LIDAR) is an RS technology that measures distance by illuminating a target with a laser and analyzing the reflected light.

MODIS The Moderate Resolution Imaging Spectroradiometer (MODIS) is a key instrument aboard the Terra (EOS AM) and Aqua (EOS PM) satellites.

Pixel In digital imaging, a pixel is a physical point in a raster image, or the smallest addressable element in an all-points-addressable display device; so it is the smallest controllable element of a picture represented on the image.

Reflectance spectra Reflectance spectroscopy is the study of light as a function of wavelength that has been reflected or scattered from a solid, liquid, or gas. As photons enter a mineral, some are reflected from grain surfaces, some pass through the grain, and some are absorbed. Those photons that are reflected from grain surfaces or refracted through a particle are said to be scattered. Scattered photons may encounter another grain or be scattered away from the surface so they may be detected and measured.

SAR Synthetic aperture radar (SAR) instruments are active systems that transmit a beam of radiation in the microwave region of the electromagnetic spectrum. SAR can provide day-and-night imagery of the earth.

SBAS The small baseline subset (SBAS) processing technique generates deformation time series starting from SAR data.

Sentinel missions The sentinel missions are a new family of missions called sentinels specifically for the operational needs of the Copernicus program. Each sentinel mission is based on a constellation of two satellites to fulfil revisit and coverage requirements, providing robust datasets for Copernicus Services. These missions carry a range of technologies, such as radar and multispectral imaging instruments for land, ocean, and atmospheric monitoring.

SMA Spectral mixture analysis (SMA) is based on a physical representation of spectrally mixed pixels as areal fractions of spectral end-members. SMA simulates the physical process of mixed radiance measurement in the IFOV with a linear mixture model. Inverting the linear mixture model yields quantitative estimates of the areal abundance (fractions) of specific land cover types (end-members) within each pixel.

REFERENCES

Adams, J.B., Smith, M.O., Johnson, P.E., 1986. Spectral mixture modeling: a new analysis of rock and soil types at the Viking Lander 1 site. J. Geophys. Res. 91 (B8), 8098–8112.

Armaroli, C., Grottoli, E., Harley, M.D., Ciavola, P., 2013. Beach morphodynamics and types of foredune erosion generated by storms along the Emilia-Romagna coastline, Italy. Geomorphology 199, 22–35.

Balouin, Y., Ciavola, P., Michel, D., 2006. Support of subtidal tracer studies to quantify the complex morphodynamics of a river outlet: the Bevano, NE Italy. J. Coastal Res. 602–606.

Berardino, P., Fornaro, G., Lanari, R., Sansosti, E., 2002. A new algorithm for surface deformation monitoring based on small baseline differential SAR interferograms. Geosci. Remote Sens., IEEE Trans. 40 (11), 2375–2383.

Bian, L., 1997. Multiscale nature of spatial data in scaling up environmental models. In: Quattrochi, D.A., Goodchild, M.F. (Eds.), Scale in Remote Sensing and GIS, Raton Lewis Publishers, Boca Raton, FL, pp. 13–26.

Bjornsson, H., Venegas, S., 1997. A Manual for EOF and SVD Analyses of Climatic Data. CCGCR Report, 97(1).

Chorley, R.J., 1962. Geomorphology and General Systems Theory. US Government Printing Office, Washington, DC.

Clauset, A., Shalizi, C.R., Newman, M.E., 2009. Power-law distributions in empirical data. SIAM Rev. 51 (4), 661–703.

Donoghue, D.N., Mironnet, N., 2002. Development of an integrated geographical information system prototype for coastal habitat monitoring. Comput. Geosci. 28 (1), 129–141.

Eastwood, J.A., Yates, M.G., Thomson, A.G., Fuller, R.M., 1997. The reliability of vegetation indices for monitoring saltmarsh vegetation cover. Int. J. Remote Sens. 18 (18), 3901–3907.

Elmore, A.J., Mustard, J.F., Manning, S.J., Lobell, D.B., 2000. Quantifying vegetation change in semiarid environments: precision and accuracy of spectral mixture analysis and the normalized difference vegetation index. Remote Sens. Environ. 73, 87–102.

Forman, R.T.T., Gordron, M., 1986. Landscape Ecology. Wiley, New York.

Green, R., Buchanan, G., Almond, R., 2012. What Do Conservation Practitioners Want from Remote Sensing? Cambridge Conservation Initiative (summary of 5 workshops).

Hannachi, A., Jolliffe, I., Stephenson, D., 2007. Empirical orthogonal functions and related techniques in atmospheric science: a review. Int. J. Climatol. 27 (9), 1119–1152.

Kéfi, S., Rietkerk, M., Alados, C.L., Pueyo, Y., Papanastasis, V.P., ElAich, A., De Ruiter, P.C., 2007. Spatial vegetation patterns and imminent desertification in Mediterranean arid ecosystems. Nature 449 (7159), 213–217.

van de Koppel, J., Herman, P.M., Thoolen, P., Heip, C.H., 2001. Do alternate stable states occur in natural ecosystems? evidence from a tidal flat. Ecology 82 (12), 3449–3461.

Lengyel, S., Déri, E., Varga, Z., Horváth, R., Tóthmérész, B., Henry, P.-Y., Kobler, A., Kutnar, L., Babij, V., Seliˇskar, A., Christia, C., Papastergiadou, E., Gruber, B., Henle, K., 2008. Habitat monitoring in Europe: a description of current practices. Biodivers. Conserv. 17, 3327–3339.

Lorenz, E.N., 1956. Empirical Orthogonal Functions and Statistical Weather Prediction. Report no.1, Statistical forecasting project. MIT Massachusetts Institute of Technology, Department of Meteorology, Cambridge, Massachusetts.

Lucas, N.S., Shanmugam, S., Barnsley, M., 2002. Sub-pixel habitat mapping of a costal dune ecosystem. Appl. Geogr. 22 (3), 253–270.

Maiti, S., Bhattacharya, A.K., 2009. Shoreline change analysis and its application to prediction: a remote sensing and statistics based approach. Mar. Geol. 257 (1), 11–23.

McDermid, G.J., Franklin, S.E., LeDrew, E.F., 2005. Remote sensing for large-area habitat mapping. Prog. Phys. Geogr. 29 (4), 449–474.

McDermid, G.J., & Franklin, S.E., 1995. Remote sensing and geomorphometric discrimination of slope processes. Zeits. Geomorph. Supp. 165–185.

Melesse, A.M., Weng, Q., Thenkabail, P.S., Senay, G.B., 2007. Remote sensing sensors and applications in environmental resources mapping and modelling. Sensors 7 (12), 3209–3241.

Nagendra, H., 2001. Using remote sensing to assess biodiversity. Int. J. Remote Sens. 22, 2377–2400.

Nagendra, H., Lucas, R., Honrado, J.P., Jongman, R.H.G., Tarantino, C., Adamo, M., Mairota, P., 2013. Remote sensing for conservation monitoring: assessing protected areas, habitat extent, habitat condition, species diversity, and threats. Ecol. Indic. 33, 45–59.

Newman, M.E., 2005. Power laws, Pareto distributions and Zipf's law. Contemp. Phys. 46 (5), 323–351.

Rietkerk, M., van de Koppel, J., 2008. Regular pattern formation in real ecosystems. Trends Ecol. Evol. 23 (3), 169–175.

Rietkerk, M., Dekker, S.C., de Ruiter, P.C., van de Koppel, J., 2004. Self-Organized patchiness and catastrophic shifts in ecosystems. Science 24, 1926–1929.

Roughgarden, J., Running, S.W., Matson, P.A., 1991. What does remote sensing do for ecology? Ecology 72, 1918–1922.

Scheffer, M., Bascompte, J., Brock, J.A., Brovkin, V., Carpenter, S.R., Dakos, V., Held, H., van Nes, E.H., Rietkerk, M., Sugihara, G., 2009. Early-warning signals for critical transitions. Nature 461, 53–59.

Scheidegger, A.E., 1994. Hazards: singularities in geomorphic systems. Geomorphology 10, 19–25.

Small, C., 2004. The Landsat ETM+ spectral mixing space. Remote Sens. Environ. 93 (1), 1–17.

Smith, M.O., Johnson, P.E., Adams, J.B., 1985. Quantitative determination of mineral types and abundances from reflectance spectra using principal components analysis. J. Geophys. Res.: Solid Earth (1978–2012) 90 (S02), C797–C804.

Taramelli, A., Melelli, L., 2009. Map of deep-seated gravitational slope deformations susceptibility in central Italy derived from SRTM DEM and spectral mixing analysis of the Landsat ETM+ data. Int. J. Remote Sens. 30 (2), 357–387.

Taramelli, A., Pasqui, M., Barbour, J., Kirschbaum, D., Bottai, L., Busillo, C., Calastrini, F., Guarnieri, F., Small, C., 2012. Spatial and temporal dust source variability in northern China identified using advanced remote sensing analysis. Earth Surf. Processes Landforms 38, 793–809. http://dx.doi.org/10.1002/esp.337174.

Taramelli, A., Valentini, E., Innocenti, C., Cappucci, S., 2013. FHYL: field spectral libraries, airborne hyperspectral images and topographic and bathymetric LiDAR data for complex coastal mapping. In: Geoscience and Remote Sensing Symposium (IGARSS). IEEE International, 21–26 July 2013, pp. 2270–2273. http://dx.doi.org/10.1109/IGARSS.2013.6723270.

Taramelli, A., Valentini, E., Cornacchia, L., Mandrone, S., Monbaliu, J., Hoggart, S.P.G., Thompson, R.C., Zanuttigh, B., 2014a. Modeling uncertainty in estuarine system by means of combined approach of optical and radar remote sensing. Coastal Eng. 87, 77–96.

Taramelli, A., DiMatteo, L., Ciavola, P., Guadagnano, F., Tolomei, C., 2014b. Temporal evolution of patterns and processes related to subsidence of the coastal area surrounding the Bevano River mouth (Northern Adriatic) – Italy. Ocean Coastal Manage. http://dx.doi.org/10.1016/j.ocecoaman.2014.06.021.

Tebbens, S.F., Burroughs, S.M., Nelson, E.E., 2002. Wavelet analysis of shoreline change on the Outer Banks of North Carolina: an example of complexity in the marine sciences. Proc. Natl. Acad. Sci. 99, 2554.

U.S. Geological Survey, 2003. Landsat: A Global Land-observing Program. Fact Sheet 023-03 http://egsc.usgs.gov/isb/pubs/factsheets/fs02303.html.

Valentini, E., 2013. A new paradigm in coastal ecosystem assessment: linking ecology and geomorphology implementing the FHyL approach. PhD Thesis at Tuscia University, Viterbo (Italy)

Valentini, E., Taramelli, A., Filipponi, F., Giulio, S., 2015. An effective procedure for EUNIS and Natura 2000 habitat type mapping in estuarine ecosystems integrating ecological knowledge and remote sensing analysis. Ocean Coastal Manage. http://dx.doi.org/10.1016/j.ocecoaman.2014.07.015.

Vanden Borre, J.V., Paelinckx, D., Mücher, C.A., Kooistra, L., Haest, B., De Blust, G., Schmidt, A.M., 2011. Integrating remote sensing in Natura 2000 habitat monitoring: prospects on the way forward. J. Nat. Conserv. 19 (2), 116–125.

Visser, J.M., Sasser, C.E., Chabreck, R.H., Linscombe, R.G., 2002. The impact of a severe drought on the vegetation of a subtropical estuary. Estuaries 25, 1184–1195.

Vandenbruwaene, W., Bouma, T.J., Meire, P., Temmerman, S., 2012. Bio-geomorphic effects on tidal channel evolution: impact of vegetation establishment and tidal prism change. Earth Surf. Processes Landforms 38 (2), 122–132.

Vicinanza, D., Ciavola, P., Biagi, S., 2009. Field experiment to control coastline subsidence: a unique case study at Lido Adriano (Italy). J. Coastal Res. 1105–1109.

VV. AA., 2009. Foce Bevano: l'area naturale protetta e l'intervento di salvaguardia. Servizio Difesa del Suolo della Costa e Bonifica, regione Emilia-Romagna.

Wang, C., Temmerman, S., 2013. Does biogeomorphic feedback lead to abrupt shifts between alternative landscape states? an empirical study on intertidal flats and marshes. J. Geophys. Res.: Earth Surf. 118 (1), 229–240.

Weerman, E.J., Van Belzen, J., Rietkerk, M., Temmerman, S., Kéfi, S., Herman, P., Van de Koppel, J., 2012. Changes in diatom patch-size distribution and degradation in a spatially self-organized intertidal mudflat ecosystem. Ecology 93, 608–618.

Weerman, E.J., Herman, P.M.J., Van de Koppel, J., 2011. Top-down control regulates self-organization on a patterned intertidal flat. Ecology 92, 487–495.

Wickland, D.E., 1991. Mission to planet Earth: the ecological perspective. Ecology 72, 1923–1933.

Making the Link

By Liette Vasseur

Remote Sensing Solutions to Monitor Biotic and Abiotic Dynamics in Coastal Ecosystems ⟺ **Managing Adaptation to Changing Climate in Coastal Zones**

Andrea Taramelli, Emiliana Valentini and Loreta Cornacchia

Daniel Lane, Colleen Mercer Clarke, John D. Clarke, Michelle Mycoo and Judith Gobin

Understanding the transitional coastal zone can be daunting. Many components of the ecosystem cannot be measured one by one at the local scale due to their required financial, human, and time resources. Remote sensing can provide a great tool to analyze the changing patterns of such complex systems. Taramelli et al. have summarized in their chapter the usefulness of combining remote sensing with LIDAR and other geospatial techniques to examine coastal changes. Through case studies, they have demonstrated how changes can be detected over time. Using these techniques, they can show changes in vegetation patterns, which are sensitive to changes in environmental conditions such as salinity. The advancement of these new geospatial techniques represents another opportunity for communities to document changes over historical to present times, especially as related to sea-level rise and impacts on coastal ecosystems.

In the next chapter, Lane et al. introduce a large international project targeting several coastal communities in Canada and the Caribbean. They describe the many challenges faced by communities from the environment, policy, and research viewpoints. Their research challenges connect very well with the previous chapter, as both chapters show the need to acquire baseline data and information with which communities can discuss their experiences and concerns.

The adage "a picture is worth a thousand words" reflects well the experience presented in both chapters as well as several previous ones in this book (e.g., Chapters 2 and 3): through computational techniques or drawings from local people, changes of coastal zones can be documented. The next chapter also reemphasizes the need to connect researchers and communities to find more lasting solutions that can be readily acceptable for local communities.

Chapter 9

Managing Adaptation to Changing Climate in Coastal Zones

Daniel E. Lane[1], Colleen Mercer Clarke[1], John D. Clarke[1], Michelle Mycoo[2], Judith Gobin[2]
[1]Telfer School of Management, University of Ottawa, Ottawa, ON, Canada; [2]The University of the West Indies, St. Augustine Campus, Trinidad and Tobago

Chapter Outline

INTRODUCTION

Coastal zones are the most biologically and economically productive regions in the world. Over 40% of the world's population lives within 150 km of the shore (United Nations Atlas of the Oceans, 2010), and that figure is growing. In Canada, approximately 38% of Canadians live within only 50 km of one of three surrounding oceans—the Atlantic, Pacific, or Arctic Oceans—or one of the Great Lakes. In the Caribbean region, coastal populations in 28 independent territories and island states are generally clustered along thin bands of land in close proximity to the shore. An estimated 60% of the Caribbean's total population of approximately 40 million people lives within less than 100 km from the coast, and approximately 40% of the population resides within a mere 2 km of the coast.

These coastal zones, where land and water interact, are key landscapes when considering (1) the environmental challenges faced by human societies and (2) the impact of values and priorities on the development of policies and the implementation of effective practices for coastal adaptation.

This chapter focuses on the red-flag challenges facing coastal communities in Canada and the Caribbean. Preparing for environmental change, shifts in ecosystem services, and impacts on changing demographics place both immediate and long-term pressures on coastal community resources. These pressures threaten both prosperity and sustainability, and represent the preeminent challenges of the twenty-first century.

The chapter contains the following subsections:

The C-Change Project: Description of the coastal community research project for managing adaptation to environmental change in selected Canadian and Caribbean coastal communities.
Challenges: Description of the significant environmental changes affecting coastal residents and coastal services, and the related policy and research challenges for coastal communities in Canada and the Caribbean.
Solutions: Examination of adaptation policies and best practices to raise awareness within the coastal community and to prepare it for the environmental changes in the coastal zone.
Conclusions: Discussion of environmental change in coastal communities.

THE C-CHANGE PROJECT

The C-Change Project, *Managing Adaptation to Environmental Change in Coastal Communities: Canada and the Caribbean,* is a collaborative International Community-University Research Alliance (ICURA) initiative whose goal is to assist participating coastal communities in sharing experiences and tools that aid adaptation to pending changes in their physical and social coastal environments. These environmental changes include (1) flooding associated with storm surge, (2) coastal erosion, (3) rising sea levels, (4) seawater intrusion in coastal aquifers, (5) coastal and sea pollution caused by wastewaters and solid wastes that have been treated ineffectively, and (6) impacts associated with increasingly extreme weather events (Lane et al., 2013).

C-Change is a multi-year (2009–2015) community-based research project funded by the Social Sciences and Humanities Research Council of Canada (SSHRCC) and the International Development Research Centre (IDRC). Project coordination and administration offices are located at the Sir Arthur Lewis Institute for Social and Economic Studies (SALISES), University of the West Indies in St. Augustine, Trinidad and Tobago, and at the Telfer School of Management of the University of Ottawa in Ottawa, Canada.

The principal research objectives of C-Change community-university research alliance are to (1) *document and share global research and adaptation strategy evaluation* in the local context; (2) *profile local community vulnerabilities and risks* and declare priorities through understanding local spatial and demographic data in the preparation of community climate guidelines and action plans; (3) *build local capacity for managing adaptation* by promoting new institutional arrangements, training, and introducing tools for evaluation of adaptation strategies; and (4) *develop interdisciplinary curricula* for undergraduate and graduate university programs and local community schools to raise awareness and train new generations of young people to learn to evaluate opportunities for their coastal communities subject to environmental change.

C-Change Communities and Partners

The C-Change study sites across Canada and the Caribbean region consist of eight coastal communities including five island communities. C-Change study sites were selected based on anticipated similarities between four sets of paired communities: (1) the low-lying regional capital cities of Charlottetown, Prince Edward Island, and Georgetown, Guyana; (2) the indigenous communities of Iqaluit, Nunavut, and San Pedro, Ambergris Caye, Belize; (3) the eco-touristic, resort communities of Gibsons, British Columbia, and Grande Riviere, Trinidad and Tobago; and (4) the small island communities of Isle Madame, Nova Scotia, and the Grenadine island of Bequia, St. Vincent, and the Grenadines. Figure 9.1 maps the locations of the communities. Table 9.1 provides further details on the distinctiveness, threats, and C-Change partnerships of these coastal communities.

Together, C-Change university-based researchers, and community partners process local feedback, knowledge, contexts, insights, and priorities toward improving local and regional policy planning and strategic response, and act as "community champions" toward developing community adaptation actions. The C-Change project has been active in taking up the challenges of community adaptation by establishing "communities of practice" among the community leaders. These challenges are discussed in further detail in the next section.

ENVIRONMENTAL, POLICY, AND RESEARCH CHALLENGES

The challenges of adapting to the hazards posed by climate change in the coastal landscape are complex, interactive, and require the attention of multiple disciplines and sectors. These challenges are categorized as follows:

1. *Environmental challenges* include the uncertainty of severe weather events, physical contexts, and vulnerabilities of communities; degradation and damage from natural sources; and changes/shifts in resource abundance.
2. *Policy challenges* represent the pressures of the changing coastal environment on local communities, and the need to take steps to prepare community asset bases and citizens for extreme events.
3. *Research challenges* include the need to engage the community to prioritize preparedness measures in the face of predicted extreme events, and to develop and implement strategic interdisciplinary adaptation strategies for the local context.

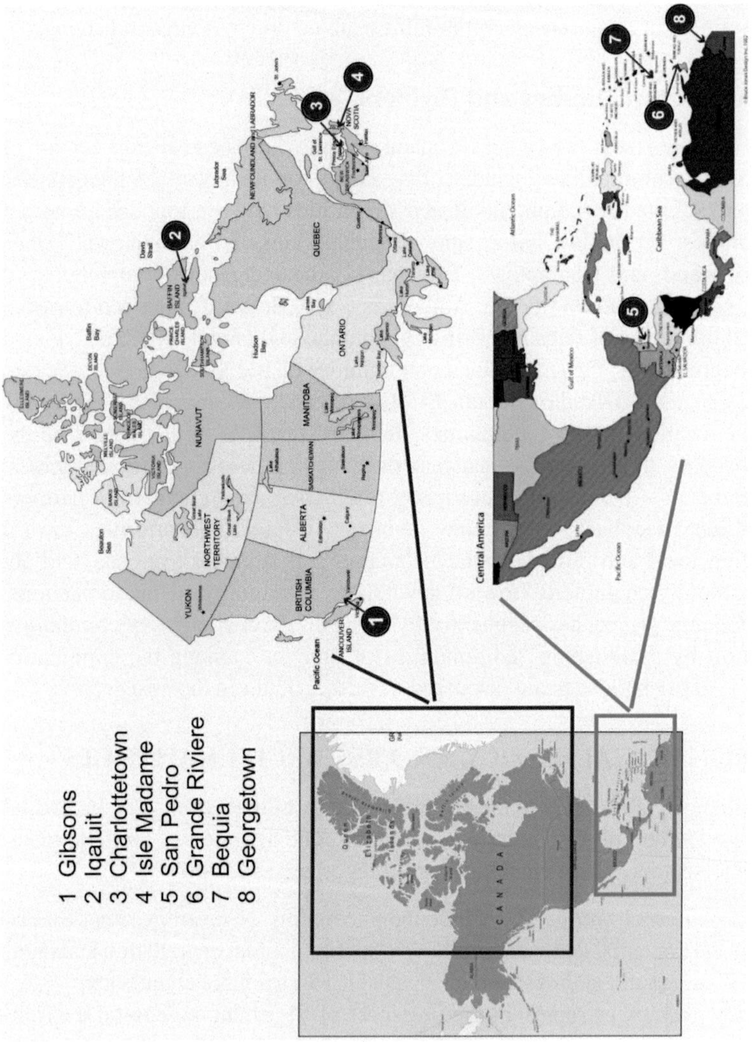

FIGURE 9.1 C-Change communities in Canada and the Caribbean region. *From Bruce Jones Design, Inc. 1992.*

TABLE 9.1 C-Change Canada–Caribbean Communities: Threats and Partnerships

Community	Profile	Partnerships and Alliances	Coastal Climate Change Vulnerabilities
Georgetown, Guyana	National capital, port; population 215,000; industrial center; large areas located below sea level	Central government, central housing and planning authority	Breaching of historic dykes; flooding of urban core saltwater; contamination of drinking water
Charlottetown, Prince Edward Island, Canada	Provincial capital city, coastal port; population 60,000; industrial activity; historic shorefront district	City manager, planning officer; sustainability coordinator	Flooding infrastructure, national historic sites; harbor sea-ice; salinity intrusion to groundwater
San Pedro, Ambergris Caye, Belize	Village on barrier reef, UNESCO World Heritage site; tourism and fishing; nearby shipping lanes	Local contacts, graduate students	Coral bleaching; coastal flooding; tourism sector at risk; fish & shellfish fisheries affected
Iqaluit, Territory of Nunavut, Canada	Remote Arctic community; air/water shipping; growing territorial capital; population 7000; traditional subsistence	Director of sustainability engineering, territorial government, climate change coordinator	Instability in ice-rich permafrost; high tidal range; urban development pressure, subsistence infrastructure threatened
Grande Riviere, Trinidad & Tobago	Subsistence fishing village, small crop farmers, ecotourism from leatherbacks; nearby shipping lanes	Local tourism environmental group, Turtle Village Trust	Beach erosion from river discharge; loss of nesting sites for turtles with effects on local ecotourism
Gibsons, British Columbia, Canada	Coastal town on Pacific shoreline, near Vancouver; access to coast by ferry, popular ecotourism resort, retirement town	Town council, Town planner, local contacts with tourism & environmental groups	Pressure for waterfront development; groundwater exposure to salinization; seasonal droughts threats to water supply
Isle Madame, Nova Scotia, Canada	Small island community; traditional fishing economy; aging population	Municipality, mayor, CAO, director of tourism, development association, university researchers	Coastal erosion; roads, causeway transportation links threatened; water infrastructure damage; stormwater management
Bequia, St. Vincent & the Grenadines	Volcanic island; coral reefs; resort development; marine ecotourism boating area; nearby shipping lanes	Local government officials, NGOs	Damage to transportation links, marine infrastructure, water-supply challenges from regional drought

Modified from Lane et al. (2013), Mercer Clarke et al. (2015).

Environmental Challenges

Environmental events affect the coastal ecosystem and have detrimental consequences on coastal communities. In the context of the changing coastal climate, there are three major effects on the physical environment that alter the provision of ecosystem services and instigate detrimental consequences with specific impacts on the coastal zone. These are as follows:

1. *Warming trends.* These include the warming of temperatures on air, land surface, and in the oceans generally attributed to global warming trends from the increasing presence of greenhouse gases in the atmosphere.
2. *Increased frequency and severity of coastal storms.* Extreme precipitation, higher winds and storminess, as well as extended periods of dryness and drought increase the frequency and severity of the storms.
3. *Increased human activity in the coastal zone.* This results in higher levels of emissions and pollution, and maladaptation to coastal changes.

Table 9.2 itemizes the major environmental effects and corresponding expression of impacts with examples from the C-Change communities. These examples mirror the concerns of the latest IPCC Area Report 5 (IPCC, 2013), and include: (1) the contamination of potable water sources due to saltwater intrusion to wells and the community water system; (2) disruption of food supplies due to poor growing seasons interrupted by extreme precipitation combined with long periods of drought; (3) interference with transportation systems (flooded roads, bridges); (4) damage to coastal facilities including wharves and ports; and (5) impaired wastewater collection, water treatment, and disposal. Inundation of lower-lying areas can result in loss of life, impairment of well-being, and economic damage at a scale most communities cannot absorb sustainably over repeated severe storm events (Lu, 2013). Among many coastal communities in Canada and the Caribbean, the hardest hit are often those that are the most disadvantaged economically and socially, for example, Georgetown, Guyana (Leung, 2010).

Policy Challenges

The physical effects of the changing coastal environment present acute policy challenges for coastal communities worldwide and threaten community sustainability. In general, top-down and centralized pyramid systems of governance that proceed from the national to the regional and finally, to the local municipality are ineffective in prioritizing action on local impacts. It is a major political challenge for communities facing severe climate impacts to invert the governance pyramid and decentralize policymaking and resource allocation in order to provide local communities with the capabilities needed to deal directly with the local changes. This authority and responsibility assume with it the capability and need of the community to plan strategically and beyond the short-term political horizon that characterizes current top-down decision-making. The following itemized list of policy challenges are developed from evidence in the C-Change communities in

TABLE 9.2 Environmental Challenges: Events, Impacts, and C-Change Examples

Environmental Interventions	Coastal Effects	Observed Impacts	C-Change Examples
Warming temperatures Air, surface, oceans	Reduced ice cover (esp. Canada)	Increased wave energy and storm surge; reduction in primary production	Iqaluit: Instability in ice-rich permafrost, exposure to enhanced wave energy (Hatcher et al., 2011; James et al., 2011) Charlottetown: Harbor sea-ice (Forbes et al., 2013)
	Thermal expansion	Decline in primary production	Ambergris Caye: Fish & shellfish fisheries decline; coral bleaching, reef destruction Isle Madame: Groundfish stocks collapse (Lane, 2011) Grande Riviere: Loss of nesting sites for turtles (Seeram, 2011)
	Sea-level rise	Saltwater intrusion	Georgetown: Breaching of dykes (Leung, 2010) Isle Madame: Salinity intrusion to groundwater coastal flooding (Pakdel, 2011) Gibsons: Groundwater exposure to salinization (Vadeboncoeur, 2014)
Increased frequency of storms and storm severity	Erosion and subsidence	Landward retreat of coastline	Grande Riviere: Beach erosion from river discharge (Seeram, 2011; Grant, 2010) Iqaluit: High tidal range (Hatcher et al., 2011; James et al., 2011); urban development pressure: subsistence infrastructure threatened (Forbes et al., 2013; Forbes, 2011)
	Extreme events: intense precipitation and extended dry periods	Flooding, droughts from storm surge	Charlottetown: Hurricane storm surge floods urban core (Hartt, 2014, 2011) Isle Madame: Severed transportation links, fish processing (Pakdel, 2011; Mostofi Camare, 2011) Bequia: Water shortage, extended dry periods, no well-water supply Gibsons: Water shortage, extended dry period, unknown limit of aquifers (Vadeboncoeur, 2014) Georgetown: Floods residential parishes (Gossai, 2010)
	Destruction of infrastructure		Charlottetown: Coastal infrastructure flooding (Liu, 2014) Isle Madame: Stormwater threatens capacity of water treatment (Barghi, 2013)
Increased human activity in the coastal zone	Development	Building infrastructure, housing, tourism, wharves, port facilities; sediment extraction	Charlottetown: Pressure on waterfront development (Hartt, 2014, 2011) Grande Riviere: Beach hotels for tourists (Sutherland, 2011) Isle Madame: Beach rock, sand removal, breakwater failures (Mostofi Camare, 2011)
	Pollution	Industrial and stormwater systems effluent release	Charlottetown: Damage to transportation links, marine infrastructure (Liu, 2014) Isle Madame: Aging water infrastructure and insufficient capacity for stormwater and sewage control (Barghi, 2013)
	Maladapation	Inappropriate armoring	Georgetown: Patchwork dykes due to failure of existing sea walls (Leung, 2012) Isle Madame: Protection of transportation links (Lu, 2013; Mostofi Camare, 2011)

Canada and the Caribbean region, and hinge on resourcing these communities to enable strategic adaptation action at the local level.

1. *Identify community priorities.* An underlying need for policy requires the profiling of community vulnerability, resilience, and adaptive capacity. The exercise of profiling community status is preliminary to identifying community priorities, especially in the face of stochastic climate changes. Resources are limited and local communities tend to be severely under-resourced (Klein, 2014). Nevertheless, the burden for implementation of change too often falls to local governments already constrained by available resources, by conflicting economic pressures, and by limited capacity to implement new instruments for coastal development. Under these circumstances, it is even more important for local communities to define their relevant trade-offs and priorities, for example, between community subpopulations (youth vs the aged), between community assets (coastal infrastructure vs transportation links), and between community social and cultural instruments (community centres vs faith-based educational supports) (Lane et al., 2013).
2. *Cede authority to local communities.* Authority is a fundamental characteristic of effective decision making and supportive implementation. With authority comes the responsibility for engagement and collaboration among multiple community partners, enhanced awareness and communication, and evidence of improved stewardship among all community members. This key policy challenge is revolutionary under current top-down systems that tend to exclude local considerations for the good of the overall regional or national policy change. Local authority and responsibility require participatory, traditional, indigenous, and direct communication to inform, sensitize, and create ownership within the local population. It is incumbent on municipal governments acting with authority to develop efficient, dynamic governance systems to prioritize under-resourced environmental changes by developing local adaptation strategies for sustainability, and to extend the planning time frame from short-term political horizons to long-term strategic considerations (Klein, 2014; Lane, 2012).
3. *Measure, track, and exercise preparedness.* Identification of community priorities (1, above) is linked to the policy challenge of targeting community objectives and moving continuously toward achieving these objectives. Clearly specified community objectives reflect the environmental, socioeconomic, cultural, and institutional priorities of the community. These objectives are expressed as measurable indicators of community preparedness that permit the ongoing monitoring of the preparedness status of the community. It is a policy challenge for coastal communities to exercise their ability to think strategically by running regular tabletop or "What If?" workshops with key emergency measures, to enable and inform community participants about the eventuality of severe environmental events of specific concern to the community (Chung, 2014).

4. *Implement the precautionary approach.* Decision making for sustainability in the coastal zone involves (1) natural resource allocation, for example, commercial and artisanal fisheries, aquaculture, and forestry, and use of agricultural lands; (2) development, for example, for residential and commercial enterprises; and (3) ecotourism, for example, hotels, and recreational and commercial use of greenspace. In the presence of the changing climate, decisions on these issues require enhanced risk analysis and a more cautious approach to interventions deemed to be sustainable. The precautionary approach implies a more closely monitored and more conservative level of natural resource exploitation, spatial allocation, and management, for example, for fisheries and aquaculture in the marine ecosystem, for local energy production and supply, for land use in the agriculture sector, and for managed degradation of coastal marine environments under tempered demands for coastal development, and sustainable coastal tourism (Mohammadi, 2014; Lane, 2012, Benidickson et al., 2005).
5. *Plan strategically.* Action planning is a significant part of global initiatives in adaptation management, for example, Snover et al. (2007), FCM (2011), UNEP (2011), CARE International (2010). It is an important policy challenge to impress upon all communities the need to prepare strategic plans. For example, the C-Change communities of Charlottetown, Gibsons, and Iqaluit are members of the Federation of Canadian Municipalities (FCM) program "Partners for Climate Protection," a network of Canadian municipalities committed to reducing greenhouse gases and taking action toward adapting to climate change. The PCP program is the Canadian component of ICLEI's global Cities for Climate Protection (CCP) network, involving more than 900 global communities. This challenge calls for the development of formal plans, for example, Business Continuity Plans (BCPs) and Emergency Operations Plans (EOPs), which involve operational (short term, i.e., current to 3 years) and strategic (long term, i.e., 5–25 years) planning periods. For example, in 2013 Canadian communities were tasked to present Municipality Climate Change Action Plans (MCCAPs) (Fisher, 2011). MCCAPs address changing coastal environmental realities and the implementation of proactive planning instruments for conflicts with existing development practices, limited institutional capacities, and established values and expectations of the rights of access to and use of coastal resources. Adaptive planning addresses the challenge to improve local preparedness, through assessment of adaptation options to "protect," "accommodate," or "retreat," and critical examination of disaster response mechanisms (Mercer Clarke et al., 2015). Adaptive planning also addresses the challenge of bringing science and public education into the decision-making process, effectively mobilizing knowledge for application at the local level.
6. *Establish an education legacy.* A significant policy challenge is related to outreach to the local schools as a fundamental means of disseminating social and cultural values throughout the community as a legacy for community sustainability. This outreach infuses information about the local environment

into elementary, high school, and university institutions. The results instill the uniqueness of local knowledge and pride of place while developing respect and environmental stewardship among all community members. The links to learning integrate local knowledge into research and improve the decision-making process acceptability and information flow back to community members through school participants and their families (Clarke, 2012). Climate change is no longer the discipline of meteorologists, climatologists, and natural scientists. Rather, there is an increasing need for interdisciplinary analysis embracing all disciplines toward meeting this policy challenge. For example, through C-Change, university engineering students at the University of New Brunswick are exposed to marine policy, law, and administration, linking geomatics tools to support adaptation planning and governance issues. At the University of Ottawa, business students are examining the effects of simulated dynamics of coastal storms and sea-level rise on community socioeconomic systems. Researchers at UWI (St. Augustine) have prepared courses on the socioeconomic impacts of climate change through applied spatial information management courses for social science students.

These policy challenges are derived from the experience of the C-Change project communities in Canada and the Caribbean region, and are presented here in general terms across all communities. It is understood that these policy challenges are more widely applicable to coastal zone challenges for the twenty-first century around the world. The following subsection builds on these challenges to present the corresponding research challenges.

Research Challenges

The research objectives of the C-Change project seek to document and share global research on climate change in the local context. This is accomplished by researching and documenting local community vulnerabilities and risks, managing local adaptation by promoting tools for adaptation strategy evaluation, and defining new institutional arrangements and educational curricula. Research in these topics presents particular challenges, as itemized below.

1. *Improve baseline information.* To understand the local community vulnerabilities, risks, priorities, and capabilities, considerable local data gathering is required. In every community, the availability of data varies considerably from data-rich communities (typically urban, suburban communities, or ecotourism centers, e.g., Charlottetown, Gibsons, Grande Riviere) to data-poor communities (small island, remote communities, e.g., Bequia, Iqaluit). Lack of data availability presents research challenges that necessitate closer community engagement and implementation of systems that enable increased community participation for collecting and validating unique local data. For example, in the ecotourism resort communities of Gibsons, and Grande Riviere, C-Change researchers undertook community surveys to gauge the position of community members on their priorities and challenges due to their changing coastal climates

(Vadeboncoeur, 2014; Hosein, 2011). Local elevation data were also collected in these communities that were used to project local flooding impacts (Seeram, 2011). In Isle Madame, global positioning system (GPS) surveys of selected vulnerable coastal areas of the archipelago were conducted to provide accurate coastal elevation data not previously available. In Georgetown, questionnaires were administered to households in city wards. While all social classes and urban environments were potentially exposed to flooding, impacts were influenced by income, location, and dwelling construction style (Leung, 2010; Gossai, 2010). A Public Participatory GIS (PPGIS) facilitated data collection via the participation of community members through a Volunteer Geographical Information (VGI) interface (Tienaah, 2011) that facilitated visualization of historical richness of communities such as that provided for Isle Madame (Boudreau and Thériault, 2012). Contributions of C-Change community partners to the project's companion social networking site also provides an active source of local information that can be linked to the VGI database initiative (C-Change, 2015).

2. *Profile community vulnerability, resilience, and adaptive capacity.* Research is required into profiling the status of coastal communities with respect to their environmental, economic, social, and cultural characteristics. For example, vulnerability analysis in Georgetown was developed through questionnaires of residents and profiles of case study wards and the comparative application of impact categories based on the Parry et al. (2007) report. The Charlottetown profile was modeled using a geographic information system linked to a system dynamics model of the sustainability pillars (environmental, economic, social, and cultural layers) to examine the impact of simulated storms on the community (Beigzadeh, 2014; Hartt, 2011; Crabbé, 2015; Lane et al. WP). Computer-generated flood simulation scenarios were developed based on LiDAR-derived topographic for Charlottetown Harbor (Forbes et al., 2013; Webster and Forbes, 2006). Hatcher et al. (2011) assessed and mapped the flood probability and hazards for Iqaluit, including evolving exposure to waves and storm surges, sea-level trends, and changes in sea-ice breakup or freeze-up dates. A failed breakwater on Isle Madame was examined in terms of the village's priorities and vulnerability toward evaluating the value of alternative adaptive strategies (Chung, 2014; Pakdel, 2011). The challenging studies provide evidence of the uniqueness of each community and the need for research into understanding community priorities, vulnerabilities, and capabilities to adapt.
3. *Evaluate opportunities for strategic adaptation.* The development of strategic, interdisciplinary adaptation strategies in the context of local problems presents an important research challenge. There are well-known techniques of participative multicriteria decision-support systems that are directly applicable to evaluate community adaptation options (Lane, 2007). However, the application of these methods needs to be made readily and easily accessible at the community level to facilitate informed decision making. Mostofi Camare (2011) developed an analysis of the multicriteria status of the community and evaluates vulnerability with respect to alternative adap-

tation strategies for Isle Madame. Such analyses underscore the significant opportunity cost of inaction compared to longer-term adaptation strategies. C-Change teams incorporate the alternatives to adaptation policy developed by international and national organizations and summarized as "Protect," "Accommodate," "Retreat," or "Status Quo" (the "do nothing" option) (Pilkey and Young, 2009; IOC, 2009; Boateng, 2008).

4. *Develop indicators of preparedness.* In the pending urgency of extreme weather events, communities are challenged to be prepared. Research is required to define community preparedness measures that take into account the uncertainties of extreme coastal storm events and their physical, socio-economic, and cultural human impacts. Preparedness indicators include the community's regular practice of assembling Emergency Operations Center (EOC) personnel to engage in simulation sessions known as tabletop exercises that play out the community's response to the emergency event. Research into developing tabletop exercises is another occasion to engage and inform community members. Chung (2014) orchestrated a tabletop exercise for the breakwater failure in the Isle Madame community of Little Anse.
5. *Define effective institutional arrangements.* Research is needed to assist communities in developing new governance and institutional arrangements to deal with the preparation, emergency planning, and decision analysis related to coastal environmental change. New governance includes the formation of new community positions such as the Sustainability Engineer in Iqaluit, and the Sustainability Coordinator in Charlottetown, whose responsibilities are related to community preparedness and emergency planning. New institutions include the formation of a Committee of the Whole to deal with strategic planning and emergency preparedness issues with dedicated funding and administrative resources. C-Change provides local governance and institutional advice comprised of the documentation of cases and best practices in institutional arrangements (Young, 2013), and the integration of multiparticipant, multidisciplinary "communities of practice" to communicate and share formally community resources and capacities to deal with extreme climate events. The community of practice links professionals from all coastal communities and advances their capacity to deliver tools as practical aids in local planning for adaptation. The community of practice builds on the existing connections among the project academic and community partners, and ensures continuance of the linkages with the wider global community addressing coastal adaptation to climate change (Mercer Clarke et al., 2015; Wenger et al., 2007; Buysse et al., 2003).
6. *Improve university-community links.* Severe weather events and increasingly frequent extreme storm surges have exemplified the need to improve working relationships to better connect local communities with the existing expertise that is vested within universities; national, regional, and provincial governments; and the economic sectors (Dyer and Andrews, 2012). Knowledge mobilization—the translation of what is known into information useful for practical application—remains a challenge. Clearly, when scientists work

closely with local stakeholders and decision makers, the two-way exchange of knowledge and experience greatly enhances the focus of research efforts, and improves the usefulness of research deliverables in application to the early planning for and adaptation to climate change challenges. It is therefore a fundamental challenge to coastal community research to integrate all stakeholders in the research process. Communicating research findings to at-risk communities helps build adaptive capacity (Mycoo and Gobin, 2013; Clarke et al., 2010; Mercer Clarke et al., 2015; Lane et al., 2013).

Finally, in acknowledging the research challenges, it is incumbent to refer explicitly to those gaps that could undermine research initiatives. These gaps include:

- *Lack of technical capacity.* Increased training and support to community leaders needs to be urgently addressed to improve the communications of their findings to policymakers.
- *Lack of sufficient funding opportunities.* Funds for community climate change focus mainly on mitigation, each with its own rules and governance. A shift in supporting community adaptation strategies must be improved.
- *Lack of effective communication.* Research findings need to be communicated to communities in an effective and timely manner to decision makers. This requires reducing the technical jargon and making research results more applicable and implementable to coastal communities.
- *Lack of understanding of adaptation.* Communities and funders need to be persuaded about the need for long-term strategic adaptation as a priority item for community sustainability. The alternative—no adaptation—may simply be devastating to the ability of the community to survive the pending environmental changes.

SOLUTIONS

There are a number of areas identified as potential solutions, priorities, and implementation of practices and policies that address the twenty-first century climate challenges and mark the potential for success in developing effective adaptation of coastal communities. These solutions areas are delineated as follows.

Solution 1. *Focus on the sustainability of coastal communities.* Improved understanding of the context and uniqueness of each community, and its expressed priorities and objectives for strategic management, are enablers for sustainability. Coastal communities are characterized by their physical environment and their economic, social, and cultural assets. These also define community vulnerabilities and threats as prioritized by community members, and they have the capabilities to evaluate strategic adaptation options. Community sustainability is ensured through stewardship that comes from active participation and engagement of community members. Sustainable communities are prepared and they regularly exercise their preparedness through emergency planning and simulation exercises that keep community leaders on alert.

Solution 2. *Develop local governance and decision making.* Increased authority, resources, and the ability to decide on participative community futures include the capacity to evaluate and analyze adaptation options through effective strategic planning while exercising, monitoring, and tracking measures of the impacts and outcomes of adaptation alternatives.

Solution 3. *Promote collaborative policymaking.* Participation and engagement of all community members are necessary for community collaboration and coherence. This is evident in the ability of community members to communicate formally and informally through established social networking activities. Collaborative practices achieve the integration of local and traditional indigenous knowledge. Collaborative policy ensures that decisions are compatible with the core cultural values of the coastal communities. Collaborative policy also implies the union of natural science data with social science information and contexts. Collaboration is realized through sharing among community members and "communities of practice." Collaboration also requires making contact with local media (e.g., MacPhail, 2011). Academic connections enable community members to participate and to provide their message to international conferences.

Solution 4. *Renew legislation and regulations.* Application of the precautionary approach in development planning, resource allocation, and exploitation is fundamental to deal with the increased uncertainty of extreme coastal weather events. This solution requires improved and reinforced legal frameworks controlling coastal activities, for example, setbacks, and includes enforced auditing of these regulations and measures of preparedness as desirable targets and preset standards to be achieved by the community. Regulations should include the need for communities to regularly test and exercise their Emergency Operation Centers.

Solution 5. *Link adaptation to education.* Develop a regime of local schools curriculum on local climate and environmental challenges and solutions, and encourage the construction of new multidisciplinary university curricula in the natural and the social sciences. Knowledge must be shared, promoted, and used in order to aid communities in developing a critical approach, exercise pressure on policymakers, and develop realistic, sustainable, and acceptable adaptation policies for implementation. Instilling the issues into the local community education system results in the best means of promoting public awareness and stewardship of community values in the coastal resources and ecosystems as a legacy solution.

Solution 6. *Carry out research in the challenge areas.* Encourage the efficient process of local information gathering, enhanced decision analysis and evaluation, and new technologies for adaptation, energy, and water use. The dissemination of this research for improved monitoring and forecasting systems with indicators of community preparedness will promote local adaptation and stewardship.

TABLE 9.3 Solution Areas for Success in Coastal Zone Adaptation

Solution Areas	Corresponding Policy Challenges	Corresponding Research Challenges
Sustainability of coastal communities	Identify community priorities Cede authority to local communities Measure and track preparedness	Improve information baseline Profile community vulnerability, resilience, and adaptive capacity Develop indicators of preparedness
Local governance and decision making	Cede authority to local communities Implement the precautionary approach Plan strategically	Profile community vulnerability, resilience, and adaptive capacity Evaluate opportunities for strategic adaptation Define effective institutional arrangements
Collaborative policymaking	Identify community priorities Cede authority to local communities Measure and track preparedness Implement the precautionary approach Plan strategically	Improve information baseline Profile community vulnerability, resilience, and adaptive capacity Improve university-community links
Legislation and regulation	Cede authority to local communities Implement the precautionary approach	Define effective institutional arrangements Profile community vulnerability, resilience, and adaptive capacity Evaluate opportunities for strategic adaptation
Education	Identify community priorities Establish an education legacy	Improve information baseline Profile community vulnerability, resilience, and adaptive capacity Improve university-community links
Research in challenge areas	Measure and track preparedness Implement the precautionary approach Plan strategically	Improve information baseline Profile community vulnerability, resilience, and adaptive capacity Evaluate opportunities for strategic adaptation Develop Indicators of preparedness Define effective institutional arrangements Improve university-community links

These solution areas are reflected in the policy and research challenges of the coastal zone as indicated above and summarized in Table 9.3 in correspondence with the solution areas. The ability to address the policy and research challenges translates into solutions for sustainability and adaptation in coastal communities.

CONCLUSIONS

Coastal communities are unique and defy generalization. The evidence of the C-Change project demonstrates that coastal communities are generally under-resourced and over-committed to immediate concerns (e.g., water and sewage infrastructure, food security, drought management, social stability, and commerce) that tend to make strategic (long-term) planning secondary. Coastal communities struggle with the dichotomy between immediate coastal development (e.g., commercialization, tourism) and longer-term resource conservation (e.g., waterfront, shorelines). Effective community-university alliance for information sharing requires ongoing engagement and communication with local governments, community professionals, and community social groups. While climate change may not be among the highest of local community priorities, community resilience can be enhanced through strategies and renewed governance to mainstream climate change considerations into the broad spectrum of community decision making.

The challenges faced by coastal areas are a compilation of multiple stressors, which—like climate change—are mostly anthropogenic in source, or have been amplified by human activities that clearly go beyond reasonable limits. Initiatives proposed to assist coastal communities toward enhanced sustainably must integrate the limits of the planet and act in union with increased authority at the local level to develop latent coastal community stewardship and improved sustainability.

GLOSSARY

ICURA International Community-University Research Alliance
IDRC International Development Research Center
MUN Memorial University of Newfoundland, St. John's, Newfoundland and Labrador, Canada
SSHRCC Social Sciences and Humanities Research Council of Canada
UBC The University of British Columbia, Vancouver, British Columbia, Canada
UNB University of New Brunswick, Fredericton, New Brunswick, Canada
UOttawa University of Ottawa, Ottawa, Ontario, Canada
UWI University of the West Indies, St. Augustine Campus, Trinidad and Tobago

ACKNOWLEDGMENTS

The C-Change research teams in Canada and in the Caribbean gratefully acknowledge the support of the Canadian funding agencies, the Social Sciences and Humanities Research Council (SSHRC) and the International Development Research Centre (IDRC). C-Change also acknowledges the host institutes at the University of the West Indies, St. Augustine

Campus, Trinidad & Tobago, and the Telfer School of Management of the University of Ottawa in Ottawa, Canada, as well as the support of research teams from The University of British Columbia, the University of New Brunswick, and Memorial University of Newfoundland. Support is also acknowledged from ArcticNet and the Networks of Centers of Excellence Canada. C-Change is deeply appreciative of the time and effort spent by municipal staff and community members of the C-Change partner communities. It is thanks to their valued participation, commitment, and contribution to managing adaptation that this research has relevance.

REFERENCES

Barghi, S., 2013. Water Management Modeling in the Simulation of Water Systems in Coastal Communities (MSc thesis, Systems Science). University of Ottawa, Ottawa. Accessed 08.02.15 at: http://coastalchange.ca/download_files/Barghi_Final_Thesis_Submission_July_2013.pdf.

Beigzadeh, S., 2014. System Dynamics Modeling of the Impacts of Maximum Water Level: The Case of Charlottetown, P.E.I., Canada (MSc thesis, Systems Science). University of Ottawa, Ottawa. Accessed 08.02.15 at: http://www.coastalchange.ca/images/stories/Documents_Tab/shima_proposal_30april2013.pdf.

Benidickson, J., Chalifour, N., Prévost, Y., Chandler, J., Dabrowski, A., Findlay, S., Déziel, A., McLeod-Kilmurray, H., Lane, D., 2005. Practicing Precaution and Adaptive Management: Legal, Institutional and Procedural Dimensions of Scientific Uncertainty. SSHRC and the Law Commission of Canada, 2005. SSRN Electronic Journal 07/2005; pp. D-85–D-131, http://dx.doi.org/10.2139/ssrn.2272684.

Boateng, I., 2008. Integrating Sea-level Rise Adaptation into Planning Policies in the Coastal Zone. Page 22, Integrating Generations: FIG Working Week 2008: TS 3F. Coastal Zone Administration, Stockholm, Sweden.

Boudreau, A., Thériault, M., 2012. Isle Madame Historical Documentation and Storm Monitoring Project 2011-2012. Final report to C-Change. Marine Research Institute, Université Sainte-Anne, Petit-de-Grat, Nova Scotia, Canada.

Buysse, V., Sparkman, K.L., Wesley, P.W., 2003. Communities of practice: connecting what we know with what we do. Exceptional Child. 69, 263–277.

CARE International, 2010. Community-based Adaptation Toolkit. Accessed 19.04.11 at: http://www.careclimatechange.org/tk/cba/en/.

C-Change, 2015. C-Change Facebook Group Site. Accessed 16.02.15 at: https://www.facebook.com/coastalchange.

Clarke, A.J., 2012. Canada-Caribbean Communities' Schools Curricula Project: Developing C-change Lesson Plans. C-Change Working Paper 39. Accessed 30.09.12 at: http://www.coastalchange.ca/images/stories/Documents_Tab/workingpaper39_lesson_planning_project_2012.pdf.

Clarke, J., Mercer Clarke, C., Lane, D.E., 2010. Proactive adaptation to climate change: building bridges between science and local government. In: Proceedings BIOECON 12th Annual Conference, from the Wealth of Nations to the Wealth of Nature. Rethinking Economic Growth, Venice. Accessed 08.02.15 at: http://www.coastalchange.ca/images/stories/Documents_Tab/workingpaper6_clarke_etal_2010.pdf.

Chung, A., 2014. Emergency Management Planning: A Value Based Approach to Preparing Coastal Communities for Sea Level Rise (MSc thesis, Systems Science). University of Ottawa, Ottawa. Accessed 08.02.15 at: http://www.coastalchange.ca/images/stories/Documents_Tab/alex_chung_m.sc._thesis_proposal.pdf.

Crabbé, P., 2015. Conceptual Framework for a Systems Dynamics Adaptation Model to Climate Change for Charlottetown, P.E.I. C-Change Working Paper 45. 48p + appendices. Accessed 04.06.15 at http://coastalchange.ca/download_files/WorkingPaper45_Crabbe_2015.pdf.

Dyer, G., Andrews, J., 2012. Higher education's role in adapting to a changing climate. A report prepared by Second Nature: Education for Sustainability for the American College & University Presidents' Climate Commitment. Boston. 33p. Accessed 04.06.15 at: http://www.presidentsclimatecommitment.org/files/documents/higher-ed-adaptation.pdf.

FCM, 2011. PCP: A Five-milestone Framework. Federation of Canadian Municipalities, Ottawa. Accessed 14.04.11 at: http://www.sustainablecommunities.fcm.ca/Partners-for-Climate-Protection/PCP_Milestone.asp.

Fisher, G., 2011. Municipal Climate Change Action Guidebook. Province of Nova Scotia, Services Nova Scotia and Municipal Relations. Canada, Halifax.

Forbes, D.L. (Ed.), 2011. State of the Arctic Coast 2010—Scientific Review and Outlook. International Arctic Science Committee, Land-ocean Interactions in the Coastal Zone. Arctic Monitoring and Assessment Programme, International Permafrost Association, Helmholtz-Zentrum, Geesthacht, Germany. Accessed 18.09.12 at: http://arcticcoasts.org.

Forbes, D.L., Webster, T.L., MacDonald, C., 2013. Projecting and Visualizing Future Extreme Water Levels for Climate Change Adaptation in Charlottetown Harbour, Prince Edward Island, Canada. C-Change Working Paper 43 (to appear at http://www.coastalchange.ca/).

Gossai, B., 2010. Vulnerability to Sea Level Rise in an Urban Centre of a Developing Country: A Case Study of Georgetown, Guyana (MSc thesis, Economics), University of the West Indies, St Augustine, Trinidad & Tobago.

Grant, S., 2010. Assessing the Socioeconomic Impact of Sea Level Rise on Coastal Communities. Geomatics Engineering and Land Management. University of the West Indies, St. Augustine, Trinidad & Tobago. Accessed 29.09.12 at: http://coastalchange.ca/images/stories/Documents_Tab/Grant_2009_FinalProject.pdf.

Hartt, M., 2014. An innovative technique for modelling impacts of coastal storm damage. Reg. Stud. Reg. Sci. 1 (1), 240–247.

Hartt M., 2011. Geographic Information Systems and System Dynamics: Modeling the Impacts of Storm Damage on Coastal Communities (MSc thesis, Systems Science). University of Ottawa, Ottawa. Accessed 30.09.12 at: http://www.coastalchange.ca/images/stories/Documents_Tab/Hartt_FinalThesis_2011.pdf.

Hatcher, S.V., Forbes, D.L., Manson, G.K., 2011. Coastal Hazard Assessment for Adaptation Planning in an Expanding Arctic Municipality. C-Change Working Paper 23. Accessed 30.09.12 at: http://www.coastalchange.ca/images/stories/Documents_Tab/workingpaper23_hatcherforbes_manson_2011.pdf.

Hosein, F.A., 2011. Assessment and Validation of the Sea Level Rise Threat to Grande Riviere, Trinidad. Geomatics Engineering & Land Management, University of the West Indies, St Augustine, Trinidad & Tobago. 91 pp. Accessed 29.09.12 at: http://www.coastalchange.ca/images/stories/Documents_Tab/Hosein_2011_SpecialInvestigativeProject.pdf.

IOC, 2009. Hazard Awareness and Risk Mitigation in Integrated Coastal Area Management. Intergovernmental Oceanographic Commission, UNESCO Headquarters, Paris. Manuals and Guides no 50, ICAM Dossier no 5.

IPCC, 2013. Climate change 2013: the physical science basis. Cambridge University Press, Newyork and United Kingdom. Intergovernmental Panel on Climate change, Working Group I Contribution to the Fifth Assessment Report of the Intergovernmental Panel on Climate Change (AR5).

James, T.S., Simon, K.M., Forbes, D.L., Dyke, A.S., Mate, D.J., 2011. Sea-level projections for five pilot communities of the Nunavut climate change partnership. In: Geological Survey of Canada, Ottawa Open File 6715.

Klein, N., 2014. This Changes Everything: Capitalism vs the Climate. Alfred A. Knopf, Toronto. 566 pp.

Lane, D.E., June 2012. Canada's commercial fishery: share the wealth or create prosperity?" Optimum Online J. Public Sect. Manage. 42 (2), 1–18.

Lane, D.E., 2011. Toward adaptive fisheries management: Is the current fisheries management toolbox sufficient to address climate change? In: International Workshop on "The Economics of Adapting Fisheries to Climate Change", OECD Committee for Fisheries and the Fisheries Policies Division, Trade and Agriculture Directorate, Fisheries Committee, 10–11 June, 2010 in Busan, Korea, p. 37. Accessed 16.02.15 at: http://www.coastalchange.ca/images/stories/Documents_Tab/workingpaper5_lane2009.pdf.

Lane, D.E., 2007. Planning in fisheries related systems: multicriteria models for decision support. In: Bjorndal, T. (Ed.), Handbook on Operations Research in Natural Resources. International Series in Operations Research and Management Science. Springer, pp. 237–272.

Lane, D.E., Mercer Clarke, C., Forbes, D., Watson, P., July 2013. The gathering storm: managing adaptation to environmental change in coastal communities and small islands. Sustainability Science 8 (3), 469–489.

Leung, K.L., 2010. Adaptation Measures to Climate Change and Sea Level in Georgetown, Guyana (MSc thesis, Urban and Regional Planning). University of the West Indies, St. Augustine, Trinidad & Tobago. Accessed 29.09.12 at: http://coastalchange.ca/download_files/community_documents/Kira%20Lise%20C-Change%20Georgetown%20Case%20Study.pdf.

Liu, M. 2014. Supply Chain Management in Humanitarian Aid and Disaster Relief (MSc thesis, Systems Science). University of Ottawa, Ottawa. Accessed 08.02.15 at: http://www.coastalchange.ca/images/stories/Documents_Tab/lml-thesis%20proposal-final%20version.pdf.

Lu, M., 2013. Coastal Community Climate Change Adaptation Framework Development and Implementation (MSc thesis, Systems Science). University of Ottawa, Ottawa. Accessed 08.02.15 at: http://coastalchange.ca/download_files/Mingliang-final-M.pdf.

MacPhail, D., December 2011. Researchers provide update of coastal study. Port Hawkesbury Rep. 34 (48), 10.

Mercer Clarke, C., Clarke, J., Lane, D., Forbes, D., Watson, P., Edinboro, R., 2015. Weathering the Storm: Community planning for adaptation to coastal climate change. In: Smith, L. (Ed.), The Caribbean in a Changing World: Surveying the Past, Mapping the Future. Cambridge Scholars Publishing, Cambridge UK. In Press.

Mohammadi, S., 2014. Risk Analysis in Coastal Communities Decision Making (MSc thesis, Systems Science). University of Ottawa, Ottawa. Accessed 08.02.15 at: http://www.coastalchange.ca/images/stories/Documents_Tab/sara_thesis_proposal_oct_2013.pdf.

Mostofi Camare, H., 2011. Multicriteria Decision Evaluation of Adaptation Strategies for Vulnerable Coastal Communities (MSc thesis, Systems Science). University of Ottawa, Ottawa. Accessed 30.09.12 at: http://www.coastalchange.ca/images/stories/Documents_Tab/Mostofi_FinalThesis_2011.pdf.

Mycoo, M., Gobin, J., 2013. Coastal management, climate change adaptation and sustainability in small coastal communities: leatherback turtles and beach loss. Sustainability Science 8(3), 1–13.

Pakdel, S., 2011. Spatial–temporal Modelling for Estimating Impacts of Storm Surge and Sea Level Rise on Coastal Communities: The Case of Isle Madame in Cape Breton, Nova Scotia, Canada (MSc thesis, Systems Science). University of Ottawa, Ottawa. Accessed 08.02.15 at: http://www.coastalchange.ca/images/stories/Documents_Tab/Pakdel_FinalThesis_2010.pdf.

Parry, M.L., Canziani, O.F., Palutikof, J.P., van der Linden, P.J., Hanson, C.E., 2007. Climate Change 2007: Impacts, Adaptation and Vulnerability. Cambridge University Press, Cambridge, UK. Contribution of Working Group II to the Fourth Assessment Report of the Intergovernmental Panel on Climate Change.

Pilkey, O., Young, R., 2009. The rising sea. A Shearwater Book. Island Press, Washington, D.C.

Seeram, A., 2011. Developing a Predictive GIS Model of Sea Level Rise for a Community: Grande Riviere, Trinidad & Tobago. LAP Lambert Academic Publishing, Saarbrucken.

Snover, A., Whitely Binder, L., Lopez, J., Willmott, E., Kay, J., Howell, D., Simmonds, J., 2007. Preparing for Climate Change: A Guidebook for Local, Regional, and State Governments. In association with and published by ICLEI – Local Governments for Sustainability, Oakland, CA. 172p. Accessed 04.06.15 at: http://archive.iclei.org/fileadmin/user_upload/documents/Global/Progams/CCP/Adaptation/ICLEI-Guidebook-Adaptation.pdf.

Sutherland, M., 2011. Sea level rise modelling in a Caribbean small island developing state. Hydro Int. 15 (6). Accessed 13.01.12 at: http://www.hydro-international.com/issues/articles/id1308-Sea_Level_Rise_Modelling.html.

Tienaah, T., 2011. Design and Implementation of a Coastal Collaborative GIS to Support Sea Level Rise and Storm Surge Adaptation Strategies. Geodesy and Geomatics Engineering, University of New Brunswick, Fredericton, NB. Technical Report No. 276. Accessed 08.02.14 at: http://coastalchange.ca/download_files/Student_Thesis_UNB_Titus_Tienaah.pdf.

UNEP, 2011. Mainstreaming Climate Change Adaptation into Development Planning: A Guide for Practitioners. United Nations Development Programme (UNDP) and United Nations Environment Programme (UNEP) Poverty-Environment Initiative (PEI).

United Nations Atlas of the Oceans, 2010. USES Worldview: Human Settlements on the Coast. Accessed 04.06.15 at: http://www.oceansatlas.org/servlet/CDSServlet?status=ND0xODc3JjY9ZW4mMzM9KiYzNz1rb3M~.

Vadeboncoeur, N., 2014. Knowing Climate Change: Modeling, Understanding and Managing Risk (Ph.D. thesis). Faculty of Graduate and Postdoctoral Studies (Resource management and Environmental Studies), The University of British Columbia, Vancouver, B.C. 310pp. Accessed 02.02.15 at: https://circle.ubc.ca/handle/2429/50777.

Webster, T.L., Forbes, D.L., 2006. Airborne laser altimetry for predictive modeling of coastal storm-surge flooding. In: Richardson, L.L., LeDrew, E.F. (Eds.), Remote Sensing of Aquatic Coastal Ecosystem Processes: Science and Management Applications. Springer, Dordrecht, pp. 157–182.

Wenger, E., McDermott, R., Snyder, W., 2007. Cultivating Communities of Practice: A Guide to Managing Knowledge. Harvard Business School Press, Cambridge MA.

Young, O., 2013. On Environmental Governance: Sustainability, Efficiency, and Equity. Paradigm Publishers, Boulder, CO.

Making the Link

By Jean-Paul Vanderlinden

Managing Adaptation to Changing Climate in Coastal Zones	⟺	**Sustainability of Artificial Coasts: The Barcelona Coast Case**
Daniel E. Lane, Colleen Mercer Clarke, John D. Clarke, Michelle Mycoo and Judith Gobin		*A. Sánchez-Arcilla, M. García-León and V. Gràcia*

Using examples from Canadian and Caribbean coastal communities, Lane et al. have presented the scope of options that are available in order to adapt to the impacts of climate change. They stress the necessity of taking into account the priorities of the communities. Yet taking into account communities' priorities involves establishing top-down and bottom-up collaborative processes. Once priorities are identified, stemming sometimes from historical choices that seriously limit the options available, what can be the use of science?

Such a situation may be found in the following chapter authored by Sánchez-Arcilla and colleagues. The Catalan coastline has been through a long history of human usage and influence. Retreat is not an option, doing nothing is risky at best, and accommodation remains as the sole option. Sánchez-Arcilla and colleagues explore a robust analysis of accommodation options in terms of sustainability. Novel "green" coastal interventions are analyzed for different types of exposed coastlines.

Chapter 10

Sustainability of Artificial Coasts: The Barcelona Coast Case

A. Sánchez-Arcilla, M. García-León, V. Gràcia
Universitat Politècnica de Catalunya, Laboratori d'Enginyeria Maritima LIM/UPC, International Centre Coastal Resources Research, CIIRC, Barcelona, Spain

Chapter Outline

INTRODUCTION

Coasts in developed countries have experienced a decrease in sediment availability together with a simultaneous increase of socioeconomic pressures so that (under present climate and socioeconomic conditions) the level of conflicts in many European Union coasts has reached dangerous levels, close to unsustainability (Hinkel et al., 2013). Under future conditions the situation will aggravate due to mean sea-level rise (Jevrejeva et al., 2014) and changes in storminess (Lionello et al., 2008b).

The "conflict," between human development and natural processes and not between land and sea, stems from the excessive pressure of uses affecting the coastal territory: extensive urbanization and heavily regulated rivers are representative examples (Sierra et al., 2003). The sediment starvation conflict appears more explicitly in narrow coastal fringes, where roads, railways, and buildings occupy (Figure 10.1) the former back beach, "freezing" the sand-dune deposits, which become no longer available for coastal sand transport.

This situation, well illustrated by Mediterranean beaches and here by the Catalan coast (Figure 10.2), exemplifies the conflict between human uses and infrastructures, which are essentially static, and the natural dynamics of the coast that require enough sediment and space available for morphodynamic evolution at various scales.

FIGURE 10.1 Rigidized coastal fringe in the Spanish Mediterranean, showing how an alongshore revetment, a railway line, and other infrastructures have frozen the back-beach sand deposits and prevent the natural coastal dynamics.

FIGURE 10.2 Example of rigidized and out-of-equilibrium coast from the Spanish Mediterranean. The perpendicular groins and jetties produce embayments that erode in the middle and loose sediment through the lateral barriers, failing to provide a wide, stable beach. The coastal interventions performed exemplify the conflict between human infrastructures and coastal natural dynamics.

The expected climatic changes in waves (Lionello et al., 2008a), sea level plus circulation patterns (Marcos and Tsimplis, 2008), and sea-level rise (Jevrejeva et al., 2014) will increase the mismatch between coastal uses and geomorphic resources, making many of our present beaches unstable and even "artificial" (requiring permanent maintenance). Global-scale projections will not allow robust or even socially acceptable decisions for a given coastal region. The necessary downscaling of climate parameters requires a significant amount of work and a number of hurdles to be circumvented, particularly for semi-enclosed seas such as the Mediterranean (Casas Prat and Sierra, 2013). In this case, projected future waves will fall (by 2100), depending on the season, within a ±10%

interval for the average and a ±20% interval for the 50-year return-period values. The observed mean sea-level rise during the last decades (1970–2008) has been about 2.7 mm/year (Jevrejeva et al., 2014), while future scenarios (Conte and Lionello, 2013) present variations on the order of 5% for both positive and negative storm surges. These values, combined with land subsidence, will increase coastal conflicts (and these lead to higher risk levels) arising from flooding and erosion for Mediterranean and, in general, developed coasts.

The sustainability (capacity to maintain a given coastal state, by natural or artificially acceptable means) of artificial coasts by means of beach nourishment or hard coastal protection may lead to modified sediment fluxes, and at the mid-range or long-range time scales, enhanced erosion, aggravating the conflicts in beaches like those in the Mediterranean and posing more difficult challenges for coastal managers and policymakers (Sánchez-Arcilla et al., 2011). More efficient alternatives should be based on bed vegetation (Paul et al., 2014; Feagin et al., 2011; Folkart, 2005), transient defense measures (Sánchez-Arcilla et al., 2012, 2014), or enhanced natural accretion mechanisms (e.g., due to controlled flooding). The efficiency of such environmentally friendly solutions is difficult to quantify and has only been recently approached with large-scale laboratory data that offer reliable information on their efficiency without undue scale-distortion effects (Manca, 2010; Koftis et al., 2013; Anderson and Smith, 2014).

The main aim of this chapter is therefore to analyze the sustainability of artificial coasts by means of novel "green" coastal interventions such as sea grass meadows or artificial (limited) nourishment. The keyword here is "limited," which means it would be a small volume of sediment lasting longer than present nourishments. The analysis (for present and future climate conditions) is based on numerical simulations (Sánchez-Arcilla et al., 2013), applied to two vulnerability hotspots in the Catalan coast (Spanish Mediterranean): the Llobregat Delta (southern beach) and the Maresme coast (northern beach), illustrated by Figures 10.3 and 10.4.

FIGURE 10.3 Vulnerable coastal case in the Llobregat Delta, south of Barcelona city, showing a typical urban beach.

FIGURE 10.4 Heavily anthropogenic coast in the Spanish Mediterranean illustrating the effect of structural and episodic erosion events and the threat for sustainability under future climates.

THE BARCELONA COAST

The Catalan coast presents a high level of diversity in about 700 km, of which 250 km correspond to sandy beaches, 150 km to urban beaches, and 75 km to open beaches. Out of this total, 208 km are cliffs and about 40 km are artificially armored coasts. The general orientation is north–south (northern part) and north-east to south-west (rest). The Barcelona littoral, with a high level of pressures and conflicts, features relatively coarse sediments (D_{50} 500 μm) except in deltaic areas, such as the Llobregat Delta where Barcelona Airport is located, where the sediment size is about $D_{50} = 250$ μm. The yearly average significant wave height is about 0.8 m, with peak periods around 6 s. During storm events (Sánchez-Arcilla et al., 2008a), values six times higher than mean conditions can be easily reached, corresponding to eastern wind events (longest fetch). The astronomical tidal range is less than 0.40 m, with storm surges up to 0.9 m.

The southern beach of this chapter represents a typical urban coastal stretch backed by a seafront promenade and buildings. Due to its orientation, the area is sheltered (Figure 10.5) from eastern waves but exposed to southern waves. The sediment size ranges from 200 to 350 μm, the beach width from 34 to 100 m (60 m average), and the average berm height is 0.8 m, within a gentle and dissipative profile.

The northern beach (Figure 10.5), exposed to eastern and southern waves, has been selected as representative of conditions downstream of small harbors that act as barriers for sediment transport and generate an erosive gradient. The beach width goes from 10 m at the northern-most part (right by the harbor barrier) up to 76 m at the southern part, where the beach width has recovered part of its initial shape. The sediment is coarse sand with a medium grain size of about 500 μm, leading to a steep and reflective profile. Storms are able to reach the existing alongshore revetment by overtopping, producing a modification of the upper part of the beach (CIIRC, 2010). Both cases illustrate artificial beaches requiring maintenance.

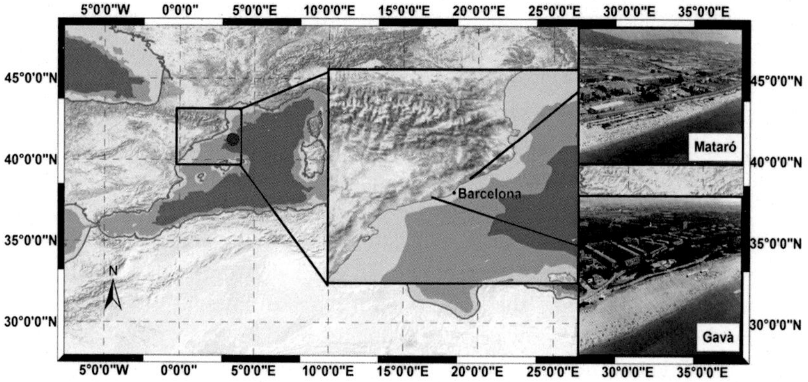

FIGURE 10.5 General map of the Western Mediterranean Sea, where the area of study is located. The two insets correspond to the northern and southern beaches analyzed in depth (Mataró and Gavà municipalities, respectively).

COASTAL STRATEGIES

Future scenarios of meteo-oceanographic conditions can enhance present coastal hazards, forcing a new approach to the traditional defend-adapt-retreat strategies. The uncertainty in future climate and socioeconomic scenarios may also alter the balance between these strategies, conceived for a set of hydrodynamic drivers and coastal responses that will likely be surpassed under future conditions. Taking the Spanish Mediterranean Coast as an example, there are two main options within a "maintain" strategy (defend or adapt), since the retreat pathway is not easily acceptable for a scarce territory and a coastal society not used to that approach:

1. Permanent infrastructures with "hard" reinforcements (e.g., higher free boards, longer revetments), associated with to a "rigid" maintenance of a given shoreline.
2. Soft or "green" solutions (e.g., vegetation, sand groins) that would maintain a given "space" but allow a certain level of dynamics for the shoreline.

Since the main aim of this chapter is the sustainability of artificial beaches under present and future conditions, we shall favor the more flexible solutions based on limited beach nourishment and sea bed vegetation, offering lower costs (in monetary and energetic terms) and preventing irreversible modifications of natural fluxes. Limited nourishments will have to be included in any other solution strategy due to the present sediment starvation of developed coastal zones to mitigate the impact of present and future storms, requiring maintenance (in monetary and energetic terms) that may render this type of strategy unsustainable in the near future, particularly for artificial beaches. This limited nourishment can also include a transient sediment buffer defense (such as a buffer dune or a buffer groin) that will dissipate part of the incoming wave energy while at the same time providing a surplus of sediment to the down-drift beaches. However, this type of solution strategy will only be feasible if there is enough sediment volume available. For instance, for the beaches located north of Barcelona, there is not enough space to deploy a buffer dune to limit beach erosion and inundation; while the southern beaches certainly offer room for that alternative (Figure 10.6).

Bed vegetation will be an element of adaptive strategies that require a certain time interval for the plant cover to become effective. It may cover the emerged and/or submerged parts of the profile (vegetated dunes or sea grass meadows), resulting in a smaller mobility than beach areas without biota in flume tests (Figure 10.7). We have considered two possible densities (50 stems/m^2 and 100 stems/m^2), maintaining the same type of plant and height, which is 0.5 m, based on the existing meadows off the Catalan Coast.

The possible coastal strategies for sustainability of the studied beaches will therefore consider (1) limited nourishments (e.g., buffer dune where there is room for that solution), (2) stabilizing vegetation, and (3) combinations of both elements. Both solution types are flexible and require a periodic control within a more proactive type of coastal management than the one we are used to.

SUSTAINABILITY ASSESSMENT

Approach

Sustainability has been evaluated for present and future storms, with return periods of 5, 10, and 15 years. This is based on the design life (20–25 years) of coastal engineering interventions and on the acceptable risk levels for coastal structures in many Mediterranean countries. The wave data to calculate the extreme probability distributions come from the two directional-wave buoys (XIOM network) covering the period from 1984 to 2013 and deployed at depths of 45 and 75 m. They characterize the wave climate for the northern (predominance of NE and SW) and southern beaches (predominance of E, S, and W). Future projections for the A1B scenario (a medium scenario from IPCC Assessment Report number 4), conveniently downscaled for this area (Cases Prat and Sierra, 2013), indicate that during the winter season, eastern waves will become less dominant while southeasterly and southwesterly waves will become more frequent. This is consistent with the projected increase of NAO (North Atlantic Oscillation climatic index) during the winter (Donat et al., 2010) and the northward shift of storm tracks (Lionello et al., 2008a).

The performance of the proposed coastal interventions has been evaluated for two different time-slices. The first one corresponds to the year 2030, which represents a near future for climate projections but implies a relatively long term for coastal management. The second time-slice corresponds to the year 2050, which represents the mid-future in climatic terms and a long-term horizon for coastal management. The directional shift in incoming wave power has been introduced by modifying the directional coefficients of the significant wave height probability functions. The basic distribution has been taken as a three-parameter Weibull distribution (Sánchez-Arcilla et al., 2008b). A moderate (±2%) change in wave

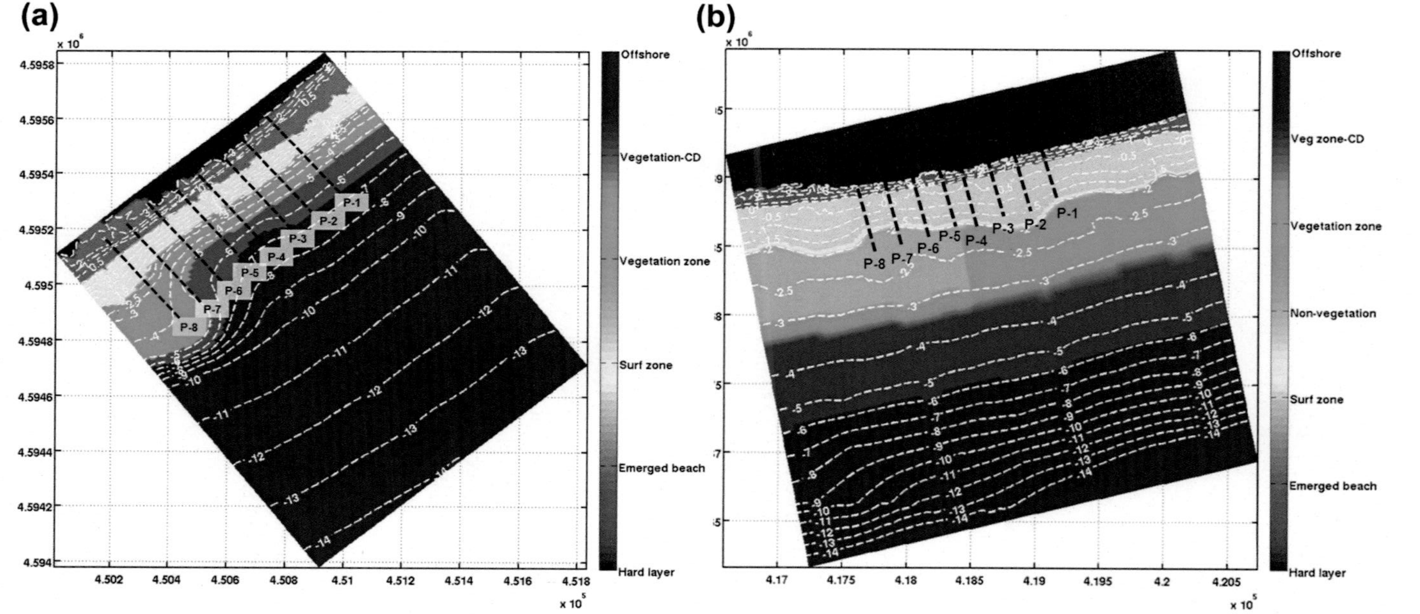

FIGURE 10.6 Plan view for the areas (a) north of Barcelona without space for a buffer dune and (b) south of Barcelona offering enough room for such an alternative. The figure shows the depth (in blue) and topographic (in yellow) contours together with the back-beach area (in burgundy). P stands for the profile numbering in the surveying.

FIGURE 10.7 Large- scale hydraulic tests carried out at the CIEM flume in LIM/UPC Barcelona, to analyze the effect of a plant canopy on hydrodynamics and morphodynamics.

storminess has been considered for 2030, while a higher variation (±8%) has been selected for 2050, in accordance with the available climate projections.

Sea-level scenarios have been derived from IPCC-AR5 projections (IPCC, 2013). A high-end representative concentration pathway (RCP 8.5) has been selected as an upper bound for the Mediterranean, where the evidence is that sea-level rise will be (at least initially) milder than for other areas in the planet. The projection for the northern beach (time-slices 2030 and 2050) reaches values around 0.12 and 0.24 m for a coastal stretch without subsidence. For the southern beach, the values proposed with deltaic subsidence (2.5 mm per year, an upper bound for the Llobregat Delta) are 0.16 and 0.33 m for the same time-slices. These global projections should be applied with care to regional or local analyses. However, for the new IPCC-AR5 projections, no regional downscaling exists yet, so the only alternative is a sensitivity study with global values, such as performed here. The estimate of 2.5 mm per year is from a larger nearby Mediterranean delta (Ibáñez at al., 2010).

The impact simulations are based on a storm with an average duration of 24 h, assuming a trapezoidal distribution for the wave height and mean sea level, that combines linearly the climatic rise with a storm surge peak of 0.35 m, based on present values (Sánchez-Arcilla et al., 2008b; Bolaños et al., 2009). These drivers have been used to evaluate the performance of the proposed solutions for two time-slices (2030 and 2050) with three storm-return periods (5, 10, and 15 years) and two main directions. This results in six possible combinations for every time-slice and climatic mean sea level (3 return periods × 2 directions), as summarized in Table 10.1.

Modeling

Morphodynamic impacts have been obtained with a locally adjusted version (PREMOS) of the X-BEACH model (Roelvink et al., 2009), based on the action balance equation for waves (Booij et al., 1999) and a 2DH Navier–Stokes shallow water solver for the currents (Stelling and Duinmeijer, 2003). These

TABLE 10.1 Summary of Hydrodynamic Drivers for Impact Projection

Hydrodynamic Variable	Magnitude	
SLR (m) [mean SLR with subsidence]	2030: [0.12; 0.16]	2050: [0.24; 0.33]
Significant wave height (m) [5,10,15] return periods	E [4.31, 4.69, 4.9], SSE [2.50, 2.72, 2.84] (north beach) S [3.14; 3.42; 3.58], SSW [3.75, 4.08, 4.26] (south beach)	
Associated peak period (s) [5,10,15]	E [12.2,12.7,13], SSE [9.2, 9.7, 9.9] (north beach) S [9.8; 10.2; 10.5], SSW [10.7, 10.2, 11.4] (south beach)	
Wave direction (°)	E, SSE (north beach); S, SSW (south beach)	

hydrodynamic terms feed an advection–diffusion equation calculating sediment concentration, and from that erosion/accretion based on a sediment mass conservation law. The bed concentration comes from a Van Rijn (2007) formulation, whose performance along the Catalan coast has been tested (Sánchez-Arcilla et al., 2013). Vegetation damping comes from Myrhaug and Holmedal (2011).

Mean sea level and wave spectra at the offshore and lateral boundary nodes were updated every 30 min, with a computational time step of 0.4 s. The southern beach (3580 m alongshore and maximum depth 19 m) grid was 26 × 21 m, while the northern beach (150 m alongshore and maximum depth 14 m) grid was 11 × 13.5 m. The sediment has been taken as uniform in size, with $D_{50} = 250\,\mu m$ for the southern beach and $D_{50} = 500\,\mu m$ for the northern beach.

The two beach domains have been subdivided into a number of subzones: (1) a hard layer with non-erodible coverage corresponding to infrastructures (railway, buildings, etc.); (2) an emerged beach layer extending down to the backwash limit; (3) a submerged beach layer covering the surf zone; and (4) a sea grass meadow layer going down to the closure depth (6.35 m for the southern case, with milder waves; and 6.9 m for the northern case, with more energetic waves).

Assessment

The sustainability of artificial beaches under present and future conditions has been determined from (1) hydrodynamics drivers and (2) beach responses. Regarding hydrodynamics, we have considered three main variables, the significant wave height before the initiation of breaking, the spatially and temporally averaged near shore circulation "strength," and the run-up. The values obtained for future scenarios correspond to an ensemble of storms with return periods of 5, 10, and 15 years plus a sea-level rise for the two selected time-slices (2030 and 2050). The calculated sedimentary volumes correspond to the different studied layers, while the strength of the near-shore circulation corresponds to the time and depth averaged module of the current; the wave height variation corresponds to the averaged time series. The run-up represents the spatial average of the maximum run-up for each of the studied beach profiles. Ratios below 1 indicate a decrease in future scenarios with respect to present conditions, and ratios above 1 indicate an increase with respect to present conditions.

The significant wave height (Hs) for the northern beach shows a reduction of 25% for the surf zone and 10% for the vegetated zone in 2030. These percentages diminish to 15% for the surf zone and 7% for the vegetated zone for 2050, representing the smaller efficiency of vegetation for the larger waves and mean sea level expected by 2050. The reduction in the emerged beach is, however, much larger, going up to 70%. This behavior is expected of narrow surf zone reflective beaches. The southern beach, wider and more dissipative, presents already important natural dissipation capabilities and thus smaller reduction due to vegetation (about 10% in the surf zone and 20% in the rest of the submerged beach). Regarding the average surf zone circulation, the southern beach shows a reduction between 15% and 20%, while the northern beach presents an increase due to the blocking effect of the plant canopy that forces the flow to pass through a narrower section. Because of that, due to the continuity equation, some undesired increases in near-shore currents will appear.

Run-up (a critical variable for erosion and flooding) presents for the northern beach a limited reduction due to vegetation between 15% and 5%, attributed to the limitations of bed vegetation for large incoming waves (close to 5 m for the 15-year storm on this beach). For some wave conditions (SSE waves and time-slice 2050), the run-up even appears to increase with respect to the case of no vegetation. This has been attributed to the enhanced wave damping that may produce an increase in set-up and a wider surf zone, leading to higher set-up and thus run-up. For the southern beach (Tables 10.2 and 10.3) the reduction in run-up is of about 25%, due to the wider dissipative beach.

The horizontal run-up action (Figure 10.10) describes the horizontal extent of the beach profile zone subject to run-up and run-down. If subsidence is taken into account (southern beach, Tables 10.2 and 10.3), run-up increases moderately for 2050 and so do erosion and flooding, with hardly any effect of plant density. For 2030 the dissipative nature of the beach dominates and there is a smaller effect. For the northern beach (not shown), the effects are less clear, depending on the relative weight of water depth, surf zone width, and plant canopy blockage. Bed vegetation also reduces run-up variability, with steadier conditions that should facilitate the management of beach deposits and hinterland infrastructure.

The plant canopy increases beach resilience since (Figure 10.10) the flooding and erosion of the upper part of the beach start at hour 12, that is near the middle of the storm duration, with a functional limit of the plant canopy to bound the incoming wave energy and associated circulatory system. The beach response (Figures 10.8 and 10.9) can be assessed from the projected beach

TABLE 10.2 Southern Beach Run-up Calculated Considering SLR Only

	SSW (202.5°)			S (180°)		
	50 stems/m^2	100 stems/m^2	Ref. (m)	50 stems/m^2	100 stems/m^2	Ref. (m)
2030	0.74	0.74	1.58	0.91	0.91	1.19
2050	0.73	0.73	1.86	0.76	0.76	1.64

Ratios indicate vegetated beach value versus beach nourishment without vegetation. The results correspond to two wave directions, two time horizons, and two plant densities (stems per square meter).

TABLE 10.3 Southern Beach Run-up Calculated Considering SLR + Subsidence

	SSW (202.5°)			S (180°)		
	50 stems/m^2	100 stems/m^2	Ref. (m)	50 stems/m^2	100 stems/m^2	Ref. (m)
2030	0.74	0.74	1.70	0.91	0.91	1.23
2050	0.75	0.75	1.87	0.76	0.76	1.79

Ratios indicate vegetated beach value versus beach nourishment without vegetation. The results correspond to two wave directions, two time horizons, and two plant densities (stems per square meter).

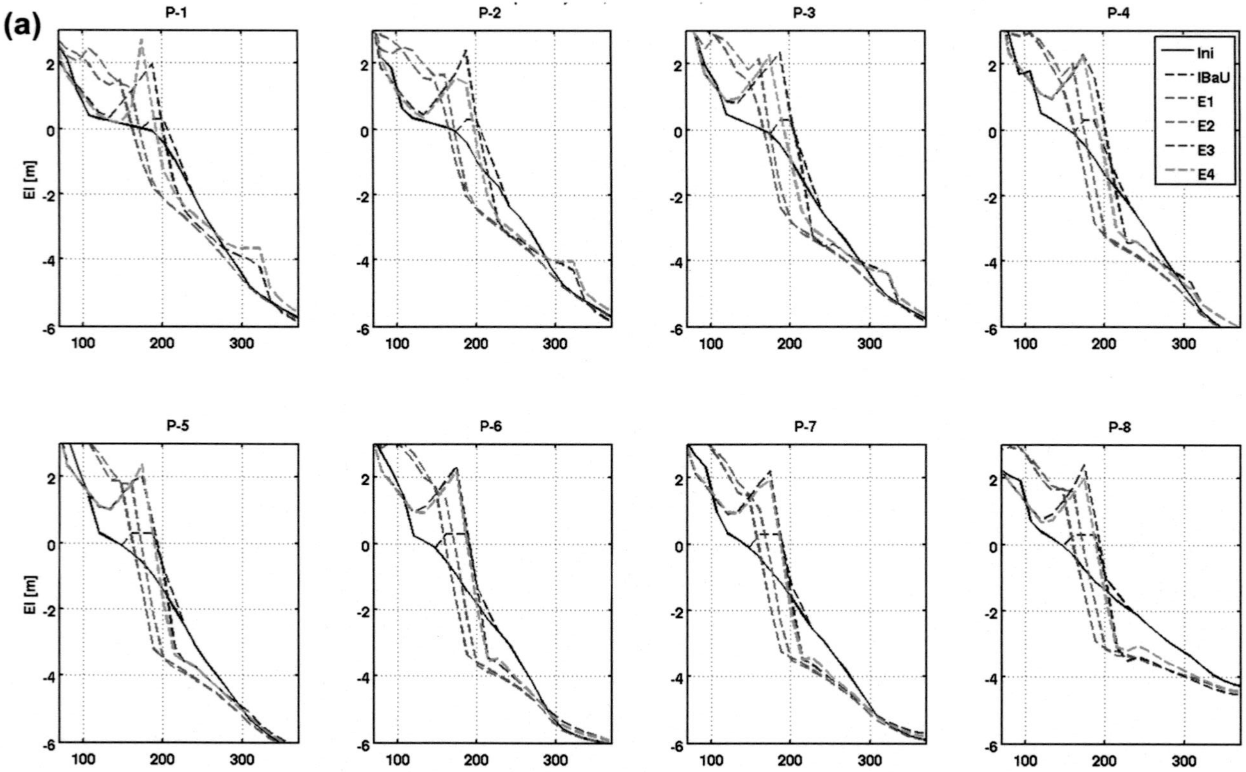

FIGURE 10.8 Northern beach profiles. Initial (solid black line for the prognostic scenario and dotted black line for the alternative) and final (colored dotted line) beach profiles for the simulated storm at 2050. The hydrodynamic drivers consist of the mean wave field with a return period of 15 years and the projected mean sea-level rise of 0.24 m above present conditions. Each profile represents the beach behavior for different intervention types: do-nothing alternative (red, E1); nourishment (magenta, E2); nourishment and sea-grass meadows with 50 stems/m^2 (blue, E3); and the same combination but with 100 stems/m^2 (green, E4). The x-axis represents the cross-shore distance in meters and the y-axis the elevation relative to mean sea level. Note how the morphodynamic response changes for different hydrodynamic scenarios. Panel (a) incident wave direction = 90° (east); panel (b) wave direction = 157.5° (south-southeast).

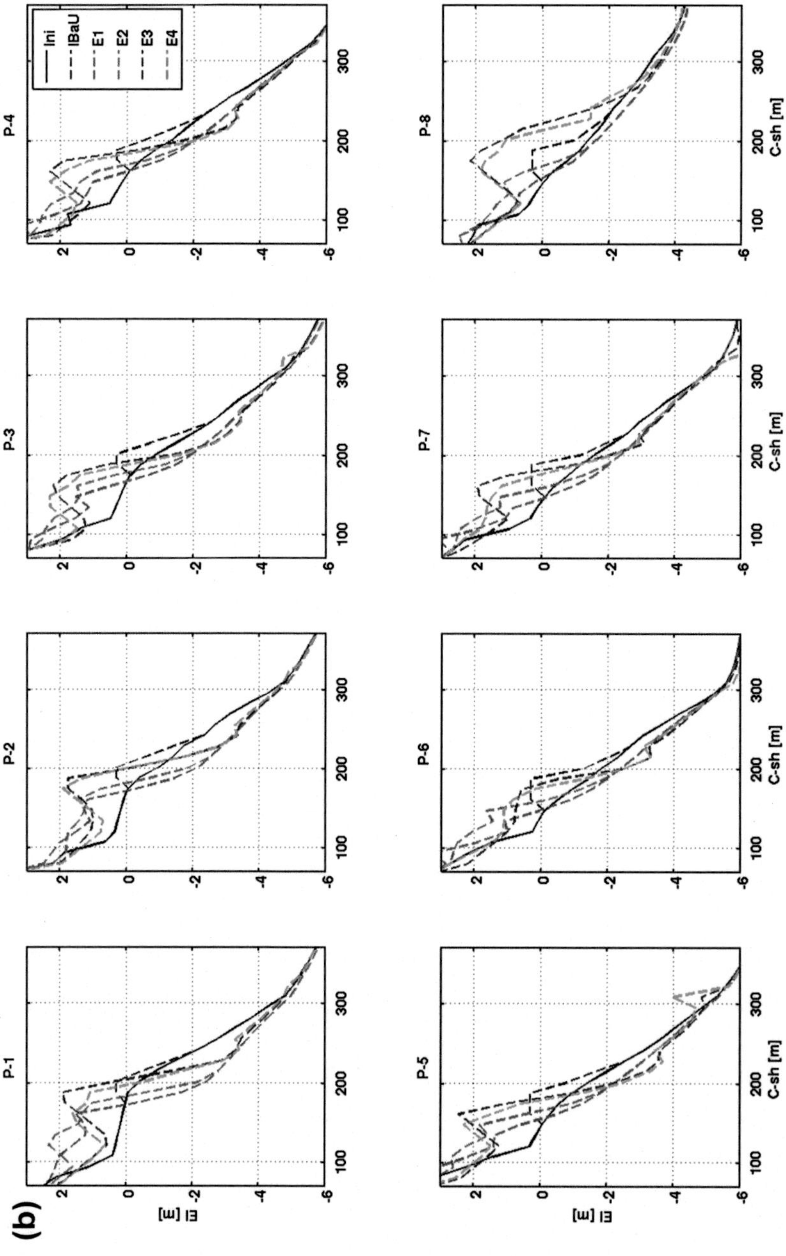

FIGURE 10.8 Continued

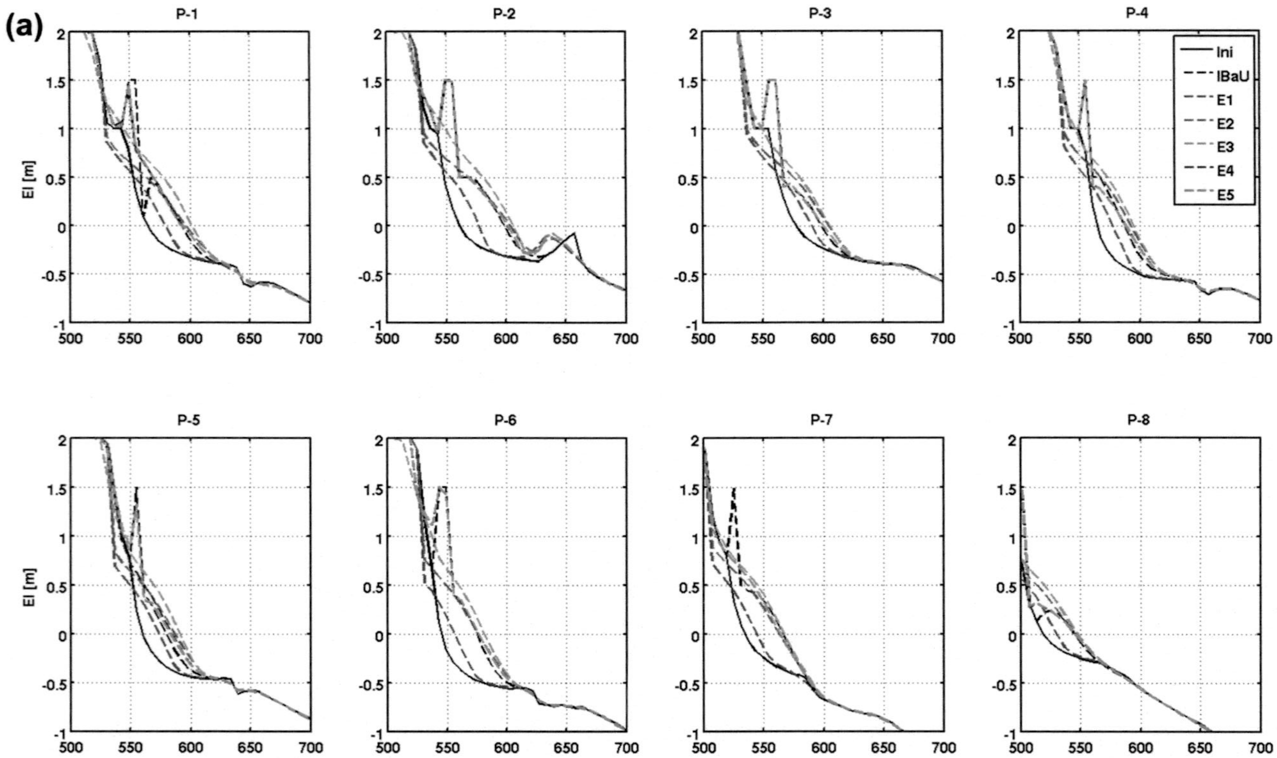

FIGURE 10.9 Southern beach profiles. Initial (solid black line for the prognostic scenario and dotted black line for the alternatives) and final (colored dotted line) beach profiles for the simulated storm at 2050. The hydrodynamic drivers consist of the mean wave field with a return period of 15 years and the projected mean sea-level rise of 0.33 m above present conditions, combining climatic sea-level rise with subsidence. Each final profile represents the beach behavior for different intervention types: do-nothing scenario (red, E1); nourishment (magenta, E2); nourishment and a dune (cyan, E3); nourishment, consumable dune, and sea-grass meadows with 50 stems/m^2 (blue, E4); and the same combination but with 100 stems/m^2 (green, E5). The x-axis represents the cross-shore distance in meters and the y-axis the elevation relative to mean sea level. Note how the morphodynamic response changes for different hydrodynamic scenarios. Panel (a) incident wave direction = 180° (south); panel (b) wave direction = 202.5° (south-southwest).

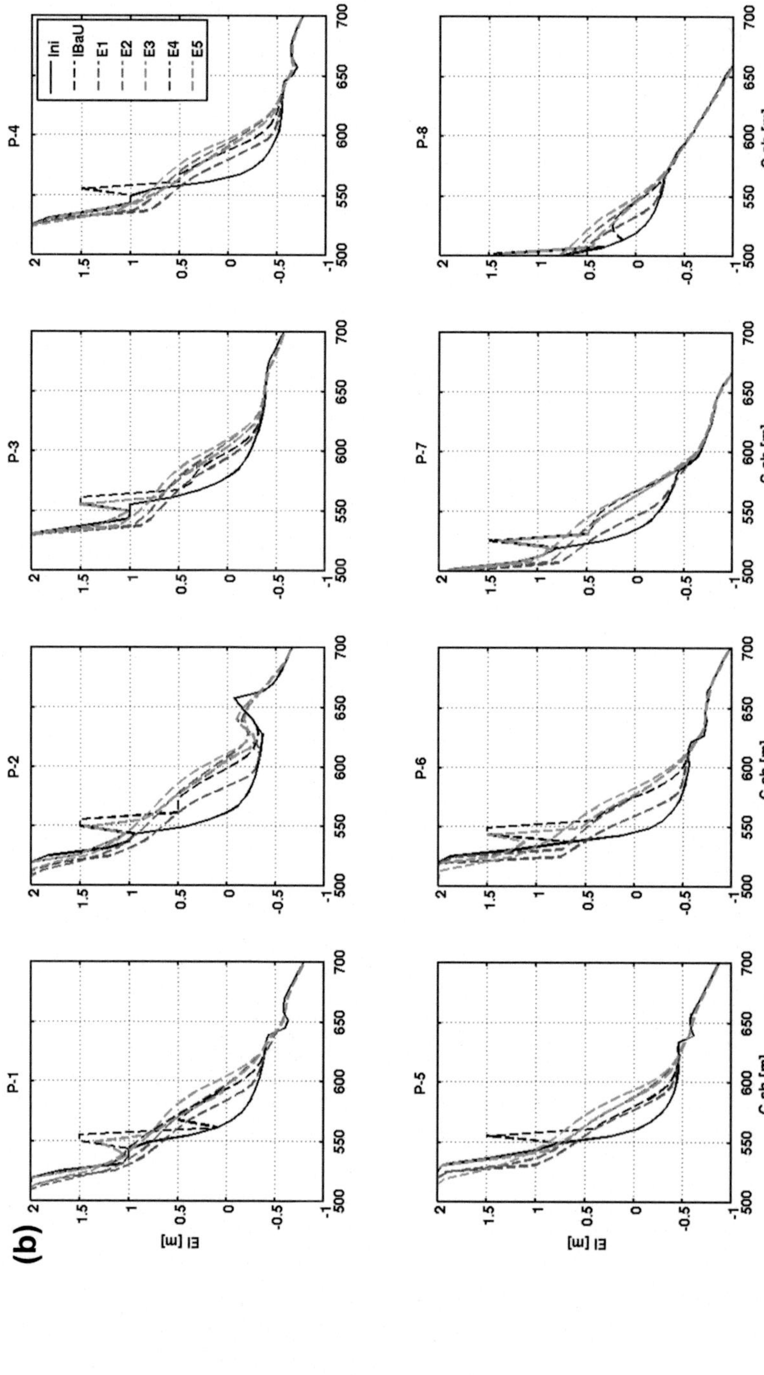

FIGURE 10.9 Continued

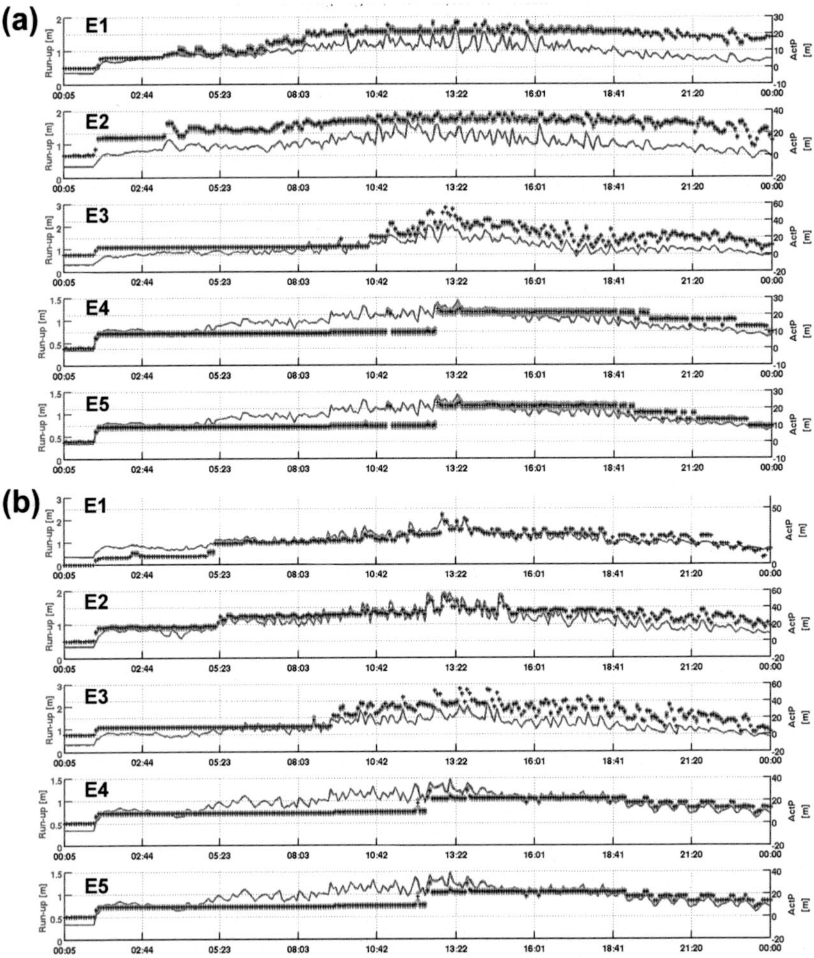

FIGURE 10.10 Run-up (vertical) and run-up action point (horizontal) displacement for the different alternatives at the southern beach. The hydrodynamic drivers consist of the same mean wave field as before, with a return period of 15 years. The mean sea level has been set at 0.33 m, including subsidence. The vertical axis on the left indicates the spatially averaged run-up (solid red line) in meters, together with the 95% (dashed magenta line) and 99% (dashed green line) run-up confidence interval. The vertical right axis indicates the spatial-averaged run-up horizontal action point displacement (dotted blue line). Positive values denote the horizontal extent of shoreline retreat and negative values represent an offshore displacement of the shoreline. The 95% (dotted magenta line) and 99% (dotted green line) action point confidence interval is also shown. Panel (a) wave direction = 180°; panel (b) wave direction = 202.5°.

evolution and eroded volumes (ratio of future to present conditions). For the southern beach there is a large reduction in eroded volumes for the emerged and vegetated areas, which goes up to 60%. The reduction in the submerged beach without vegetation is also significant (up to 20%). This reduction is larger for the time-slice 2050 (larger waves) and the SSW condition, where there is also a longer fetch and more energetic waves. The effect of vegetation on shoreline retreat results in erosion starting at about half the storm duration (i.e., 12 h after the start of the storm; see Figures 10.8–10.10) and the "consumable" dune does not disappear throughout the erosive event, whereas the same dune was indeed consumed by the erosive power of the waves in the case without vegetation.

The performance of the solution (limited nourishment, vegetation, or combinations) will thus depend on the relative weight of the various competing mechanisms, not all of which are represented in the numerical model, and the considered area (emerged or submerged parts of the beach).

DISCUSSION

More reflective beaches, exposed to eastern waves such as the north beach in this chapter, experience a larger percentage reduction in incoming wave height. The corresponding more dissipative beaches (exemplified by the south beach in this chapter) experience a smaller percentage reduction due to vegetation since the natural dissipative character of the profile already exerts a strong control on the near shore hydrodynamics.

Because of that it can be said that although in percentage terms vegetation may appear to be more efficient, in absolute terms vegetation will be a more robust type of solution for dissipative profiles with a wider near-shore zone. As a consequence for dissipative beaches the percentage reduction in erosion is smaller but the total effect on near-shore morphodynamics is clearer. This means that the circulation, the incoming turbulent wave energy, and the resulting beach dynamics are easier to control within a dissipative profile where the effect of vegetation will in all cases suppose a reduction in incoming wave height and strength of near-shore circulatory systems, leading to a reduction in erosion.

Regarding flooding the run-up, the reduction in reflective profiles is again smaller due to increased set-up because vegetation prompts wave breaking at a larger distance from the shoreline and therefore the reduction of wave run-up can be offset by the increase in set-up. However, for dissipative beaches, vegetation produces a significant reduction in run-up and nearly no change in set-up, which results in a decrease of flooding. Associated with this, the horizontal extent of beach affected by run-up is easier to control in dissipative beaches while in more reflective profiles the horizontal extent affected by run-up is a combined function of all of the hydrodynamic and morphodynamic parameters involved (water depth at the initial breaking and secondary breakings, surf zone width, blockage effect of plants, etc.).

In summary, although vegetation reduces in all cases the hydrodynamic variability and postpones the onset of erosion and flooding at more acceptable levels, for dissipative profiles the effect of vegetation appears to be more robust and therefore easier to include within a coastal engineering or management program. As a consequence, from the perspective of beach vegetation, dissipative beaches appear to be more resilient. In all cases the effect of vegetation, regardless of the type of beach morphology, reduces mobility and increases resilience, moving the onset of unacceptable erosion and flooding to nearly half of the storm duration for the studied Mediterranean cases.

CONCLUSIONS

The sustainability of artificial beaches is linked to the sediment volumes and fluxes that occur under present and future climatic conditions. From the analysis performed in this chapter, it is apparent that reflective beaches (illustrated by the northern beach) are more sensitive to the prevailing mean sea-level conditions, including here the mean sea level due to climate warming plus storm surges and set-up associated with storms. They also exhibit a smaller shoreline retreat (higher resilience) with sea-grass meadows and occasional increases of the near-shore circulatory system (due to the blockage effect of the plant canopy). This increase in currents, together with a higher sensitivity to water fluxes linked to sea-level rise, make (for a given sediment size) reflective beaches more sensitive to future climates.

For dissipative beaches, particularly under the more energetic wave conditions, there is a wider *surf* zone, where the attenuation due to vegetation can be more clearly perceived. This results in a weaker near-shore circulatory system and reduced erosion and accretion rates.

The sediment trapping effect of sea-bed vegetation, although not modeled in our simulations, is expected to act as a sediment buffer for near-shore morphodynamics that will prevent further erosion under storm action. This is considered to be another important asset for sea-bed vegetation, since the occurrence of storm sequences is one of the main responsible factors for cumulative damages on beaches and coastal zones.

The sustainability of dissipative beaches appears, therefore, to be more clearly enhanced by the proposed interventions than would be the case for reflective beaches. However, the coarser sediments normally found in these latter beaches result in smaller transport rates and thus a reduced mobility.

The existence of a large "enough" sediment volume (for the required functions of the beach), and the space for the natural beach dynamics to act, will also enhance the capacity of the beach to adjust to climate variability and change, resulting in greater sustainability.

ACKNOWLEDGMENTS

This work has been supported by the research project RISES-AM (contract FP7-ENV-2013-two-stage-603396). It has also benefited from the iCoast project (DG-ECHO, contract ECHO/SUB/2013/661009). The second author acknowledges the PhD scholarship from the Government of Catalonia (DGR FI-AGAUR-14).

REFERENCES

Anderson, M., Smith, J., 2014. Wave attenuation by flexible, idealized salt marsh vegetation. Coastal Eng. 83, 82–92.

Bolaños, R., Jordà, G., Cateura, J., López, J., Puigdefabregas, J., Gómez, J., Espino, M., 2009. The XIOM: 20 years of a regional observatory in the Spanish Catalan coast. J. Mar. Syst. 77, 237–260.

Booij, N., Ris, R., Holthuijsen, L., 1999. A third-generation wave model for coastal regions, part I, model description and validation. J. Geophys. Res. 104 (C4), 7649–7666.

Casas-Prat, M., Sierra, J., 2013. Projected future wave climate in the NW mediterranean sea. J. Geophys. Res. Oceans 118, 3548–3568.

CIIRC, 2010. LlibreVerd: Estat de la zonacostanera a Catalunya. Generalitat de Catalunya.

Conte, D., Lionello, P., 2013. Characteristics of large positive and negative surges in the Mediterranean Sea and their attenuation in future climate scenarios. Glob. Planet. Change 111, 159–173.

Donat, M.G., Leckebusch, G.C., Pinto, J.G., Ulbrich, U., 2010. European storminess and associated circulation weather types: future changes deduced from a multimodel ensemble of GCM simulations. Clim. Res. 42 (1), 27–43.

Feagin, R., Irish, J., Moller, I., Williams, A., Colon-Rivera, R., Mousavi, M., 2011. Short communication: engineering properties of wetland plants with application to wave attenuation. Coastal Eng. 58, 251–255.

Folkard, A.M., 2005. Hydrodynamics of model Posidonia oceanica patches in shallow water. Limnol. Oceanogr. 50 (5), 1592–1600.

Hinkel, J., Nicholls, R., Tol, R., Wang, Z., Hamilton, J., Boot, G., Vafeidis, A., McFadden, L., Ganopolski, A., Klein, R., 2013. A global analysis of erosion of sandy beaches and sea-level rise: an application of DIVA. Glob. Planet. Change 111, 150–158.

Ibáñez, C., Sharpe, P.J., Day, J.W., Day, J.N., Prat, N., 2010. Vertical accretion and relative sea level rise in the Ebro Delta wetlands (Catalonia, Spain). Wetlands 30, 979–988.

Jevrejeva, S., Moore, J., Grinsted, A., Matthews, A., Spada, G., 2014. Trends and acceleration in global and regional sea levels since 1807. Glob. Planet. Change 113, 11–22.

Koftis, T., Prinos, P., Stratigaki, V., 2013. Wave damping over artificial Posidonia oceanica meadow: a large-scale experimental study. Coastal Eng. 73, 71–83.

Lionello, P., Boldrin, U., Giorgi, F., 2008a. Future changes in cyclone climatology over Europe as inferred from a regional climate simulation. Clim. Dyn. 30, 657–671.

Lionello, P., Cogo, S., Galati, M., Sanna, A., 2008b. The Mediterranean surface wave climate inferred from future scenario simulations. Glob. Planet. Change 63, 152–162.

Manca, E., 2010. Effects of Posidonia Oceanica Seagrass on Nearshore Waves and Wave Induced Flows (Ph.D. thesis), University of Southampton.

Marcos, M., Tsimplis, M., 2008. Comparison of AOGCMs in the Mediterranean Sea during the 21st century. J. Geophys. Res. 113, C12028.

Myrhaug, D., Holmedal, L.E., 2011. Drag force on a vegetation field due to long-crested and short-crested nonlinear random waves. Coastal Eng. 58, 562–566.

Paul, M., Henry, P.-Y., Thomas, R., 2014. Geometrical and mechanical properties of four species of northern European brown macroalgae. Coastal Eng. 84, 73–80.

Roelvink, J., Reniers, A., Dongeren, A.V., van Thiel de Vries, J., McCall, R., Lescinski, J., 2009. Modeling storm impacts on beaches, dunes and barrier islands. Coastal Eng. 56 (11), 1133–1152.

Sánchez-Arcilla, A., Gónzalez-Marco, D., Bolaños, R., 2008a. A review of wave climate and predition along the Spanish Mediterranean coast. Nat. Hazard. Earth Syst. Sci. 8 (6), 1217–1228.

Sánchez-Arcilla, A., Jiménez, J., Valdemoro, H.I., Gràcia, V., 2008b. Implications of climatic change on spanish mediterranean low-lying coasts: the Ebro Delta case. J. Coastal Res. 24 (2), 306–316.

Sánchez-Arcilla, A., Jiménez, J., Marchand, M., 2011. Managing coastal evolution in a more sustainable manner. The conscience approach. Ocean Coastal Manage. 54, 951–955.

Sánchez-Arcilla, A., Gràcia, V., Grifoll, M., García, M., Pallarés, E., 2012. Operational forecast of beach morphodynamics. Reability and predictions limits. In: 33rd International Conference in Coastal Engineering (2012).

Sánchez-Arcilla, A., Gràcia, V., Solé, J.M., García, M., Sairouni, A., 2013. Forecasting beach morphodynamics: the shoreline border as a control for flooding risks. In: Klijn, Schweckendiek (Eds.), Comprehensive Flood Risk Management. Taylor and Francis Group.

Sánchez-Arcilla, A., García, M., Gràcia, V., 2014. Hydro Morphodynamic Modelling in Mediterranean Storms: Errors and Uncertainties under Sharp Gradients. Natural Hazards and Earth System Sciences 2 (2), 1693–1728.

Sierra, J.P., Sánchez-Arcilla, A., Figueras, P.A., González Del Río, J., Rassmussen, E.K., Mösso, C., 2003. Effects of discharge reductions on salt wedge dynamics of the Ebro River. River Res. Appl. 20.

Stelling, G., Duinmeijer, S., 2003. A staggered conservative scheme for every Froude number in rapidly varied shallow water flows. Int. J. Numer. Methods Fluids 43, 1329–1354.

van Rijn, L., 2007. Unified view of sediment transport by current and waves, part I, II, III and IV. J. Hydraul. Eng. 133 (6,7) 649–689 (part I & part II), 761–793 (part III & part IV).

Making the Link

By Omer Chouinard

Sustainability of Artificial Coasts: The Barcelona Coast Case	⟺	**Protected Shores Contaminated with Plastic: From Knowledge to Action**
A. Sánchez-Arcilla, M. García-León and V. Gràcia		*Juan Baztan* et al.

On the Spanish Mediterranean Coast, Sánchez-Arcilla et al. developed a model to assess sustainable approaches for adapting the coast to uncertainty in future climate and socioeconomic scenarios. A model was built to understand the more efficient adaptation strategies for the reduction of coast erosion. The choices were (1) maintaining the permanent infrastructures with "hard" reinforcements associated with rigid maintenance of a given shoreline, or (2) soft or "green" solutions that would maintain a given space but allow a certain level of dynamics for the shoreline. For different reasons, the second option appears to be a more robust solution and would lead to a reduction of erosion.

Baztan et al. have developed initiatives to protect shores contaminated with plastic. Their field work started with a group of dedicated people in Lanzarote, part of the Canary Islands, interested in "the conservation of biological and cultural diversity and economic and social development through partnerships between people and nature." To answer the question, "how can we truly solve the plastic problem?" they recognized the importance of a local to regional and global approach through a participatory process. To address the question of implementation, their solution is the creation of the Communities-Based Observatories Tackling Marine Litter (COASTAL) initiative. Its focus is on working with stakeholders to improve ocean literacy and enhance social engagement about the issue of marine litter.

While differing greatly in content, these two chapters both emphasize the importance of flexibility in the solutions they offer and recognition of the dynamic contexts within which they are operating.

Chapter 11

Protected Shores Contaminated with Plastic: From Knowledge to Action

Juan Baztan[1,2], Bethany Jorgensen[3,2], Jean-Paul Vanderlinden[1,2], Sabine Pahl[4], Richard Thompson[4], Ana Carrasco[5], Aquilino Miguelez[5], Thierry Huck[6,2], Joaquim Garrabou[7], Elisabetta Broglio[7,2], Omer Chouinard[8,2], Céline Surette[8,2], Philippe Soudant[9], Arnaud Huvet[10], François Galgani[11], Ika Paul-Pont[9]

[1]Université de Versailles Saint-Quentin-en-Yvelines, OVSQ, CEARC, Guyancourt, France; [2]Marine Sciences for Society, www.marine-sciences-for-society.org; [3]The University of Maine, Orono, Maine, USA; [4]Plymouth University, Drake Circus, Plymouth, UK; [5]Observatorio Reserva de Biosfera, Cabildo de Lanzarote, Arrecife, Spain; [6]UBO-CNRS-LPO, UFR Sciences F308, Brest, France; [7]Institut de Ciències del Mar, CSIC, Barcelona, Spain; [8]Université de Moncton, Moncton, New Brunswick, Canada; [9]IUEM, CNRS/UBO, Laboratoire des Sciences de l'Environnement Marin, Plouzané, France; [10]IFREMER, Centre de Brest, Laboratoire Physiologie des Invertébrés, Plouzané, France; [11]IFREMER, Centre de Corse, Laboratoire Environnement Ressources PAC/Corse Imm Agostini, ZI Furiani, Bastia, France

Chapter Outline

INTRODUCTION

Since the first use of the term "protected area" in 1933, at the International Conference for the Protection of Fauna and Flora in London, the number of protected areas has increased exponentially throughout the world. In early 2015, there were more than 209,000 designated marine and terrestrial protected areas. When taken together, these areas span over 30 million km^2 of the earth's surface (Deguignet et al., 2014).

In parallel with efforts for environmental protection, the production is increasing of materials that if improperly disposed of are potentially harmful. Plastic production, for example, follows this trend, with annual production in

2012 reaching nearly 300 million metric tons (Plastics Europe, 2013). From its initial production until the moment it is discarded as waste, plastic's life cycle has advantages and disadvantages for our societies and the ecosystems in which we live.

Five years ago, a working group formed to focus on the question of plastics contaminating the shores of protected areas in the Canary Islands. In this chapter, we will share the story of that working group's collaborative efforts to date, and also describe future directions in the shape of a larger regional working group: Communities-Based Observatories Tackling Marine Litter (COASTAL). As an entry point into the key elements presented in this chapter, we offer a brief case study highlighting work unfolding on the island of Lanzarote, followed by a concise summary of the plastic pollution problem, and a description of the aims and vision proposed by the COASTAL working group.

PLASTIC DEBRIS THREATENS PROTECTED AREAS

In 1993, Lanzarote was designated a Biosphere Reserve by the Man and Biosphere UNESCO program, and it is a founding member of the World Network of Islands and Coastal Zones, created in 2012. Lanzarote exemplifies issues related to marine debris for several reasons: (1) from the oceanographic perspective, it is strategically positioned in the Canary Current; (2) from a conservation perspective, it includes natural and national parks, and is one of the biggest Marine Protected Areas in Europe; (3) socially, its residents actively and enthusiastically support an ongoing plastic debris awareness campaign in schools, they spontaneously organize beach cleanings, they seek out scientific and technical support, and they take questions into their own hands and develop locally relevant solutions; and (4) politically, local administrations identify tackling plastic pollution as one of their priorities for the coming years, they support local initiatives, and look for ways to connect local efforts with regional and global endeavors.

For these reasons, we begin at this place and with these people who "seek to reconcile conservation of biological and cultural diversity and economic and social development through partnerships between people and nature" (www.unesco.org 2015). From here, our work has grown into COASTAL, at the North Atlantic regional scale, aiming to build solutions between locations in the oceanic circulation system in Figure 11.1, from sources to sinks.

During our preliminary fieldwork session in Lanzarote, from December 2008 to January 2009, we were astonished by the amount of accumulated plastic on the northern shores of the three visited Canary Islands. One of the first pictures we took of Famara Beach (Lanzarote) shows the overwhelming extent of this contamination, most of which is transported via the ocean from abroad and does not come from the island's inhabitants or visitors. We quickly realized that even if local solutions could be implemented, an

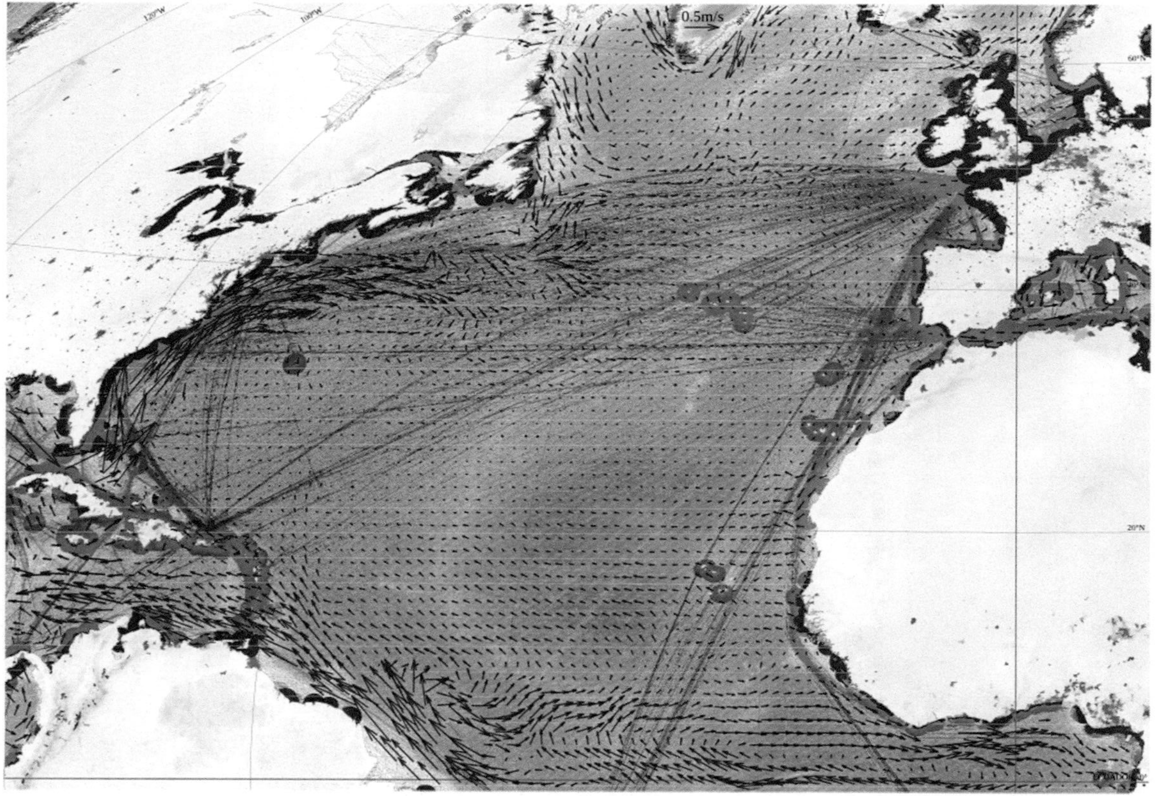

FIGURE 11.1 North Atlantic–Mediterranean system: We illustrate here the combination of (1) the oceanographic conditions, (2) the anthropogenic impact at sea (Halpern et al., 2008), and (3) population density (GPWv3, 2011).

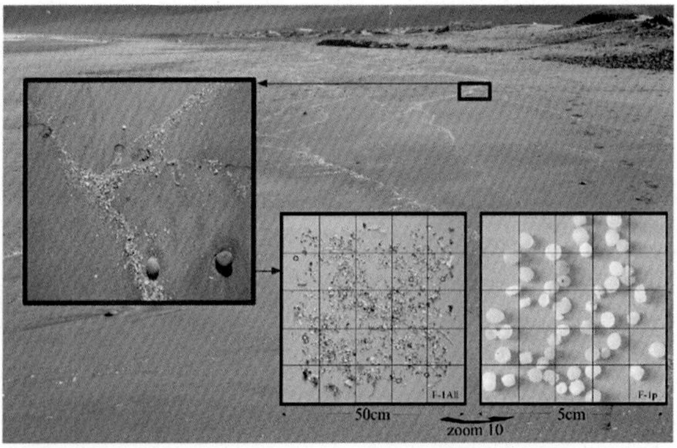

FIGURE 11.2 Microplastic deposition on Famara Beach. The background image shows extensive white lines of microplastics deposited on the beach by successive waves. The left box shows a closer view of the sampled area. The middle box displays the plastic extracted from the sample, and the right box shows the pellets found.

integrated regional approach would be necessary to truly solve the problem (Figure 11.2).

From here, we began a participatory strategizing process with local stakeholders at different levels: the fishermen's institute, municipal governments, the national park, the natural park, the marine reserve, and the biosphere reserve. We built a coordination team, which is managed at the island level by the biosphere reserve through an "inaugural contract." The contract helps dissipate the fuzziness inherent in the interdisciplinary approach (Blanchard and Vanderlinden, 2010). Once plastic pollution was identified as a top priority for these stakeholders, the first move was to transition from the identification of plastics, one of the most salient concerns at the local level, to selecting and implementing solutions at the regional scale.

As an initial step toward this goal, in January and February of 2013, we completed extensive microplastic sampling on every beach of Lanzarote and two other islands sharing its volcanic platform: Fuerteventura and la Graciosa. We found that despite being located in highly protected natural areas, all beaches on these three islands are substantially affected by microplastic pollution. On some beaches, pollution levels reached concentrations greater than 100 g of plastic per liter of sediment (Baztan et al., 2014). These results show that the plastic debris problem is shared by at least three of the islands in the Canary Islands archipelago. Moreover, it reflects conditions of extensive contamination in our oceans on a global scale (e.g., Thompson et al., 2009a,b; Law and Thompson, 2014).

To bring the public's attention to plastic pollution on Lanzarote and highlight plastic's accumulation in coastal zones around the world, in February 2013 the working group started a social media group (https://www.facebook.com/Aguitaconelplastico).This group continues to gain active followers who post news, alerts, questions, and local initiatives. Some posts have been read by over 25,000 people, and as of March 2015, there are more than 7000 participating members.

This social media method of raising awareness is one of many approaches taken to inform the public of the ongoing campaign. The message has also been spread via radio, newspapers, and television. In March 2013, a program from Spain's national television network focused two of its episodes on plastic pollution and the work being done on Lanzarote to combat it. The episodes were first aired in June 2013, and then again in September 2013, February 2014, and July 2014. They were shown during prime time, and with each broadcast, participation in the social media group jumped. After the last broadcast, 1000 new participants joined the social media group within the span of one week. This increase in membership appears to be associated with an increase in the diversity and richness of the information participants share with each other, as well as the pertinence and legitimacy of their shared messages.

In January 2014, the Spanish National Research Council (CSIC) connected with the working group through the Institute of Marine Sciences (ICM), and launched a program called "Plastic 0" This program introduced a focus on plastic pollution into ICM's Seawatchers initiative (http://www.seawatchers.org), a web-based platform where citizens and institutions can share geo-referenced observations or pictures of plastic pollution in marine and coastal areas. The goal is to construct an open database of common facts that help users visualize the extent of this global problem. As such, beach walkers, divers, sailors, and others who enjoy marine recreation can participate actively in refining the global picture of plastic pollution.

These media efforts have been complemented by (1) our participation in more than five international conferences; (2) an awareness campaign using posters on public transportation and in markets, small shops, and shopping centers; and (3) an educational campaign in more than 50 schools. These shared efforts behind the local "Plastic 0" campaign were recognized in 2014, when it won the Social Innovation Prize from Laguna University.

At this point, after five years of work, the working group has arrived at the question: What is the best way to continue efforts and participate in constructing solutions to solve the problem locally and at the regional scale?

As we consider this question, we draw upon lessons learned in the first five years of this process. We recognize the need to (1) connect with other regions in the North Atlantic and Mediterranean system; (2) formalize and synchronize a working group at this scale; (3) reach political implications through scientific and social engagement; and (4) work for the sustainability of the process politically, socially, and scientifically.

These four points are the basis for a new regional-level working group, Communities-Based Observatories Tackling Marine Litter (COASTAL). The core of the working group is composed of researchers from Versailles University, Plymouth University, CNRS, IFREMER, ICM-CSIC, Moncton University, and UNESCO-MaB. It is open to other colleagues as an initiative of the research network Marine Sciences for Society. From our case study in Lanzarote, we now shift to a brief summary of the available literature on the plastic debris problem, followed by an overview of the suggested solutions we are beginning to implement through the COASTAL working group.

WHY PLASTIC POLLUTION?

In the 1970s, reports clearly documenting plastic pollution in oceans and coastal zones began to be published (e.g., Carpenter and Smith, 1972; Wong et al., 1974). It took several years, and many more reports, before this work was seen as a red flag. Meanwhile, plastic pollution spread and has become established throughout the marine environment. Currently, its range extends to coastal areas, both poles, throughout the water column, and the open ocean, sea floor, and ocean surface (e.g., Day and Shaw, 1987; Shomura and Godfrey, 1989; Galgani et al., 2000; Thompson et al., 2004; Eriksen et al., 2014; Law and Thompson, 2014). It is estimated that plastics constitute the majority of marine litter (Derraik, 2002), and in some places microplastics make up over 80% of intertidal plastic debris (Browne et al., 2007). This material is ubiquitous, and can even be found washed up on the remote island beaches of highly protected natural areas (e.g., Barnes et al., 2009; Baztan et al., 2014).

Plastic contamination is a proxy for other types of contamination as well. When viewed this way, it seems we have reached an ecological tipping point, and must be ready for unanticipated consequences and changes (Rockström et al., 2009; Steffen et al., 2015). This is why it is crucial to bring attention to the plastic debris problem on coasts and in oceans. We must work to better understand the extent of the problem, how it is changing over time and across scales, and the short- and long-term consequences of marine plastic pollution. Alongside ongoing research, we must use the knowledge we already have to inform political, civic, and industrial responsibilities and solutions.

As global rates of plastic production continue to increase, so too does our understanding of marine litter's detrimental impact on marine and coastal zones and the wildlife that live there. Given the unsolved question of plastic's fate once it reaches the ocean, the little we know about effectively removing it from marine environments, and the need for further investigation into its potential environmental effects, the plastic debris problem is more relevant today than ever (Figure 11.3).

Cumulative knowledge about plastic pollution is expanding rapidly. In the past 10 years, hundreds of papers have been published, including works like the special issue of *Philosophical Transactions of the Royal Society* entitled "Plastics, the environment and human health," compiled by Thompson et al. (2009a). This issue contains the review paper, "Plastics, the environment and human health: Current consensus and future trends," which concisely summarizes perspectives on plastic pollution from many natural scientists'

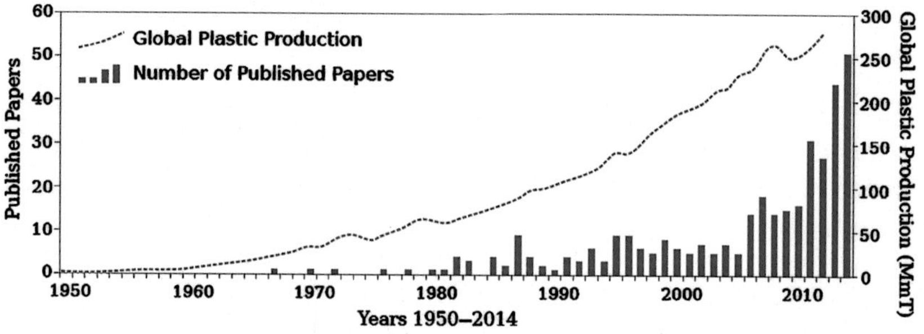

FIGURE 11.3 This graph shows global plastic production (red dotted line) and (in blue) the number of papers published by Elsevier between 1950 and 2014 with "plastic pollution" in the title, abstract, or keywords. This figure raises the question: How do we move from knowledge to action?

points of view. The synthesis of these perspectives is reflected in this quote: "Plastics offer considerable benefits for the future, but it is evident that our current approaches to production, use and disposal are not sustainable and present concerns for wildlife and human health" (Thompson et al., 2009b). Since 2009, hundreds more scientific articles have contributed to our understanding of the issue (e.g., Galgani et al., 2010; Cole, 2012; Law and Thompson, 2014; Gall and Thompson, 2015; Moore, 2014). All of these works address the same concern: even if it is one of the most useful materials in our societies, plastic represents an explicit hazard if we continue to treat it the same way we do now.

Additionally, the role of human decisions and behavior is increasingly recognized within the context of the plastic debris problem. The Lanzarote case study and the EU-funded MARLISCO project (www.marlisco.eu) are only two examples of many ongoing efforts working to integrate different stakeholder perspectives into collaborative solutions. These approaches focus on a systematic understanding of different actors in this complex problem, and take into account perceptions and behavioral patterns, using social science methods and interventions to bring about change.

In January 2014, more than 100 specialists in microplastic pollution came together in Plouzané, France for the international workshop "Fate and Impacts of Microplastics in Marine Ecosystems" (see http://micro2014.sciencesconf.org/). Participants were surveyed for their opinions about the challenges facing the scientific community, the commitments they were willing to make related to plastic, and their suggestions for solutions to the plastic debris problem. We grouped their responses under two broad questions: "What are we struggling with?" and "How will we solve the plastic problem?" Most of the suggested solutions were related to three themes: alternative materials, public awareness, and regulation of plastic's use and disposal or reuse. Many participants made commitments to continue researching the effects of microplastics on human health and the environment. Participants also committed to sharing data, monitoring strategies, and protocols. With these insights in mind, we transition into the next section of this chapter.

MOVING FROM KNOWLEDGE TO ACTION TO FACE THE PLASTIC DEBRIS CHALLENGE

From 2008 to 2012, our community-based work on the plastic debris problem in Lanzarote was extremely local, focused on, and for the island. We recognized, however, that as critical as this local effort is, it is clearly not enough. Plastic debris is a complex human problem, and to address it, we must connect different scales of action—local, regional, and global—while reconciling environmental, political, and industrial priorities. Additionally, we must weave together the web of available knowledge to inform and be informed by sustainable development actions.

But how?

Drawing from collaborative experiences, stakeholder theory, interdisciplinary frameworks, perspectives on risk perception and communication (e.g., Morgan et al., 2001), and behavior change (e.g., Darnton, 2008), we offer the following steps toward a reflexive research/action program.

- Establish a collaborative working group that includes world experts on marine litter and experts on other complementary and cross-cutting topics (e.g., governance, policy, behavior change, communication) to work together toward shared goals.
- Define a geographic context in which stakeholders share similar challenges and in which solutions could be implemented.
- Start with pilot study sites to develop protocols, apply methodologies, and test solutions.
- Gradually expand the working group to include stakeholders to enrich and legitimize deliberation and action processes.
- Work with policymakers, industry leaders, and the general public to implement and monitor solutions.
- Develop local, regional, and global actions that reduce new inputs of plastic debris while respecting community values and individuals.

These steps should be taken in tandem with the efforts already set in motion by thousands of people working on this problem—from volunteers to professionals, using spontaneous approaches or highly sophisticated protocols. For all the diverse work on the plastic debris problem, we often hear the following two solutions: (1) put a stop to plastic debris entering the oceans; and (2) where feasible, clean up the "sinks" where plastic has accumulated.

These steps sound deceptively simple.

Plastic products are widely used because they are convenient and inexpensive. Their production is increasing exponentially for these reasons. In this sense, we must consider the roles played by consumers, producers, recyclers, waste managers, and policymakers; in other words, those whose choices drive trends in plastic manufacturing, use, and disposal. To be successful, we must consider the perspectives of these individuals and find integrative solutions while being mindful of the barriers.

For example, since the ascription of responsibility is associated with taking action (e.g., Dietz et al., 1999), who shoulders the responsibility for solving the plastic debris problem?

People differ on who should ultimately take ownership of such a diffused problem. The most popular choice is to externalize the blame: "Someone other than me is responsible."

Here the waters swirling around the question of responsibility become even muddier. Although it is widely recognized that plastic debris poses a global environmental and waste management problem, perceptions and representations of the plastic debris problem vary. For example, some argue that the products are not the problem—the challenge is in their disposal. To make progress, we must first reconcile the various perspectives relating to this problem.

Once we calibrate our perceptions of the problem, there is the question of how to solve it. For a consumer who uses plastic because it is the most cost-effective option for his or her household, it may seem that solutions need to be supplied by the plastic industry, who have the means to make their products more "environmentally friendly." But, from the industry's side, they have responsibilities to their shareholders, investors, employees, etc., and need to base their decisions on staying competitive. So, does the responsibility then fall back on the consumers to use their purchasing power to drive change in manufacturing practice? Or is it on voters to get their elected officials to pass legislation that regulates manufacturers and recycling? And what of citizens without enough genuine political agency to influence national policy?

This list of questions could go on and on, but in the end our question is: How do we implement solutions when we may not share the same values and motivations and we operate within different yet geographically overlapping and linked contexts?

Our approach to addressing this question is the COASTAL initiative. Its focus is on compiling existing knowledge about marine litter and working with and for stakeholders to improve ocean literacy and enhance social engagement around the issue of marine litter. Our end goal is to move from awareness to action through social engagement and integrated stakeholder collaboration, and to continue using what we learn from our experiences to inform our efforts.

To do this, COASTAL is taking the basic steps outlined earlier, and proposes the following work streams to realize our goals.

- Social engagement with tourists to create geographical connections between (potentially remote) causes and consequences.
- Recreational and social activities as elements of community building and encouraging participation from diverse stakeholders.
- Open dialog with the media to help raise awareness.
- Sea and land roaming actions.
- Creation of an open-source Atlas of Marine Litter.
- Development of the Knowledge and Action Observatory.
- Devising and testing mutually acceptable solutions.

Between technical, scientific, and local communities, there is so much we already know. By forming a solid base through COASTAL, we hope to create opportunities to share this knowledge and foster efforts with stakeholders in affected communities to help reduce the sources and driving factors behind marine litter. In doing so, we will work toward solutions that are acceptable to a wide range of stakeholders.

Today, we conclude that (1) there is a wealth of knowledge and information on the extent and impacts of plastic debris; but (2) this knowledge is currently not used optimally to communicate with and engage stakeholders; so (3) we need to draw conclusions from, and base our actions upon, what we already know, with the explicit aim of working toward mutually acceptable solutions.

Pressures from human activities are causing unprecedented changes in the marine environment, especially in coastal zones. Our societies are responsible for identifying ways to reverse, mitigate, or adapt to these changes, which collectively pose one of the biggest challenges we face. Marine litter is an unequivocal expression of human activity, threatening wildlife, economies, and human health and well-being. The COASTAL consortium effort hopes to (1) integrate technical and scientific knowledge with that of local communities and stakeholders; (2) improve awareness through increasing ocean literacy; and (3) implement pertinent local, regional, and transnational action plans.

Let us commit to working collaboratively and actively implementing solutions to truly solve the plastic debris problem.

REFERENCES

Barnes, D.K.A., Galgani, F., Thompson, R.C., Barlaz, M., 2009. Accumulation and fragmentation of plastic debris in global environments. Philos. Trans. R. Soc. B: Biol. Sci. 364, 1985–1998.

Baztan, J., Carrasco, A., Chouinard, O., Cleaud, M., Gabaldon, J.E., Huck, T., Jaffrès, L., Jorgensen, B., Miguelez, A., Paillard, C., Vanderlinden, J.-P., 2014. Protected areas in the Atlantic facing the hazards of micro-plastic pollution: first diagnosis of three islands in the Canary Current. Mar. Pollut. Bull. 80, 302–311.

Blanchard, A., Vanderlinden, J.-P., 2010. Dissipating the fuzziness around interdisciplinarity: the case of climate change research; SAPI EN. S. Surveys Perspectives Integrating Environment and Society, 3.1, I.V.E. eds.

Browne, M.A., Galloway, T., Thompson, R., 2007. Microplastic an emerging contaminant of potential concern? Integr. Environ. Assess. Manage. 3, 559–561.

Carpenter, E.J., Smith, K.L., 1972. Plastics on the Sargasso sea surface. Science 175, 1240–1241.

Cole, M., 2012. Microplastics: small plastics, big problems? Mar. Sci. Mag. 39, 26–28.

Day, R.H., Shaw, D.G., 1987. Patterns in the abundance of pelagic plastic and tar in the north pacific ocean, 1976–1985. Mar. Pollut. Bull. 18, 311–316.

Darnton, A., 2008. Reference report: an overview of behaviour change models and their uses. GSR behaviour change knowledge review. Centre for Sustainable Development, University of Westminster.

Deguignet, M., Juffe-Bignoli, D., Harrison, J., MacSharry, B., Burgess, N., Kingston, N., 2014. 2014 United Nations List of Protected Areas. UNEP-WCMC, Cambridge, UK.

Derraik, J.G., 2002. The pollution of the marine environment by plastic debris: a review. Mar. Pollut. Bull. 44, 842–852.

Dietz, T., Abel, T., Guagnano, G.A., Kalof, L., 1999. A value-belief-norm theory of support of social movements: the case of environmentalism. Hum. Ecol. Rev. 6, 81–97.

Eriksen, M., Lebreton, L.C.M., Carson, H.S., Thiel, M., Moore, C.J., Borerro, J.C., Galgani, F., Ryan, P.G., Reisser, J., 2014. Plastic pollution in the world's oceans: more than 5 trillion plastic pieces weighing over 250,000 tons afloat at sea. PLoS One 9, e111913.

Galgani, F., Leaute, J., Moguedet, P., Souplet, A., Verin, Y., Carpentier, A., Goraguer, H., Latrouite, D., Andral, B., Cadiou, Y., Mahe, J., Poulard, J., Nerisson, P., 2000. Litter on the sea floor along European coasts. Mar. Pollut. Bull. 40, 516–527. http://dx.doi.org/10.1016/S0025-326X(99)00234-9.

Galgani, F., Fleet, D., van Franeker, J., Katsanevakis, S., Maes, T., Mouat, J., Oosterbaan, L., Poitou, I., Hanke, G., Thompson, R., Amato, E., Birkun, A., Janssen, C., 2010. Marine Strategy Framework Directive Task Team 10 Report Marine Litter. JRC (EC Joint Research Centre) Scientific and Technical Reports.

Gall, S.C., Thompson, R.C., 2015. The impact of debris on marine life. Mar. Pollut. Bull. 15, 92(1–2), 170–179.

GPWv3, 2011. Gridded Population of the World. NASA Socioeconomic Data and Applications Center (SEDAC) hosted by The Center for International Earth Science Information Network (CIESIN).

Halpern, B.S., Walbridge, S., Selkoe, K.A., Kappel, C.V., Micheli, F., D'Agrosa, C., Bruno, J.F., Casey, K.S., Ebert, C., Fox, H.E., Fujita, R., Heinemann, D., Lenihan, H.S., Madin, E.M.P., Perry, M.T., Selig, E.R., Spalding, M., Steneck, R., Watson, R., 2008. A global map of human impact on marine ecosystems. Science 319, 948–952.

Law, K.L., Thompson, R.C., 2014. Microplastics in the seas. Science 345, 144–145.

Moore, C.J., 2014. How much plastic is in the ocean? You tell me!. Mar. Pollut. Bull. 92 (1–2), 1–3.

Morgan, M.G., Fischhoff, B., Bostrom, A., Atman, C.J., 2001. Risk Communication: A Mental Models Approach. Cambridge University Press, Cambridge.

Plastics the Facts, 2013. An Analysis of European Latest Plastics Production, Demand and Waste Data. Plastics Europe, Association of Plastics Manufacturers.

Rockström, J., Steffen, W., Noone, K., Persson, A., Chapin, F.S., Lambin, E.F., Lenton, T.M., Scheffer, M., Folke, C., Schellnhuber, H.J., Nykvist, B., de Wit, C.A., Hughes, T., van der Leeuw, S., Rodhe, H., Sorlin, S., Snyder, P.K., Costanza, R., Svedin, U., Falkenmark, M., Karlberg, L., Corell, R.W., Fabry, V.J., Hansen, J., Walker, B., Liverman, D., Richardson, K., Crutzen, P., Foley, J.A., 2009. A safe operating space for humanity. Nature 461, 472–475.

Shomura, R.S., Godfrey, M.L. (Eds.), April 2–7, 1989. Proceedings of the Second International Conference on Marine Debris. US Dep. of Comm., NOAA Tech. Memo. NMFS, NOAA-TM-NMFS-SWFSC-154, Honolulu, Hawaii.

Steffen, W., Richardson, K., Rockstrom, J., Cornell, S.E., Fetzer, I., Bennett, E.M., Biggs, R., Carpenter, S.R., de Vries, W., de Wit, C.A., Folke, C., Gerten, D., Heinke, J., Mace, G.M., Persson, L.M., Ramanathan, V., Reyers, B., Sorlin, S., 2015. Planetary boundaries: guiding human development on a changing planet. Science 347, 1259855.

Thompson, R.C., Moore, C.J., vom Saal, F.S., Swan, S.H., 2009a. Plastics, the environment and human health: current consensus and future trends. Philos. Trans. R. Soc. B: Biol. Sci. 364, 2153–2166.

Thompson, R.C., Olsen, Y., Mitchell, R.P., Davis, A., Rowland, S.J., John, A.W.G., McGonigle, D., Russell, A.E., 2004. Lost at sea: where is all the plastic? Science 304, 838.

Thompson, R.C., Swan, S.H., Moore, C.J., vom Saal, F.S., 2009b. Our plastic age. Philos. Trans. R. Soc. B: Biol. Sci. 364, 1973–1976.

Wong, C.S., Green, D.R., Cretney, W.J., 1974. Quantitative tar and plastic waste distribution in the Pacific Ocean. Nature 247, 30–32.

Part III

Local Management of Common-Pool Resources

Chapter 12

Challenges to Sustainable Development along Peruvian Coastal Zones

Francisco Miranda Avalos[1], **Mariano Gutiérrez Torero**[2]
[1]*Foro Hispano Americano de Intercambio de Información sobre Temas del Mar (ONG OANNES);*
[2]*Universidad Nacional Federico Villarreal, Lima, Peru*

Chapter Outline

INTRODUCTION

Peru has a privileged geographical location; its Amazonian, Andean, and marine ecosystems are designated as world priority ecoregions for conservation (Brack, 1986), and consequently Peru has signed commitments such as the Convention on Biological Diversity (MINAM, 2012). However, Peru is also found among the most vulnerable regions to climate change effects, especially its coastal ecosystem, which extends along the Northern Region of the Humboldt Current System (Gutiérrez et al., 2011). This is one of the most productive large marine ecosystems on the planet, and hosts the world's largest single-species fishery with anchovy (*Engraulis ringens*), and many artisanal fisheries (Chávez et al., 2008).

Since the early 1990s, Peru has experienced sustained economic development mainly based on extractive activities (mining, hydrocarbons, fishing), and trade, tourism, construction, moderate rates of domestic consumption, and a relatively high saving capacity of the state, private sector, and citizens (Estela, 2001). The government's fierce fiscal discipline, along with an investment framework attractive to foreign investors and the country's reintegration into the international economic system after two decades of controlled economic management, have provided a base for economic and industrial development never before experienced in Peru's history as a republic (Hunt, 2011). Also, since the early 2000s the country has benefitted from the sustained increase in the price of "commodities" such as fishmeal and fish oil; Peru has been the world's main producer of these products for the past five decades (Paredes, 2013). This economic boom has been supported by significant improvements in the use of modern technologies in all areas of production, thus contributing to poverty reduction and better quality of life for many of Peru's citizens (CIES, 2012).

However, if we analyze the details of this economic expansion in terms of its effects on habitats, ecosystems, and coastal economies, we can see some detrimental impacts of improperly planned urban and industrial development (Valdivia, 2013). The population reduction of some wild coastal species, marine pollution, dead polluted rivers, reduced wetlands, deforestation, reduction of top-predator populations, and other negative environmental impacts are contributing to the decrease of marine ecosystem quality (PNUD, 2013).

Lack of interest and commitment from regional governments regarding environmental issues, and the incipient state of implementation of an ecosystem approach to management, have led to a negative public perception of the role of the state regarding the sustainability of marine ecosystems (MINAM, 2012).

The result is the inappropriate use of natural resources, a disordered expansion of cities and tourism, and the weakening of institutions that cannot exercise their territorial competencies due to an asymmetrical distribution of financial resources (Abusada et al., 2008). This set of conditions, despite the government's relative financial prosperity, threatens the sustainability of economic activities that depend on coastal and marine resources, and challenge the capacity of Peru's ecosystems beyond the limits of their normal processes and productivity (UNDP, 2009).

To analyze the challenge of reversing the current situation and achieving sustainable development of Peru's coastal areas, a public call was made to conduct a National Workshop in May 2012. The goal of the workshop was to produce an interdisciplinary consensus document showing what civil society could undertake on the issue of coastal zones, and also to present management alternatives that could be developed in the near future. The outcomes from this National Workshop were extended to the Peruvian Government as a contribution to the national statement presented at the United Nations Conference on Sustainable Development (Rio+20 Summit) held in June 2012 in Brazil. The workshop was entitled "Evaluating the Challenges of Coastal Areas in the XXI Century: Proposals for the Earth Summit Rio+20."

METHODOLOGY

The National Workshop was held in Lima from May 7th to 11th, 2012. The resulting report was delivered to the Peruvian government to be included with other national documents shared during the Rio+20 Summit.

The Organizing Committee proposed five main topics for discussion:

1. Climate change and its implications for marine and coastal resources.
2. Loss of habitat and biodiversity as a result of accelerated industrial and urban development and the lack of an adequate legal framework.
3. Ecosystem-based management, sustainability, and food security; resource use, and national issues including artisanal fisheries and offshore fisheries.
4. Sustainable development of aquaculture as an alternative to capture fisheries.
5. Development of infrastructure in the marine and coastal sector: hydrocarbons, ports and industrial plants.

Having defined these topics, a public call was made through the website of the NGO Foro Hispano Americano para el Intercambio de Información sobre Temas de Mar (OANNES) (www.oannes.org.pe), which has a distribution list of over 16,000 members. Through this medium, previous "statements" from entities interested in participating were requested, with the aim of making an inclusive and participatory process for discussing proposals that would achieve consensus on ten cross-thematic needs and concerns connecting the five proposed topics.

The topics were discussed during the first three days of the workshop (May 7–9, 2012). A plenary meeting to draft the report was held on the fourth day (May 10). On the last day (May 11), there was a public meeting and conference call to present the results to members of the International Working Group "Coastal Zones: Twenty-First Century Challenges."

RESULTS

Fifteen statements were received from 15 organizations that also participated in the workshop, including NGOs, universities, fishermen associations, enterprises, and regional governments. The workshop was attended by 37 representatives from 23 organizations (ministries, state agencies, NGOs, universities, fishermen associations, companies, regional governments, professional schools, and other academic institutions). An integration of these statements and participant perspectives is presented below.

Climate Change and Its Implications for Marine and Coastal Resources

The impact of climate change on marine resources and coastal zones seems to present two possible scenarios for the next 20–30 years in Peru: (1) an intensification of coastal upwelling in favor of biological productivity; or (2) the weakening of coastal winds, leading to increased temperatures in coastal areas, which would damage global ecosystem productivity (Gutiérrez et al., 2011). Both scenarios are combined with global effects that will put food security at risk, such as sea-level rise, loss of wetlands and mangroves, and changes in precipitation.

In the last two decades, Peru has made efforts to promote awareness of and implement policies on climate change. In 1993, a National Committee on Climate Change was established with the participation of government institutions and representatives of civil society. In this context, and through a participatory approach, the National Climate Change Strategy is being updated. There is also an initial National Action Plan on Adaptation and Mitigation (MINAM, 2010).

However, these efforts alone are insufficient, and more participatory and awareness-raising methods are needed, together with comprehensive policies on mitigation and adaptation to the impacts of climate change on coastal and marine ecosystems (EEA, 2011). We also must strengthen research on climate change, which includes developing appropriate diagnostics, risk analyses, and prioritization of adaptation measures. For this, the development and implementation of an ecological risk analysis that considers the positive and negative aspects of climate change would be a valuable tool for making management decisions, both on vulnerable species and for specific fisheries (Daw et al., 2009; Pecl et al., 2011). One particular aspect that requires attention is the effect of ocean acidification on bivalve mollusks and its implications for the sustainability of natural shellfish beds and aquaculture development in the long term. Peru has great potential for mollusk aquaculture, so the risk associated to acidification needs to be quantified.

Loss of Habitat and Biodiversity as a Result of Accelerated Industrial and Urban Development and Lack of an Adequate Legal Framework

In Peru, there are concerns regarding the loss of habitats and biodiversity resulting from rapid urban coastal development (Valdivia, 2013). However, the government's mid- and long-term objectives could ensure the integrity of marine ecosystems and their species at the highest possible level, but it would require a multidisciplinary and technical-scientific approach. This approach is especially needed for all aspects regarding regional management. It also requires well-informed citizens with high levels of environmental education who fully understand their role as environmental stewards.

Peru needs laws that align with the ecosystem approach to fisheries and integrated coastal zone management. Laws promoting responsible ecotourism, recreation, and best environmental practices will also be necessary. Overall, the main goal should be to have a healthy environment that is protected by a knowledgeable political system and public.

Peru has taken initiatives to reduce the loss of habitat and biodiversity. These include the creation of the Peruvian Guano Islands, Isles and Capes National Reserve (RNSIIPG) (MINAM, 2009). However, this is a set of 33 marine coastal protected areas whose dimensions are rather small; they do not yet count established buffer zones. Some wetlands have disappeared with current industrial expansion, so there is a pressing need for legal protection of these and other critical ecosystems along coasts.

The state should allocate necessary resources for monitoring ecosystems and controlling illegal activities; this is part of its international commitments. There is a need to restore damaged habitats, encourage restocking programs, and promote aquaculture of native species of fish and shellfish in coastal areas. Many species have been affected by pollution and anoxia in estuarine areas where these species spawn.

Ecosystem-Based Management, Sustainability, and Food Security; Resource Use; and National Issues Including Artisanal Fisheries and Offshore Fisheries

The concept of ecosystem fisheries management, where the preservation or conservation of a resource is understood as being essential to the existence of others, is certainly a way forward for the sustainable development of a robust fishing sector capable of generating wealth and food security in Peru and elsewhere (CEPAL, 2011). Another basic element required for effective resource management is respect for the rules on fishing selectivity (Zhou et al., 2010). We believe concepts such as conservation and fisheries management are parts of a whole, and not opposing positions.

Within this perspective, the use of goods and services provided by the marine ecosystem requires a multidisciplinary approach, including socioeconomic interests and public governance.

Along with the lack of proper supervision, monitoring, and control of economic activities in coastal areas, illegal activities such as discarding fish, destructive practices, and unreported catches have increased. The five-mile protection zone along the coast, reserved for species reproduction and exclusively for artisanal fishing, is often violated.

This situation fuels conflict. Sechura Bay (Piura Region) presents the largest number of conflicts in the coastal region, including disputes between artisanal fishermen and oil companies, illegal fishing, conflicts among fishermen from two regions (Piura and Lambayeque), and conflicts over access to the marine protected area of Lobos de Tierra Island.

There are also positive experiences. For example, in the region of San Juan de Marcona (southern Peru), a community of fishermen (COPMAR) has developed a model of successful local co-management even though it is not yet an official model of fisheries management policy. The management model, developed by the fishermen themselves, includes monitoring, quotas, and seasons for the extraction of several algae and other benthic species.

Therefore, Peru must develop an Action Plan for fisheries management of all species targeted by fisheries. Also, research on noncommercial species (top predators, sponges, starfish, jellyfish, etc.) must be promoted and supported, as these species are biological indicators of ecosystem functioning and welfare.

Since good laws alone are not enough, we must promote activities that add value to historical and ecological heritage—for example, using products and designations of origin that give personality to local production. It was also suggested to promote tourism through food fairs and the responsible use of marine resources, while promoting the formalization, payment of taxes, and overall improvement of economic conditions.

Sustainable Development of Aquaculture as an Alternative to Capture Fisheries

Global production of capture fisheries has exceeded its maximum sustainable level (FAO, 2009, 2010). This reality extends to the local context in Peru, where the main resource, anchovy, can no longer be subject to higher fishing levels (IMARPE, 2010). At the same time, the artisanal and small-scale fleets are growing in an uncontrolled manner, causing social problems as well as ecosystem stress in the restricted five-mile nautical zone off the coast, and in areas of high vulnerability and biodiversity (Alfaro et al., 2007, 2008, 2010).

Aquaculture in general, and mariculture in particular, can be developed to create value chains to produce equivalent benefits for all participants in an economic activity (Tveteras, 2012). Value chains should also provide optimal resource usage if certain actions, regulations, and incentives are implemented. Eco-labeling of fisheries and aquaculture provide the framework for creating value chains to maximize the benefits and achieve sustainability by linking responsible consumers with committed producers.

However, the National Aquaculture Development Plan (PRODUCE, 2010) does not focus on management plans or the central role of artisanal fishermen in the development of mariculture, which is a gap that should be bridged to reinforce approaches for social inclusion. Similarly, the Multi-Sector Strategic Plan (PRODUCE, 2012) does not assign a central role to artisanal fisheries in development, despite the fact that there are two main productive modes in the country: mariculture of scallops in the north and the harvest of macroalgae in the south. Both activities are largely based on the work of artisan fishermen.

An important part of aquaculture development is monitoring its impacts on the environment (UICN, 2007). For example, some aquaculture practices, such as the cultivation of fish or shellfish in tanks or ponds, create effluent by-products. In the case of shrimp aquaculture in the north, mangroves swamps have been destroyed. In other instances, cleaning accumulated marine fouling from nets does not follow waste management regulations. Fouling could be used as compost instead of being dumped in the sea. Aquaculture certification processes are being promoted by organizations like the Aquaculture Stewardship Council as a means of encouraging environmentally friendly practices.

Development of Infrastructure in the Marine and Coastal Sector: Hydrocarbon Exploration, Ports, and Industrial Plants

In the most recent decade, there has been an obvious increase in the construction of infrastructure in the coastal marine zone. This infrastructure includes piers and docks, mainly for the shipment of gas, oil, and minerals. Unfortunately, these investment and development plans have not included thorough environmental impact studies, nor have they considered potential effects on other economic activities such as the artisanal fishery or tourism (e.g., Paracas, Ica Region), which are obviously impacted (Ecoceanica, 2013). This development, which has not considered the interests of other groups, has caused social conflicts as well as damaging environmental impacts.

Proper management of marine space should be similar to the management of development in terrestrial areas. When done appropriately, such management can harmonize, or at least balance, different interests and eliminate potential sources of social conflict in the impacted zones, while controlling unwanted effects on the ecosystem. Therefore, land use decisions in coastal zones and maritime spatial planning require management tools such as Environmental Impact Assessment (EIA) or Ecological Risk Assessment (ERA). The demands from groups with the most vulnerable interests—for example, artisanal fishing communities—can be resolved using new management tools; methodologies for quantifying development impacts, such as Methods of Impact Quantification (MIQ); and compensation mechanisms (MCMs). The MCMs involve retributions to the ecosystem, not necessarily to communities directly, but could possibly include funds to support research that might lead to a higher degree of protection in surrounding areas that are targets for exploration and exploitation. Such compensation is not necessarily expressed in economic terms, but instead may involve restocking and/or the protection of a neighboring area proportional to or greater than the impacted zone. This approach can also include compensation for the loss of value of a previously existing economic activity that will be affected. These methodologies are not currently used in Peru's marine areas.

Interdisciplinary Aspects for the Management of Natural Resources in Coastal Areas

During the workshop, participants identified shared elements relevant to all the topics that were addressed. Everyone present agreed these should receive priority attention from authorities and policymakers. These issues relate primarily to the identification of problems or needs, and are described in the next sections.

Increase Scientific and Technological Research

The country dedicates few financial resources to scientific research and marine-coastal technologies despite having a large number of national and private universities; therefore, scientists' and professionals' production capacity and creativity are untapped. Also, the ratio of expected production to investment in research is increasingly unbalanced; despite increased revenue, funding for research remains stagnant.

Reform the Professional Profile for Ocean-Related Careers

Ocean studies programs need improved curricula. Without modern scientific and technological training, students in this field will not develop the research skills and knowledge needed to address the threats of the twenty-first century—like those posed by climate change, for example. Proposals such as creating a national bank of theses can guide the development of new capabilities in research. This requires access to grant funds for financing development research, which could be provided by resource industries (e.g., taxes from mining) (FOCAN).

Take Multisectoral Approaches Addressing Socioeconomic Issues

Management plans for resources like fishing, agriculture, mining, and tourism, when they exist, do not take into account the entire range of effects their resource uses have on communities or ecosystems. In the case of artisanal fisheries, the construction of gas infrastructure and mineral extraction have affected the normal distribution of coastal species in certain places, resulting in higher costs and operating times for artisanal fishermen. These setbacks include the reduction of catch quality. In other cases, some communities settled next to rivers or the sea pump or drain waste directly to rivers or the ocean, which has severely increased pollution. To counteract these impacts, all economic activities should generate funds to be used for mitigating environmental damage. This also demonstrates the need for legislation to protect communities and the environment from harm incurred by development.

Promote the Socioeconomic Development for Artisanal Fishermen and Fisheries for Human Consumption

Artisanal fishermen are key stakeholders in the country's sustainable development. They provide most of the marine food protein for Peruvians, and yet the index of malnutrition is high in Peru. These fishermen are key for increasing food security, and it is necessary to promote their environmental education and provide them with technical assistance to achieve co-management of fisheries resources. People themselves are the most valuable assets of a society. In Peru, we must confront acute problems of malnutrition and chronic child malnutrition. Peru is ranked among countries with the highest indicators in these areas, despite the fact that the country is the highest producer of marine protein. Overcoming the challenge of hunger requires the creation of new consumption habits for anchovy and giant squid, among other abundant species. These two species comprise over 90% of national catches, and are exported as fish meal and oil or a cheap food source for foreign markets. Most importantly, we must solve logistic problems related to post-harvest infrastructure: inadequate landing sites, poor or nonexistent cold-storage chains, and the lack of clean water facilities in almost all existing landings.

Strengthen the Value Chain of Artisanal Fisheries

Not only nutrition can be improved by focusing efforts on artisanal fishing. It might also be a source of welfare for the country. Artisanal fishing is a big source of employment, even temporarily. However, the management of this diverse and complex sector is almost nonexistent. While there are regulations in place, resources have not been allocated to make them effective (e.g., new fishing boats are still being built despite the fact there are legal stipulations that expressly prohibit this). Artisanal fisheries also lack the logistics for collecting and marketing through a cooperative approach between fishermen's associations. As artisanal fisheries increase their number of production units without increasing the value of production, the result is more poverty in an already impoverished sector. Transferring control from intermediaries to the Social Organizations of Artisanal Fishermen (OSPA) will require planning and the development of efficient supply chains, with properly designed ports, ice plants, and adequate collection and distribution methods.

Governmental, Civic, and Corporate Social Responsibility and Environmental Commitments

From a socioeconomic point of view, all industrial companies should focus their social responsibility toward development, not subsidies. They should contribute to the state's efforts by identifying and supporting vulnerable human groups. This could include providing proposals for solutions to reverse poverty, child malnutrition, and lack of educational opportunities, along with other challenges that are sources of chronic social exclusion in Peru.

Also, all fishing companies should support and promote the participation of their staff and crew in the collection of information relevant to scientific interest under the guidance of scientists. In a country where resources for research are scarce, we must use every avenue possible to collect data.

Increase Environmental Education and Awareness

Regional Governments (GORE) do not yet invest in Integrated Coastal Zone Management due to political misunderstandings and lack of environmental commitment. The Regional Environmental Commissions (CAR) should activate their functions; their authority exists in law, though not in practice. GOREs have responsibilities for impacted communities and individuals who rely on environmental resources for their livelihoods. Strong leading management from GOREs is needed, with linkages to the Ministry of Education (MINEDU) and the Ministry of the Environment (MINAM) to develop permanent awareness campaigns conveying concepts of sustainable development. Special attention should be given to the role of families, especially mothers, as the first educators of children.

Health, Fisheries, and Trade Certifications

The expected value increase in fish production includes the need for various certifications, which seek to ensure good practices in terms of hygiene, health, quality, and sustainability of fisheries for human consumption. It also requires the promotion of management approaches that integrate cooperatives of workers in order to ensure the fidelity necessary to add value and sustainability to an economic activity currently dominated by intermediaries and third-party operators.

Spatial Planning and Integrated Coastal Zone Management

Coastal zones should be subject to zoning. The growth of cities and the emergence of new towns (e.g., Pampa Melchorita, Lima Region) are affecting not only agricultural areas, which serve as buffer zones for pollution and climate regulators adjacent to cities, but are also directly damaging areas of high biodiversity such as coastal wetlands, river mouths, lagoons, and natural shellfish beds. An increase in the number of Marine Protected Areas (MPAs) is therefore required, which should channel industrial coastal development under sustainable standards regarding the ecosystem functioning.

Legal Reform to Eliminate Overlapping Responsibilities between Authorities

The functions and competences of various ministries are in some cases contradictory, and lead to conflicts of jurisdiction. Complaints about judges who dictate on matters that are not within their competence are common. There are also provisions for regional governments that come into direct conflict with the central government in terms of control of access to fishing. Overlapping ministerial standards in response to environmental laws are also common and problematic (Abusada et al., 2008). Lamentably, important laws do not contain sufficient information on the responsibilities of different ministries. For example, seismic surveys aimed at the exploration of gas and oil at sea are authorized by the Ministry of Energy and Mines (MINEM), and input from the Ministry of Production (PRODUCE) or the Ministry of the Environment (MINAM) are not legal requirements despite the fact that PRODUCE and MINAM are the national authorities in the case of living marine resources and coastal areas. New laws may be required, or reforms to existing ones, to provide transparency and guarantee the validity of stricter protocols that would harmonize, or at least consider, the interests of all sectors.

DISCUSSION: WHAT ARE THE MANAGEMENT PROPOSALS FOR PERUVIAN COASTAL ZONES?

A first step is the creation and implementation of an Action Plan for Mitigation and Adaptation to Climate Change in the coastal marine area that includes timelines, targets, and indicators. The Action Plan should involve the participation of national, regional, and local governments, with support from scientific institutions. In this regard, a valuable landmark is the synthesis collected by FAO on the impact of climate change on fisheries and aquaculture (FAO, 2012; Daw et al., 2009).

Another important proposal is the creation of a National Network of Marine and Coastal Impacts of Climate Change (RNIMCC) involving the participation of different users and/or stakeholders. The RNIMCC should help to collect and monitor indicators and evidence of the effects of climate change on coastal marine ecosystems, promote cooperative approaches to adapting to the impacts, and facilitate public awareness. Also, RNIMCC should be articulated with other sources of information and early warnings, especially the National Study of El Niño (ENFEN), and should contribute to the generation and transmission of information for reducing climate change impacts and to promote technology for this purpose.

A third important action is the development of a National Strategy for Risk Management (NSRM), enabling the identification of threats and opportunities arising from climate change or other potentially damaging coastal events. For example, this could include identifying areas vulnerable to tsunamis that have not yet received any special attention from regional governments. NSRM may develop an incentive program for mitigation measures, such as recovery of wetlands or mangroves to help sequester atmospheric carbon dioxide.

Introduce Adequate Legal Frameworks and Improve Environmental Education

MPAs should increase in number and size to protect wetlands and coastal zones. It is essential to implement co-management tools with artisanal fishing communities and other stakeholders in order to create the necessary legal regulations to protect exclusion areas (inside MPAs). For example, the MPA Lobos de Tierra Island, despite its protected legal status, is being affected by a number of illegal activities, including the presence of tourists arriving on large boats. The active participation of artisanal fishermen could contribute to the monitoring, conservation, and rational use of areas under co-management schemes.

Without stakeholders who are well-informed about coastal ecosystem issues, it will be very difficult to change the everyday behaviors of Peruvians that negatively impact coastal zones, or gain political traction to make the necessary policy changes we outlined earlier. Awareness campaigns currently reach a very small segment of the population.

We need intervention at all levels to share knowledge about ecological sustainability. This will require a modification of school and university curricula to include courses that reflect the principles and concepts of marine ecology, fisheries ecology, industrial ecology, ecotourism, etc., and connect them with local knowledge and values.

Develop Sustainable Aquaculture as an Alternative to Capture Fisheries

Profound reforms in the management of the artisanal fishery sector, as the main users of marine ecosystems, are required. The legendary richness of the Peruvian sea has not alleviated the poverty and prevailing conditions of socio-economic exclusion of fishermen. The prevention of major social conflicts requires coordinated action between all levels of the public sector, with contributions from all industry stakeholders through their plans for social responsibility.

A National Plan that integrates relevant state agencies linked to natural resources management is needed in order to achieve realistic and sustainable solutions to eradicate malnutrition in Peru and create lasting habits for fisheries' target populations. This will also increase the value of fisheries products.

The development of mariculture would create a large number of jobs, though it requires previous comprehensive experimental research. In this context, it would make the most sense to use native species (fish, echinoderms, bivalve and gastropod mollusks, and algae with high commercial value), concentrate efforts on a few species, and give priority, wherever possible, to those species that do not require fishmeal (FAO, 2011).

Use Development by Design Methods for Infrastructure Development in Marine and Coastal Zones

Sustained political decisions are required to lead and reconcile the requirements of economic growth with the needs of the environment in areas of high conservation value such as the Bay of Paracas. To achieve this balance, Development by Design methods can be used to create specific Portfolios of Biological Diversity (PBDs) for previously defined ecoregions for which there is a relatively high abundance of technical information.

The overall objective is to achieve a net benefit for the environment without neglecting the development of viable infrastructure projects or other economic activities. These approaches do not replace existing Environmental Impact Assessments (EIA), but complement them and provide a more comprehensive view into management. The PBDs could be used on a voluntary basis, but it could be highly convenient to incorporate them into the existing legal framework. From another point of view, these methodologies to quantify impacts and compensations can be applied in riparian areas where the coastal zones often receive the cumulative impact of pollution that is discharged by rivers (e.g., the surroundings of Callao harbor, where divers testify to the existence of "garbage carpets" all across the sea floor).

Conclusions

These proposals are methodologically feasible, inclusive, and scale to responsibilities at different levels of the organization of the state (district councils, provincial councils, regions, ministries) and private sector (agricultural, mining, fisheries, urban, oil and gas, tourism, shipment, etc.). Threats to coastal ecosystems are numerous (García and Rosenberg, 2010), and the additional need to ensure food security is worth the effort and costs of preserving the resilience of our ecosystems.

ACKNOWLEDGMENTS

We want to sincerely thank the participation of over 50 representatives of 23 institutions during the National Workshop that took place in Lima, Peru, during May 2012 before the United Nations Conference on Sustainable Development: Rio+20 Summit. The report of the workshop contained the key information include in this paper. Very especially we want to thank Bethany Jorgensen by her positive criticism and support during the drafting and editing of this document.

REFERENCES

Abusada, R., Cusato, A., Pastor, C., 2008. Eficiencia del gasto en el Perú. Instituto Peruano de Economía. 39 pp.

Alfaro-Shigueto, J., Mangel, J., Pajuelo, M., Dutton, P., Seminoff, J., Godley, B., 2010. Where small can have a large impact: structure and characterization of small-scale fisheries in Peru. Fish. Res. 106, 8–17.

Alfaro-Shigueto, J., Dutton, P.H., Van Bressem, M.F., Mangel, J., 2007. Interactions between leatherback turtles and Peruvian artisanal fisheries. Chelonian Conserv. Biol. 6, 129–134.

Alfaro-Shigueto, J., Mangel, J.C., Seminoff, J.A., Dutton, P.H., 2008. Demography of loggerhead turtles *Caretta caretta* in the southeastern Pacific ocean: fisheries based observations and implications for management. Endang. Species Res. 5, 129–135.

Brack, A., 1986. Ecología de un país complejo. In: Baca, Mejía (Ed.), La gran geografía del Perú, Tomo 2, pp. 177–319 (Edit. Manfer-Mejía Baca).

Chávez, F.P., Bertrand, A., Guevara, R., Soler, P., Csirke, J., 2008. The northern Humboldt current system: brief history, present status and a view towards the future. Prog. Oceanogr. 79, 95–105.

CIES (Consorcio de Investigación Económica y Social), 2012. Perú: Atlas de la pobreza departamental, provincial y distrital 2007–2009. Consorcio de Investigación Social (CIES) y Banco Interamericano de Desarrollo (BID), Lima.

Comisión Económica para America Latina y el Caribe (CEPAL), 2011. La sostenibilidad del desarrollo a 20 años de la Cumbre para la Tierra: Avances, brechas y lineamientos estratégicos para América Latina y el Caribe. 238 pp.

Daw, T., Adger, W.N., Brown, K., Badjeck, M.-C., 2009. Climate change and capture fisheries: potential impacts, adaptation and mitigation. In: Cochrane, K., De Young, C., Soto, D., Bahri, T. (Eds.), Climate Change Implications for Fisheries and Aquaculture: Overview of Current Scientific Knowledge, pp. 107–150. FAO Fisheries and Aquaculture Technical Paper. No. 530. Rome, FAO.

Ecoceanica, 2013. Informe del Taller de Formulación del Plan Maestro de la Reserva Nacional Sistema de Islas, Islotes y Puntas Guaneras – Islas Ballestas e Islas Chinchas (SERNANP) & Diversidad. In: Comunidad y Ecosistemas Costeros (GEF-PNUD Humboldt). 23, 24 y 25 de Abril, 2013. Paracas, Perú.

Estela, M., 2001. Ocho apuntes para el crecimiento con bienestar. Fondo Editorial del Banco Central de Reserva del Perú.

European Environment Agency (EEA), 2011. Methods for assessing coastal vulnerability to climate change. European Topic Centre on Climate Change Impacts, Vulnerability and Adaptation. 76 pp.

FAO, 2009. Fishery and Aquaculture Statistics. Roma, 107 pp.

FAO, 2010. El estado mundial de la pesca y la acuicultura (SOFIA). Roma, 114 pp.

FAO, 2011. Demand and Supply of Feed Ingredients for Farmed Fish and Crustaceans: Trends and Prospects. Roma, 102 pp.

FAO, 2012. Consecuencias del cambio climático para la pesca y la acuicultura. Roma, 246 pp.

García, S., Rosenberg, A., 2010. Food security and marine capture fisheries: characteristics, trends, drivers and future perspectives. Phil. Trans. R. Soc. B 365, 2869–2880.

Gutiérrez, D., Bertrand, A., Wosnitza-Mendo, C., Dewitte, B., Purca, S., Peña, C., Chaigneau, A., Tam, J., Graco, M., Echevin, V., Grados, C., Fréon, P., Guevara-Carrasco, R., 2011. Sensibilidad del sistema de afloramiento costero del Perú al cambio climático e implicancias ecológicas. Revista Peruana Geo-Atmosférica 3, 1–26.

Hunt, S., 2011. La formación de la economía peruana. Fondo Editorial del Banco Central de Reserva del Perú.

IMARPE, 2010. V Panel de Expertos en Evaluación de Anchoveta. Bol. Inst. Mar Perú N~25, 86 pp.

MINAM, 2012. Perú, informe país 20 años después de Río 92: gobiernos locales y regionales. Ministerio del Ambiente. 50 pp.

MINAM, 2010. Plan de Acción y Mitigación frente al Cambio Climático. Lima, Peru, 147 pp.

MINAM, 2009. Decreto supremo que aprueba el establecimiento de la Reserva Nacional Sistema de Islas, Islotes y Puntas Guaneras. DS.024-2009-MINAM.

Paredes, C., 2013. Eficiencia y equidad en la pesca peruana: La reforma y los derechos de pesca. Universidad San Martin de Porres, Instituto del Perú. 162 pp.

Pecl, G.T., Ward, T., Doubleday, Z., Clarke, S., Day, J., Dixon, C., Frusher, S., Gibbs, P., Hobday, A., Hutchinson, N., Jennings, S., Jones, K., Li, X., Spooner, D., Stoklosa, R., 2011. Risk Assessment of Impacts of Climate Change for Key Marine Species in South Eastern Australia. Part 1: Fisheries and Aquaculture Risk Assessment. Fisheries Research and Development Corporation. Project 2009/070.

PNUD, 2013. Informe sobre Desarrollo Humano Perú 2013 Cambio climático y territorio: Desafíos y respuestas para un futuro sostenible. Programa de las Naciones Unidas para el Desarrollo, Lima.

PRODUCE, 2010. Plan Nacional de Desarrollo Acuícola. 94 pp.

PRODUCE, 2012. Plan Estratégico Sectorial Multianual 2012–2016.

Tveteras, S., 2012. Price analysis of Peruvian Fishery Value Chains: Peruvian Anchovy, Shrimp, Trout and Scallop. Pontificia Universidad Católica del Perú. 51 pp.

UICN, 2007. Guía para el Desarrollo Sostenible de la Acuicultura Mediterránea. Interacciones entre la Acuicultura y el Medio Ambiente. UICN, Gland, Suiza y Málaga, España. VI + 114 paginas.

UNDP, 2009. Hacia un manejo con enfoque ecosistémico del Gran Ecosistema Marino de la Corriente Humboldt. Documento del Proyecto GEF-PNUD-Humboldt. Programa de las Naciones Unidas para el Desarrollo, Lima.

Valdivia, H., 2013. el crecimiento urbano y su relación con el deterioro de las zonas arqueológicas caso: distrito Comas-Lima, Perú. Grupo de Investigación Geográfica Ambiental. http://www.unmsm.edu.pe/iigeo/giga/articulos/arqueolog.htm.

Zhou, S., Smith, A., Punt, A., Richardson, A., Gibbs, M., Fulton, E., Pasco, S., Bulman, C., Bayliss, P., Sainsbury, K., 2010. Ecosystem-based fisheries management requires a change to the selective fishing philosophy. PNAS 107 (21), 9485–9489.

Making the Link

By Bethany Jorgensen

Challenges to Sustainable Development along Peruvian Coastal Zones	⟺	Solutions for Sustainable Coastal Lagoon Management: From Conflict to the Implementation of a Consensual Decision Tree for Artificial Opening
Francisco Miranda Avalos and Mariano Gutiérrez Torero		*D. Conde* et al.

From an effort to establish priorities and recommendations for policymakers at the national level in Peru, we go now to a successful example of integrated coastal lagoon management in Uruguay. These two chapters share some of the major themes that run through this book: seeking community input and incorporating local needs and values into the policymaking process.

In Miranda Avalos and Gutiérrez Torero's chapter, this is done through a national workshop that facilitated an inclusive and participatory process to synthesize the concerns of interested parties throughout Peru into five overarching topics and ten cross-cutting themes related to sustainable development along the Peruvian coast. Their work resulted in a report with recommendations for the Peruvian government and priorities for future work.

The example provided in Conde et al.'s chapter is particularly exciting because it tells of a successful effort to design a consensual decision model that can be easily used by community members, environmental monitors, and administrators when deciding if and when to artificially breach the sandbar of the coastal lagoon Laguna de Rocha—a designated UNESCO biosphere reserve.

Their decision tree blends hydrological, geomorphological, and local knowledge to help decision-makers determine when artificial breaching of the sandbar should be authorized, and when not.

We leave it to them to fill you in on the details.

Chapter 13

Solutions for Sustainable Coastal Lagoon Management: From Conflict to the Implementation of a Consensual Decision Tree for Artificial Opening

D. Conde[1,2,3], J. Vitancurt[1,5], L. Rodríguez-Gallego[1], D. de Álava[1], N. Verrastro[1], C. Chreties[4], S. Solari[4], L. Teixeira[4], X. Lagos[5], G. Piñeiro[6], L. Seijo[7], H. Caymaris[8], D. Panario[9]

[1]Centro Universitario Regional Este, Universidad de la República, Rocha, Uruguay; [2]Seccion Limnología, Facultad de Ciencias, Instituto de Ecología y Ciencias Ambientales, Universidad de la República, Montevideo, Uruguay; [3]Espacio Interdisciplinario, Universidad de la República, Montevideo, Uruguay; [4]IMFIA, Facultad de Ingeniería, Universidad de la República, Montevideo, Uruguay; [5]Dirección Nacional de Medio Ambiente, Ministerio de Vivenda, Ordenamiento Territorial y Medio Ambiente, Uruguay; [6]Departamento de Evolución de Cuencas, Instituto de Ciencias Geológicas, Facultad de Ciencias, Universidad de la República, Montevideo, Uruguay; [7]Oficina de Planeamiento y Presupuesto, Presidencia de la República, Montevideo, Uruguay; [8]Intendencia de Rocha, Rocha, Uruguay; [9]UNCIEP, Instituto de Ecología y Ciencias Ambientales, Facultad de Ciencias, Universidad de la República, Montevideo, Uruguay.

Chapter Outline

Coastal Zones

INTRODUCTION

Coastal Lagoons and Sandbar Dynamics

Coastal lagoons are among the most productive ecosystems in the world (Knoppers, 1994; Duck and da Silva, 2012), sustaining important environmental services such as fisheries (Pauly and Yáñez-Arancibia, 1994; Cañedo-Argüelles et al., 2012). Coastal lagoons represent over 12% of the South America coastline and coincide with densely populated areas (Gönenç and Wolflin, 2004; Esteves et al., 2008). Their importance for biodiversity conservation has been recognized extensively (Barbosa et al., 2004; Isacch, 2008; Soutullo et al., 2010). Particularly, choked coastal lagoons are inland shallow waters periodically connected with the ocean by a narrow channel that opens through a sand barrier. These lagoons are physically dominated systems with large salinity and hydrodynamic fluctuations driven by the intermittent connection with the ocean through the sandbar (Kjerfve, 1994). The connection with the ocean is probably the single most important factor governing the structure and functioning of the resident biotic communities (Smakhtin, 2004).

Despite their relevance for conservation, coastal lagoons are seriously threatened by eutrophication, pollution, urbanization, and diverse forms of modification in their watersheds, caused by human activity in the coastal zones of all continents (Esteves et al., 2008). Worldwide, one of the most common threats to coastal lagoons is the modification of their natural hydrology, particularly the artificial connection with the ocean. Among 60 intermittently open estuaries along the southeastern coastline of Australia (Pollard, 1994), more than 72% are artificially opened (Gale et al., 2007). A similar situation is observed in South Africa, Brazil, and other countries with large numbers of coastal lagoons or intermittently opens estuaries. Artificial openings are performed to manage these coastal systems for a variety of reasons, such as improving fisheries, lowering water level to avoid floods in urban or private lands, raising oxygen concentration, removing algal blooms, or reducing nutrient level (Sei et al., 1996; Griffiths, 1999; Palma-Silva et al., 2000; Thomas et al., 2005; Esteves et al., 2008).

In many coastal lagoons, the artificial opening is a cultural artisanal tradition, generally associated with fisheries (Bertotti Crippa et al., 2013), although currently most of the openings are performed mechanically and endorsed by national and local governments (Suzuki et al., 1998; Griffiths, 1999; Young and Potter, 2002).

According to ecologists, local people, and academic groups, this practice represents an inadequate short-term solution to address a long-term complex socioeconomic problem. Despite the fact that artificial openings are widely practiced in coastal lagoons (Bally, 1987; Pollard, 1994; Griffiths and West, 1999; Schallenberg et al., 2010), relatively little is known about the impact on their biological communities and ecosystem dynamics.

Ecological and Ecosystem Impacts of Artificial Openings

Most authors who have analyzed the impacts of artificial openings of coastal lagoons and intermittent open estuaries recommend strong caution concerning this management practice (Suzuki et al., 1998; Griffiths, 1999; Saad et al., 2002; dos Santos et al., 2006; Santangelo et al., 2007; Rodríguez-Gallego et al., 2010; Bertotti Crippa et al., 2013). Besides the already demonstrated impacts on hydrology and some natural communities, a major reason calling for cautiousness is the unknown and complex effects that may probably occur on other ecosystem components. The most evident effects of artificial openings are observed on water abiotic variables. Generally, sandbar breaching promotes a freshwater discharge into the ocean and a reduction of the water level, followed by the entrance of marine waters into the lagoon. This discharge is then followed by marine intrusion and the creation of steep salinity gradients, as well as nutrients and chlorophyll changes (Suzuki et al., 1998; Conde et al., 2000). In Brazilian eutrophic lagoons, the reduction of nutrients after water discharge into the ocean can be compensated by the resuspension of bottom sediments caused by water motion and by advective transport of sediment pore waters after the marine intrusion (Suzuki et al., 2002).

Macrophytes and benthic algae mortality driven by salinity and water level changes can also cause nutrient release, promoting a shift of plant-dominated lagoons to a phytoplankton prevalence and causing dystrophic crises, as observed in Brazilian and Portuguese systems (Suzuki et al., 1998; Duarte et al., 2002). Light climate in the water column is also altered due to the connection with the ocean. Marine intrusions with low organic content exhibit higher UV penetration and therefore reduction in photosynthetic rates of phytoplankton can be observed, as reported for many lagoons around the globe, including Uruguayan coastal lagoons (Conde et al., 2000, 2002). dos Santos et al. (2006) found strong impacts on macrophyte stands (*Typha dominguensis*) in a Brazilian coastal lagoon after artificial openings and the decrease of water level caused the mortality of this species in the lagoon. Nutrients were released to the water column through plant decomposition, so the expected nutrient decrease after the lagoon opening, allegedly implemented to mitigate eutrophication from urban discharges, was compensated, and a further increase in the phosphorus concentration occurred. Rodríguez-Gallego et al. (2010) showed that artificial openings of a coastal lagoon in Uruguay (Laguna de Rocha) could cause sudden submerged aquatic vegetation (SAV) decrease in biomass and richness, affecting other processes like solid particles and nutrient resuspension by wind. Moreover, sudden salinity changes driven by frequent sandbar openings may promote a highly fluctuating SAV community, with plants and macroalgae proliferation that could also alternate with phytobenthos and phytoplankton blooms (Rodríguez-Gallego et al., 2015).

Artificial openings of Brazilian lagoons have been reported to cause severe changes in zooplankton, as well as in aquatic and floodplain macroinvertebrate assemblages (Santangelo et al., 2007). A replace of typical freshwater species by brackish and marine zooplankton taxa was observed. After sandbar breaching, the main factor driving the zooplankton community was salinity, even overcoming the eutrophication effects on this assemblage. Two years after this episode, the previous zooplankton community had not yet been reestablished, denoting low resilience to this disturbance. Macroinvertebrate richness and abundance were negatively associated with salinity and were also affected by changes in the hydroperiod after artificial sandbar openings (Bertotti Crippa et al., 2013).

Effects of artificial openings of coastal lagoons on fish assemblages are dual, some being beneficial for some coastal species but detrimental for others. Generally, this practice has negative impacts on freshwater species that must migrate to less saline areas or streams, while it can have positive effects on marine and brackish species. Higher richness and recruitment of marine and estuarine fish species were observed in Australian (Griffiths, 1999), Brazilian (Saad et al., 2002), and North American (Reese et al., 2008) coastal lagoons after artificial openings. However, effects can be different depending on species life cycles, spawning, and dispersion timing, and how they synchronize with the frequency, duration, and time of year when the system is connected to the sea. Studies do not recommend artificial openings based only on fish effects until consequences on other biological communities and ecosystem processes are assessed (Santangelo et al., 2007; Bertotti Crippa et al., 2013).

Other factors, such as the location or the time of the year of the sandbar breaching, the sort of opening (manmade or mechanical), and frequency and duration can drive different and complex effects on ecological, chemical, and physical processes, but these factors have not yet been explicitly evaluated (Yañez-Arancibia et al., 2014). Moreover, effects on the flood plain vegetation, soil maintenance, sedimentation, and aquifer dynamics may also be expected. Flooding duration and frequency, and the salinity regime, may promote changes in the wetlands vegetation and therefore on the habitat and forage production for cattle-raising.

Active participation of local stakeholders (particularly fishermen and neighbors) in the decision-making process concerning the artificial control of lagoons' sandbars has been described as essential for successful and sustainable management, especially in protected areas (Pomeroy and Douvere, 2008). Although more complex, long-term processes to involve an eclectic array of stakeholders to discuss and make consensual decisions offer numerous benefits for a successful practice, based on the best available technical information

(Whitfield et al., 2008). This approach will help to improve and democratize actions concerning relevant ecosystem services, like those associated with sandbars of coastal lagoons, where dominant interests still prevail over less influential stakeholders in the decision-making process. For example, opening of lagoons with the exclusive purpose of avoiding floods in private farms could be avoided by an active participation of other stakeholders, for example, by establishing specific intersectoral committees and by scientifically designing multicriteria decision models that take into account other local interests and needs.

The ecological relevance of the connectivity of coastal lagoons and intermittent open estuaries with the ocean is not under discussion. Nevertheless, the global impact of the artificial manipulation of this natural dynamic is still not satisfactorily known, although most of the evidence suggests strong effects on communities and processes. Although this is a historical and common management practice, an increasing institutionalization and mechanization of the practice seems to be occurring worldwide.

In this chapter we analyze (1) the institutional and social demands and conflicts for the artificial opening of the sandbar of a legally protected lagoon on the Atlantic coast of Uruguay (Laguna de Rocha), (2) the hydrodynamics that drives the lagoon opening and the extent of flooding in the floodplain, (3) the natural opening and closing dynamics of the sandbar, and (4) the geological and geomorphological processes taking place at the sandbar in relation to its fragility and evolution over time. The goal of this transdisciplinary research was to implement a consensual multidimensional decision model for the sandbar management of this coastal lagoon, aimed at reducing conflicts among stakeholders and improving the long-term environmental quality of the lagoon.

A LONG CONFLICT IN A PROTECTED AREA

Laguna de Rocha is a subtropical choked lagoon located on the Atlantic coast of Uruguay (34° 35′ S–54° 17′ W), included in a series of lagoons along the Uruguayan and the Brazilian coast (Bonilla et al., 2006). This is a very shallow system with an average depth of 0.6 m, a surface area of 72 km^2, and a watershed of 1214 km^2. The population of the watershed is over 30,000 people, although only a few hundred live close to the lagoon. Among other relevant ecosystem services, the lagoon supports the most important inland fisheries of Uruguay's Atlantic coast (Fabiano and Santana, 2006). The northern area is influenced by freshwater discharge, while the southern area is influenced by the Atlantic Ocean due to the periodically breaching of the sandbar (Figure 13.1); thus salinity ranges from freshwater to marine conditions due to the intermittent connection with the sea. After freshwater discharge, the marine intrusion determines a steep salinity gradient in the lagoon, decreasing from south to north, while turbidity (Conde et al., 2000), sediment nutrients, organic matter, and granulometry (Sommaruga and Conde, 1990; Rodríguez-Gallego et al., 2010) follow the inverse pattern. Hydrology is the main driving force for

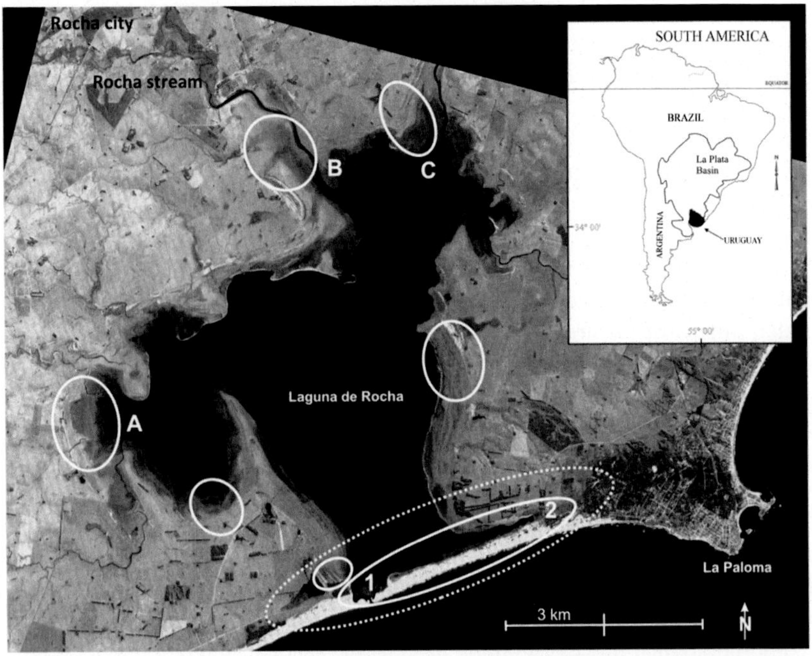

FIGURE 13.1 Overview of Laguna de Rocha and its low basin. La Paloma city is shown on the lower right margin. The provincial capital, Rocha city, is located just outside the image, on the upper left corner, where the main tributary of the lagoon (Rocha steam) can be seen. The dotted zone indicates the sandbar subsystem, where our research is focused. 1. Current opening site of the sandbar. 2. Zone called "barra vieja" (old sandbar) by local people. The circled zones indicate where the flooding areas were specifically studied. A, B, and C are the three areas referred to in Table 13.2 (Satellite image CBERS-2B-HRC 03/05/2009.)

the whole ecosystem functioning (Conde et al., 2000), and its effects on salinity gradients determine phytoplankton (Bonilla et al., 2005), phytobenthos (Conde et al., 1999), bacteria (Piccini et al., 2006), benthos (Pintos et al., 1991), and SAV (Rodríguez-Gallego et al., 2010) abundance and composition.

Phosphorus enrichment seems to be occurring in Laguna de Rocha (Aubriot et al., 2005), possibly due to land use intensification (Rodríguez-Gallego et al., 2013) during the last two decades. Concomitantly, SAV proliferation was recorded (Rodríguez-Gallego et al., 2010) and cyanobacteria blooms were observed twice (V. Hein and D. Calliari, personal communication). Afforestation with exotic trees was almost nonexistent in 1997 and increased to 8.5% of the catchment area in 2005, while agriculture increased 3% during the same time period (Rodríguez-Gallego, 2010). Agriculture is located at short distances from the lagoon and its affluents, and is changing to fertilizer-demanding crops (e.g., soybean). These quantitative and qualitative changes represent a significant increase in the input of nutrients from the catchment area to the lagoon. Moreover, the higher pressure to expand agricultural lands is restricting cattle-raising to marginal, less productive areas; therefore saline wetlands of the lagoon are now more intensively used for cattle, which in turn increases the interest of farmers to decrease the lagoon water level at the onset of flooding. On the margins of the main tributary of the lagoon (Rocha stream), Rocha city is located (about 22 km from the sandbar). This is the provincial capital, with 26,000 inhabitants, partially located on flatlands with poor pluvial evacuation and suffering frequent and short-term flooding events. In this city, there is a generalized incorrect belief that the artificial opening of the lagoon avoids or alleviates flooding; therefore during intense precipitation events, there is high pressure from people (e.g., on the media and by authorities) to accelerate the sandbar breaching process.

Laguna de Rocha constitutes a feeding and reproductive area for important fish resources (Vizziano et al., 2002; Norbis and Galli, 2004) and for resident and migratory aquatic birds (Aldabe et al., 2009; Sarroca, 2008). Therefore, it is included in the MAB-UNESCO Reserve "Bañados del Este" and since 2010 belongs to the National System of Protected Areas (SNAP) under the Protected Landscape category. Since its inclusion into SNAP, the area is co-administered by the local provincial government of Rocha (IDR) and the National Environmental Agency (DINAMA). A Director working with four rangers heads the management of the area. An Advisory Local Committee (CAE) has been installed, according to the mandate of the Protected Areas National Law, which is constituted by local and national governmental institutions, landowners, farmers, fishermen, and other local stakeholders (e.g., tourism sector). Although the CAE was formally installed in 2010, its members have been meeting for more than a decade to discuss and bring diverse environmental problems of the area to advising authorities, which represents a major advantage in the process of building consensus on the artificial opening of the sandbar. Presently, the Management Plan of the protected area is being considered for legal approval by national authorities, which includes a system for decision making for the sustainable management of the sandbar, derived from this research.

The breaching of the sandbar is performed by the local government (in coordination with DINAMA) to (1) avoid flooding of cattle-raising fields, (2) avoid floods in urban zones located on the floodplain of the lagoon and streams and in Rocha city, and (3) favor fisheries in the lagoon, which depends on the entrance of larvae and juveniles of several species from the sea. Nevertheless, there is no long-term assessment of the benefits of this management practice both on mitigating these problems and on the impact of the hydrological modification on ecosystem health. The sandbar manipulation has always been under discussion by stakeholders, but especially between two main groups: one group made up of landowners and local authorities (landowners raise cattle on the floodplain and exert pressure on authorities to open the sandbar); and the other group made up of conservationists, researchers, and some fishermen, who understand that the flooding process is part of the lagoon ecosystem dynamics and are therefore concerned about the ecological impact. Several artificial openings had apparently positive effects on fisheries (e.g., shrimp; N. Norbis, personal communication), while others had negative consequences, mainly when conducted close to summer, when the refilling of the lagoon is slower and the sediments and wetlands are exposed for several months. In many cases fishermen and researchers questioned the site selected to dig the channel on the sandbar, while on other occasions the opening was mechanically performed but insufficient water level or weather conditions closed the sandbar immediately.

More recently, discrepancy on the opening decision mechanism surfaced between both administrators of the protected area (IDR and DINAMA). Rangers pointed out that the water level to proceed to open the sandbar must lie above a certain threshold to ensure that the artificial channel dug by the machine effectively allows the discharge of water and sediments to the ocean. In other cases, the machine operator, depending on IDR, started the breaching processes without consulting the protected area director. In many cases the artificial opening was not discussed among the local and national authorities that administrate the protected area, adding confusion and tension to the management process. Furthermore, predictions of increasing risks driven by climate change adds more uncertainty to the whole discussion about the sandbar manipulation, mainly considering the potential expansion of coastal urbanization close to the sandbar of the lagoon.

DEVELOPING SOCIAL, HYDROLOGICAL, AND GEOMORPHOLOGICAL INFORMATION FOR DECISION MAKING

Sandbar Management and Stakeholder Perceptions on Sandbar Vulnerability

Laguna de Rocha is recognized for its natural and cultural values. Predominantly, its sandbar plays a central role in the socioecological balance of this environment, safeguarding its natural and cultural heritage. The intrinsic value of this fragile subsystem is crucial both from an ecological and social perspective, for example, for the sustainability of traditional fishermen communities established in the bar for many decades. The social perception of the environment is a complex mosaic of seeing, interpreting, and interacting with the environment, influenced by the particular links of people with coastal resources, the life history and permanence of these relationships, and the new knowledge they generate (Larson and Edsall, 2010). Understanding these perceptions as well as the diachronic and synchronic aspects of influential geo-historical processes, and their inclusion in management proposals for the area, will allow actions to be legitimized by local people (Shackeroff et al., 2011).

Management is an inherently human process and therefore it is essential, particularly in protected areas with a stable human population, to properly understand key socio-institutional processes and their impact on the socioecological system (Christensen and Krogman, 2012). Based on this perspective, local ecological knowledge is relevant, especially in situations where information gaps or lack of information do not allow researchers to reconstruct environmental changes (Berkes, 2009; Espinoza-Tenorio et al., 2013).

In this study, blending fishermen's ecological knowledge with the dynamics of the academic interdisciplinary work posed several challenges. For example, it was necessary to incorporate interpersonal skills that enabled a climate of communication and an open dialog to transcend disciplinary approaches and the dichotomy of academic versus local knowledge, with the aim of constructing a common knowledge focused on sustainable management (Armitage et al., 2009; Buizer et al., 2011; Hopkins et al., 2012). This approach provided higher robustness to our management proposals, which grew from diversity and cooperative reflections concerning the management of the sandbar.

Historical Strategies of Sandbar Use and Management

Given the relevance of social participation in biodiversity conservation and adaptation to climate change, especially in an area with a long history of stakeholder involvement, a systematization of sectoral interests that promote the artificial opening of the sandbar was primarily made at Laguna de Rocha, in an attempt to understand the decision process from a long-term perspective. For this, historical information on the traditional management of the sandbar made by local communities was collected, and an analysis of the current institutional mechanism of decision making was also performed. Criteria and attributes informed by stakeholders to consent to the artificial opening were also registered.

The methodological strategy consisted of a historical and documentary review on the management of the sandbar, its potential socioecological impact, and the identification of the main institutional stakeholders associated with the process. A representative sample of stakeholders was selected and 26 semistructured and structured interviews were conducted, which included most of the representatives on the CAE of the Protected Area. Field observations and open interviews with selected stakeholders were also performed. Based on this information, an analysis of the perceptions of the artificial opening was carried out, as well as geo-historical timeline analysis from 1920 to the present, describing major milestones and distinguishing transformations in the sandbar management process. We also identified diverse types of social networking between stakeholders, regarding their affinity with the opening of the sandbar and their influence on the decision-making process.

Human occupation in Laguna de Rocha area can be characterized by three stages, the first of which corresponds to the original settlement process (3000 years ago) until the colonial expansion (seventeenth century). The occupation pattern of the original colonizers corresponded to semipermanent and seasonal settlements on the paleoshore, medium and low hills, streams and coastal edge of the lagoon. Later, the colonial process imposed a clearance of the region, transforming the lagoon into an area of circulation. A second stage occurred between 1890 and 1950, when the modern settlement occurred. It was not until the early twentieth century that the permanent settlement of fishermen occurred in Los Botes port (northern area of the lagoon, on the main tributary) and in the sandbar of the lagoon. This period is characterized by the onset of agricultural and fisheries exploitation. A third stage (1950–present) is defined by the consolidation of the permanent settlement of the fishing communities and a more intense productive exploitation and housing development for tourism, mainly close to the sandbar area. A gradual process of institutionalization of the environmental management in the area started to occur later in this stage (Thompson, 2008).

Regarding the process of managing the sandbar, two periods can be distinguished, corresponding to second and third stages described above. The first is characterized by collaboration between local people, farmers, and fishermen, and scarce institutional influence. These artificial openings were executed under physical and technological conditions that produced little affect on the system, both spatially and on a temporal scale, given the time required to concretize the opening process, which better accompanied the natural processes, as well as the lack of excavating machinery in the process. In the second period, by the time that local institutions—namely IDR, the National Direction of Aquatic Resources (DINARA), and the Program for Biodiversity Conservation of Eastern Wetlands (PROBIDES)—became more actively involved in the process, the participation of fishermen in the management process was largely reduced, and the relationship between farmers and fishermen also weakened.

From that institutionalization onwards, the bottom-up character of the sandbar manipulation changed into a top-down process (Zagonari, 2008), progressively leading to an increased frequency of the artificial openings. Gradually, IDR attained more control in the decision-making process, but opening practices were still carried out without a technical monitoring and evaluation of their consequences over more than three decades. In 2011, after questioning by the CAE, a rudimentary protocol for the artificial opening, based exclusively on a water level criterion, was developed. It consisted of a virtual perimeter or limit around the lagoon, indicated by four marks in the floodplain set following an agreement between fishermen, farmers, and authorities. It represents a subjective limit between the nonflooding and the flooding condition, according to the perception of the local people. The protocol simply stated that the sandbar could be opened once the water level reaches the four marks.

Environmental Perception of Sandbar Vulnerability and Management

The perceptions of different stakeholders concerning the sandbar management were established based on three main dimensions: (1) changes and impacts (both positive and negative) generated by the artificial opening, (2) causes and conditions for an artificial opening to be made, and (3) the elements of conflict—or consensus—for developing a decision-making system for the opening (Table 13.1). Each of these dimensions is conformed by subdimensions collected from the interviews with stakeholders.

Regarding the identification of changes and impacts on the biophysical components over the last 30 years in Laguna de Rocha, some of these elements are described by most stakeholders as negative changes as a direct effect of the artificial openings, such as the location of the opening channel, the clogging of channels, and the decreased floodplain area. To a lesser extent, stakeholders also mentioned negative impacts on wetlands and pastures, as well as other changes that coincide with the intensification of artificial openings, but that cannot be

TABLE 13.1 Dimensions and Subdimensions Concerning the Management of the Sandbar of Laguna de Rocha, According to Stakeholder Perceptions

Changes and Impacts Attributed to the Artificial Opening
Decreased depth of the lagoon and channels
Changes of the breaching site
"Greening" of the bottom of the lagoon
Reduction of the floodplain area
Decrease in bird abundance
Increase of extreme weather effects
Increased grasses on dunes
Reduced size and abundance of fish species
Increased abundance of siri crab
Loss of marshes
Loss of natural pastures
Main Reasons for the Artificial Opening
Flooding: human risk
Flooding: loss of productive areas for cattle-raising
Entrance of coastal shrimp larvae
Variables Influencing the Decision-Making System for the Sandbar Opening
Arbitrariness in the decision making
Excessive bureaucracy
Lack of technical and scientific information
Protection of private property

explained straightforwardly, such as decreased bird populations or increased populations of blue crab (*Callinectes sapidus*). Most fishermen and other permanent residents, except most cattle farmers, emphasize both the impact on fisheries and on the physical structure of the lagoon (e.g., soil characteristics, depth and area of discharging channels) as consequences resulting from frequent artificial openings inadequately performed.

Concerning the causes for the artificial opening to be authorized, stakeholders mainly argue productive reasons (e.g., negative impact of flooding on livestock, or reduced shrimp harvest) when the lagoon is closed), as well as the impacts of flooding on the human population around the lagoon (i.e., urban and suburban flooding). Flooding of housing was basically the most sensitive element that justifies the artificial opening among consulted people. While the causes of urban flooding are not directly or only associated with a closed sandbar condition but also with inadequate urban planning (especially in capital Rocha city), the local perception is that the cause is based on the status of the sandbar (i.e., closed) (this topic is further addressed in Section Evolution and Functioning of the Sandbar). Notably, several stakeholders showed a preference for the lagoon and the sandbar to behave naturally from a hydrological standpoint, that is, without human intervention for artificial openings. This perception is associated with an expressed need to restore a system that has been negatively impacted by the artificial openings in the long term. An argument for this, especially among fishermen and researchers, is that a ban of the artificial opening, even partial, would facilitate the restoration of natural processes and an understanding of how the system works without intervention.

Finally, several elements were identified as key factors that promote consensus for the artificial opening decision-making system (e.g., sound information) or those that generate conflict to reach agreements (e.g., arbitrary decisions, bureaucracy in the decision process, or respect for private property). Arbitrariness is still perceived as a problematic factor in the sandbar management. On the other hand, there is full consensus that the opening decision system has to be firmly based on technical and scientific criteria. Also, it is worth noting that the CAE is unanimously recognized as the most appropriate space for building consensus and agreements to manage the artificial opening.

There was also consensus among stakeholders in supporting the initiatives of the Direction of the protected area concerning these topics. This explicit support is evident both from the interviews and from the analysis of the relationships between stakeholders and their affinity and influence concerning the artificial opening (Figure 13.2). This analysis shows that the Direction of the area has strong linkages with most of the stakeholders but particularly with all those who have less affinity for the artificial opening. Moreover, the Direction of the protected area significantly helps keeping links and permanent communication between stakeholders. On the other hand, the local government shows leadership for the artificial opening of the sandbar, both for their legal competences and for establishing stronger relationships with those stakeholders with

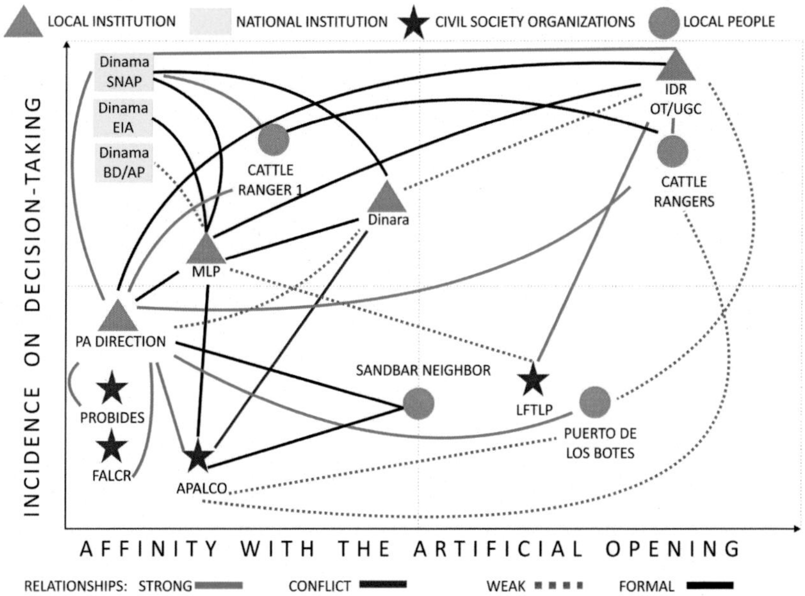

FIGURE 13.2 Laguna de Rocha stakeholder mapping, showing the individual influence on decision making (vertical axis) and affinity concerning the artificial opening of the sandbar (horizontal axis), as well as the relationships among stakeholders (indicated by different types of lines). Geometric symbols indicate stakeholder sectors. Abbreviations: DINAMA (National Environmental Agency, with three offices); SNAP (Protected Areas); EIA (Impact Assessment); BD/AP (Biodiversity); PA Direction (Laguna de Rocha protected area Director and rangers); PROBIDES (Program for Biodiversity Conservation of Eastern Wetlands); FALCR (Friends of Coastal Lagoons Foundation); MLP (Municipality of La Paloma city); APALCO (Fishermen's Association); DINARA (Aquatic Resources National Agency); LFTLP (League for Promotion and Tourism of La Paloma city); IDR OT/UGC (Rocha provincial government, with two offices; OT (Land Planning); UGC (Unit of Coastal Management). Other group stakeholders include fishermen in Puerto de los Botes settlement (PUERTO DE LOS BOTES) and cattle rangers or farmers (CATTLE RANGERS). Individuals with particular roles in the area include a cattle ranger (CATTLE RANGER 1) and neighbor (SANDBAR NEIGHBOR). (More details are given in Section Environmental Perception of the Sandbar Vulnerability and Management.)

higher affinity for the opening. This is partially balance by the role played by DINAMA, the other administrator, with low affinity for the opening but high influence on the decision making. Other stakeholders have a variety of opinions about the opening, but most of them have low influence on the decision.

Hydrological Dynamics of the Lagoon and Determination of Flooding Areas

The basin of Laguna de Rocha extends from the hilly area North of Rocha city down to the mouth in the Atlantic Ocean. The concentration time of the basin, based on the formulation of Ramser–Kirpich (Chow et al., 1994), is

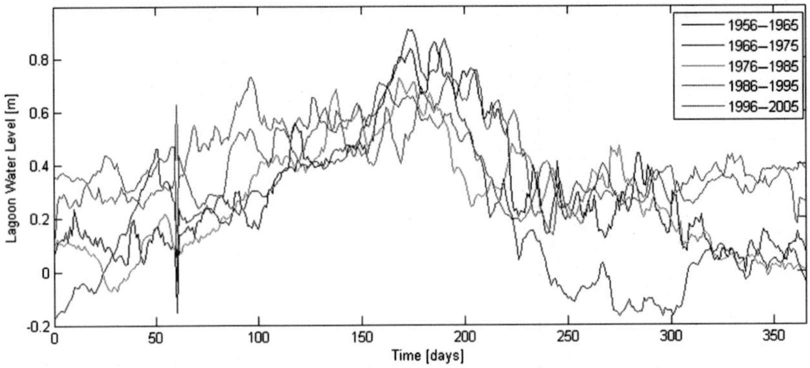

FIGURE 13.3 Mean daily water level (m) in Laguna de Rocha, estimated for five different 10-year periods from 1956 to 2005. *Data from MTOP.*

14 h. Consequently, it is expected that in less than a day the lagoon receives the water runoff generated in its basin after an extreme precipitation event. In order to characterize the hydrological dynamics of the lagoon, we analyzed the available water level data from the period 1956 to 2005 (from the Ministry of Public Works, MTOP) recorded daily every 8 h. Water level exhibited a significant inter- and intra-annual variability that essentially corresponds to the variability characteristics of Uruguay concerning monthly precipitation, but also due to the variability of the sandbar openings. Openings, both natural and artificial, occur at levels between 1.10 and 1.75 m above the Official Zero (OZ) and produce sharp declines of the water level inside the system (Figure 13.3) (The Official Zero is an official reference defined as the average water level in the Port of Montevideo).

The empirical cumulative distribution function of the water level in the lagoon is presented in Figure 13.4, highlighting the value 0.87 m (88% non-exceedance probability), which corresponds to the legal public demarcation of the lagoon limits, according to national regulations. The data show that in 72% of the 50-year series, the maximum water level commonly occurs in winter, a fact that is confirmed by applying a hydrological monthly balance (UNESCO, 2006) to the system, based on precipitation data, evaporation, and bathymetry of the lagoon under closed bar conditions. In Figure 13.3, the evolution of the average daily water level in decadal periods can also be observed. The general hydrological behavior of the system involves the superposition of two processes, one on a seasonal scale and another one on a daily scale. The seasonal process corresponds to the increase in water level of the lagoon toward the final months of summer and autumn, until during the winter the lagoon presents its highest water levels. Then, the opening of the sandbar occurs (data not shown), decreasing the water level of the lagoon. Superimposed with this process, if an extreme rainfall event can significantly increase the water level of the lagoon, the opening of the sandbar may also occur outside winter—even during summer. This process occurs on a daily scale. The proposed model describes the

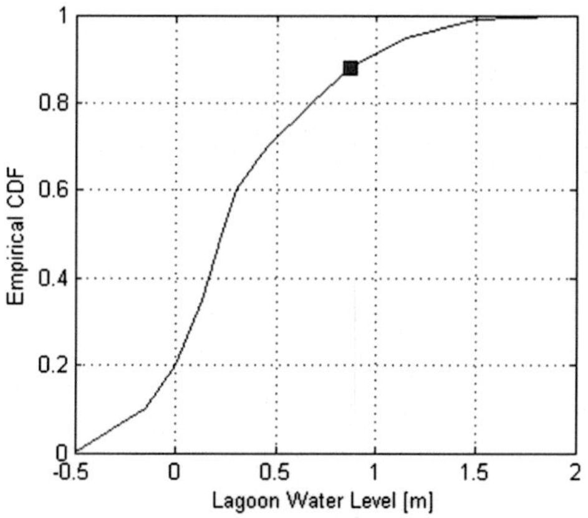

FIGURE 13.4 Empirical cumulative distribution function (CDF) of the water level (m) in Laguna de Rocha. Water level of 88% nonexceedance probability is highlighted (red square dot), corresponding to 0.87 m, that is, the limit of the legal public demarcation of the lagoon.

global behavior of the system, considering only the continental variables but not the effects of the incident sea waves on the sandbar.

In order to characterize the flooded areas surrounding the lagoon and the water retention in both public and private lands, a detailed topographic survey was conducted in selected areas (see location in Figure 13.1). With this information, a digital topography model was developed and the flooded areas were established for different water levels in the lagoon, further discriminating between public or private areas, according to the limit level of 0.87 m. The average duration of flooding in each area is given by integrating the flooded areas with the empirical cumulative distribution function of the water level in the lagoon (Figure 13.4). The results of the flooded areas for three zones of major interest are shown in Table 13.2. The result show that in all areas, the water level that can cause a natural opening floods a much higher proportion of public than private lands. For zone 1, for example, which represents one of the most critical areas because of its flatness, an opening at 1.49 m OZ floods 23% private lands, while the remaining 77% of the flooded area is public.

The effect of the water level in the lagoon on the water flow of the main tributary (Rocha stream in the Rocha city area) was preliminarily assessed based on the available hydrologic information, in particular on two bridges on the western margin of Rocha city (Estiba and Paso Real). A simplified hydrodynamic model was implemented based on the free software HEC-RAS 4.1 (U.S. Army Corps of Engineers), between the sandbar of the lagoon and a section located 5 km upstream of Rocha city. The model solves equations for free-surface flows by numerical methods based on finite differences, and in this case the modeling was performed in steady-state mode, considering the upstream flow and the level at the sandbar as the boundary conditions.

Three flow scenarios (700, 350, and 5 $m^3 s^{-1}$) were considered. These flows correspond to the maximum flow of an extreme precipitation event in the basin associated with a 100-year return period, an estimate of the base flow in the watershed, and an intermediate value of both. The selected water levels in the

TABLE 13.2 Quantification of Flooded and Nonflooded Areas for Different Water Levels in Laguna de Rocha. Results for Zones A, B, and C (Shown on Figure 13.1) Are Included, Discriminating Between Public and Private Areas, According to the Legal Public Demarcation of the Lagoon Limits (i.e., 0.87 m).

	Zone A		Zone B		Zone C	
Water Level (m)	Flooded (ha)	Nonflooded (ha)	Flooded (ha)	Nonflooded (ha)	Flooded (ha)	Nonflooded (ha)
0.35	14.78	271.6	24.4	148.5	23.0	70.6
0.87	183.8	102.6	120.9	51.9	68.6	24.9
0.93	200.7	85.6	130.8	42.0	71.7	21.9
1.15	225.7	60.6	149.1	23.65	81.6	11.8
1.49	238.8	47.5	166.8	6.0	84.6	8.9
2.0	256.2	30.1	172.8	0	87.3	6.3

TABLE 13.3 Results of the Hydrodynamic Modeling of Rocha Stream Water Levels in Two Sections (Estiba Bridge and Paso Real), According to Selected Water Levels in Laguna de Rocha (1 and 4 m). Water Levels Are Referred to the Official Zero (See Text for Explanation). Three Flow Scenarios (700, 350, and 5 $m^3 s^{-1}$) Are Modeled

Flow ($m^3 s^{-1}$)	Estiba Bridge (m)	Paso Real (m)	Lagoon Level (m)
700	10.39	9.50	1.0
700	10.39	9.50	4.0
350	9.57	8.71	1.0
350	9.57	8.71	4.0
5	6.36	5.61	1.0
5	6.36	5.61	4.0

lagoon were 1.0 and 4.0 m, the latter being a magnified value to test the hypothesis of no influence of the lagoon levels in the flooding of Rocha city. Solving the six scenarios that arise from the combination of the boundary conditions given above, the results of the water level in Estiva bridge and in Paso Real are presented in Table 13.3. Despite not having detailed cross-sectional information for this stream segment, all information led to the conclusion that there is no influence of a variation of up to 4 m of the water level of Laguna de Rocha on the water levels upstream, in both sites analyzed in the Rocha city margin.

This conclusion is relevant for sandbar management, since there is a local widespread tradition that erroneously states that when Rocha city is flooded by Rocha stream after extreme precipitation events, if the sandbar of the lagoon is closed, flooding is more intense and lengthy in the city. This has subsequently led to more intense pressure by Rocha city inhabitants on the local government to open the lagoon artificially as a way of supposedly reducing the flooding event in the city. This modeling proved that there is no reason for this conduct to be sustained in the future.

Understanding the Natural Opening and Closing Dynamics of the Sandbar

At the site of the present opening of the sandbar there are two processes that can lead to natural openings: (1) opening from the lagoon, produced when the level in the lagoon exceeds the berm crest elevation in the sandbar; and (2) opening from the ocean, caused by the combined action of high sea and heavy swell. From the available wave data based on re-analysis (Alonso, 2012) of sea level for the period 1912–2012, water levels in the lagoon (1956–2005 series from MTOP) and from DINAGUA (National Water Agency) records concerning opening and closing of the sandbar, we proceeded first to identify and characterize situations in which the opening was produced. Then, according to sea level and wave height data, we estimated the level up to which the berm at the sandbar can grow by wave effect, using the model of Horikawa and Maruo (2011). The probability distributions of the expected berm level on the short and medium term in the lagoon bar, obtained by the two above methods (DINAGUA records and Horikawa model), were very similar (Figure 13.5). In both cases the mode of the distribution lies between 1.10 and 1.49 m OZ.

Then, the effect of a wave storm over the beach profile was modeled. For this, four different berm crest elevations were used, corresponding to quantiles 10%, 50%, and 90%, and a berm crest elevation representative of a recent closure of the sandbar of the lagoon. This simulated storm event corresponds to a 72 h severe condition, with significant wave height of 4.5 m and maximum sea level of 1.1 m OZ. For the simulation, bathymetric and granulometric information of the beach area was used to construct a characteristic profile transverse to the sandbar under closed conditions. The results indicate that the sandbar can

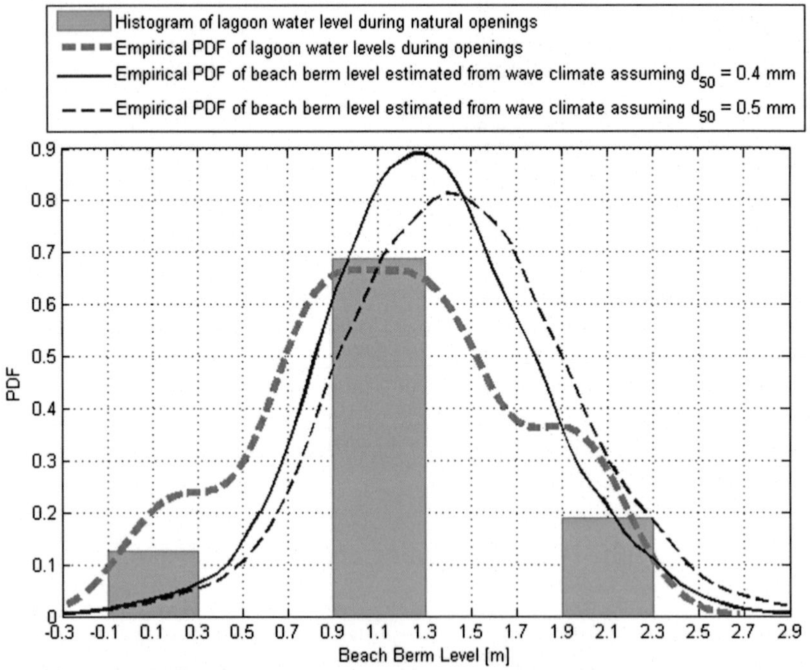

FIGURE 13.5 Empirical density function probability (PDF) estimated from berm elevation data obtained from Laguna de Rocha water level during natural openings (green) and from wave climate (blue). d_{50} is the average size of the beach sediment, used to estimate the height of the berm according to the wave climate.

be opened by sea action only if the elevation of the beach profile is very close to sea level, which is coherent with recent closing conditions.

In addition, the littoral sediment transport was estimated for the zone of the coast where the mouth of the lagoon is located. To do so, all sea states of the wave series were propagated from intermediate waters to the breaking zone, using Snell law (Dean and Dalrymple, 2002) and the breaking criteria of Thorton and Guza (1983) for irregular waves. Once the significant wave height and the breaking direction were determined, the littoral transport for each state was estimated, using the formula of CERC (U.S. Army Corps of Engineers, 2002). The net littoral transport in the sandbar area of Laguna de Rocha was zero for all practical purposes when compared to the gross longshore transport. The gross transport rate was on the order of 3 million $m^3 year^{-1}$ (i.e., 360 $m^3 h^{-1}$), so there is a strong potential to close the sandbar as soon as the discharge channel of the lagoon loses the capacity to clean sediments out of the system.

The results obtained considering the expected berm elevation, the erosion of the beach profile, and the coastal sediment transport show that the predominant opening mechanism of the sandbar occurs when the level of the lagoon exceeds the berm. A conceptual diagram of the natural opening and closing mechanism of the sandbar, based on the hydro-sedimentological behavior of the system, is presented in Figure 13.6, showing that the water level of the lagoon produces

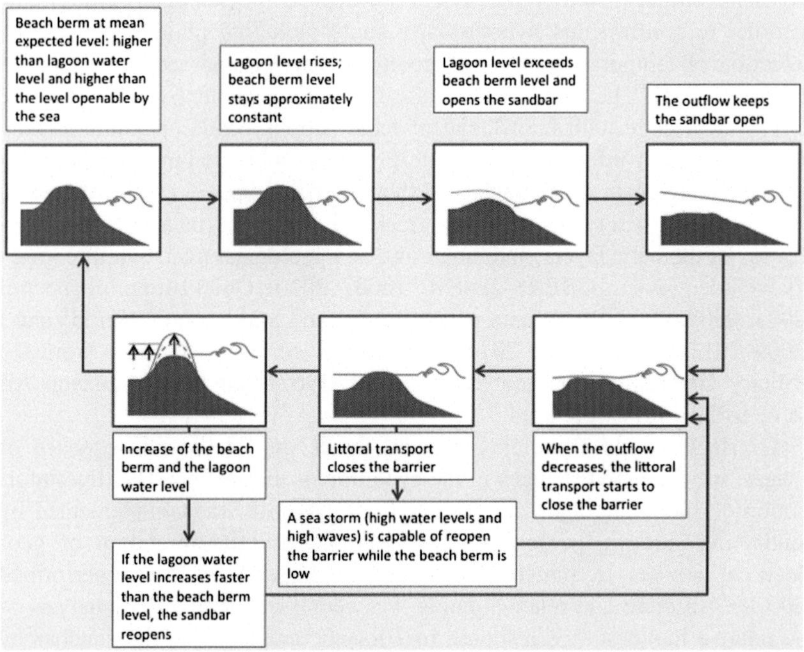

FIGURE 13.6 Conceptual diagram of the dominant opening and closing natural mechanisms operating at Laguna de Rocha sandbar (see text in Section Understanding the Natural Opening and Closing Dynamics of the Sandbar for explanation).

the breach of the berm and subsequently the water discharge to the sea occurs. After the mouth's closing by coastal sediment transport, and while the level of the berm is low, the sea can break the berm, creating temporary connections with the sea, which are again closed by the longshore transport. However, when the berm grows significantly, the sea loses its overtopping ability and under this condition only high water levels inside the lagoon can produce an opening.

Geomorphology and Vulnerability of the Sandbar

From a geological viewpoint, coastal lagoons are ephemeral environments, frequently evolving to fluvial or tidal dominated environments (Duck and da Silva, 2012), especially where human action has triggered major physical changes and deterioration. Due to their intrinsic ecological fragility, knowledge of their structure and functioning is key to facilitate risk assessments, model scenarios, and for their sustainable management. This research sought to understand the evolution of the sandbar subsystem of Laguna de Rocha (Figure 13.1) and to analyze the appropriate conditions for hypothetical artificial openings. The previous available information suggested that this sandy bar would be a relict of Pleistocene deposits shaped during the past 10,000 years (Preciozzi et al., 1985; Heinzen et al., 1986), with a relict area of closure at the Northeastern end known as "barra vieja" by local people (zone 2 in Figure 13.1). In order to confirm this hypothesis, a study including photointerpretation, detection of temporal changes, exploratory surveys, and sediment analysis was performed. The study had a transdisciplinary context by merging the expert knowledge with knowledge of local people, based on joint working sessions with key informers. Photointerpretation and digital image processing included image georeferencing and orthogonalizing, images change detection analysis (1943–2011), geomorphological and geological interpretation, digitization of thematic layers, and the design of a geological map. Satellite imagery included sensors CBERS-2B-HRC (2007–2010), QuickBird multispectral (2004, 2007, 2008, 2009), and panchromatic and multispectral WorldView I (2009–2011) and II (2010–2011). In addition, aerial photographs from the Military Geographic Service (SGM; 1943, 1967) and the Uruguayan Air Force (from 1986) were used.

Overlapping and combining spectral bands on a temporal sequence of images, we examined changes in the evolution of the sandbar. Baseline information on the stratigraphy and sedimentary sequence was complemented by conducting surveys, geological cross-sections, and interpretation of geo-electrical profiles. A mineralogical analysis of sands was also performed (600 X). Absolute and relative dating was addressed through the analysis of the relative luminescence response in 10 sand samples from the sandbar by infrared and temperature stimulation (IROSL and TL). Samples were collected in surveys ~40 cm deep, performed in the lower portions of inter-dunar areas, which are considered the oldest elements of the subaerial environment of the sandbar. Luminescence analysis was complemented by dating peat through the carbon-14 method. Bathymetric and precision altimetry data were also used. A detailed bathymetry of the lagoon and the marine area of the lagoon's mouth was also performed with centimeter accuracy.

Evolution and Functioning of the Sandbar

In general, we obtained consistent results between the remote sensing study, the mineralogy, the IROSL/TL and the carbon-14 analysis, the bathymetric and the altimetry survey, the analysis of the historical information, and the local ecological knowledge. All results indicate that the sandbar that separates the lagoon from the ocean consists of a recent sedimentary sequence formed by a series of sand ridges originated on the littoral drift and remobilized by wind above the water table. The fine sand fraction in the bar showed a more diverse mineralogy than the lacustrine samples, its luminescence was low, and its edaphic differentiation was confined to a very limited translocation of quartz silt. Survey in the sandbar also showed the presence of thin lenses of lacustrine gray muds.

We originally hypothesized that the "barra vieja" zone was a relict area where the lagoon connected with the ocean in the past. Nevertheless, this hypothesis was rejected because of the identification of areas with evidence of relatively recent ingress of sea waves into the lagoon, as well as zones of low depression (close to sea level) clearly corresponding to opening channels of the lagoon during the Holocene epoch. In particular, one depression was located about 800 m distant from the area of the current opening of the lagoon, which was registered in pictures from 1943 with evidence of hydraulic percolation (Figure 13.7(c)). Given this information, the new operating model of the sandbar system considers that while "barra vieja" served as a site of connection with the sea in the last 2660 ± 50 years (according to carbon-14), the whole connection process was governed by multiple, and probably synchronous, openings until late 1930–1940, both on the western and eastern extremes of the sandbar. This would indicate that the dynamic opening mechanism that can be observed today is very recent, that is, less than 100 years old.

Geomorphologically, the evolution of the sandbar opening mechanism can be explained by sediment transported by wind and by wave action under a regimen of multiple ocean–lagoon exchanges, where the dominance of the relationship marine/eolic energy has incremented in relation to the hydraulic lacunar energy, as major forces of change in the dynamic equilibrium of the system. This evolution toward the current situation can be explained by the interaction of three forces: eolic, marine, and hydraulic, which are described next.

The marine regression occurring during the last period of the Holocene epoch shaped large plains around Laguna de Rocha, which were frequently waterlogged. Under these conditions, the formation of peat soils by deposition of sediments and plant biomass took place. During dry periods like the one known as the "bonariensis little ice age," about 300 years ago (Rabassa et al., 1984), and probably from 3500 and 2500 years ago (Bracco et

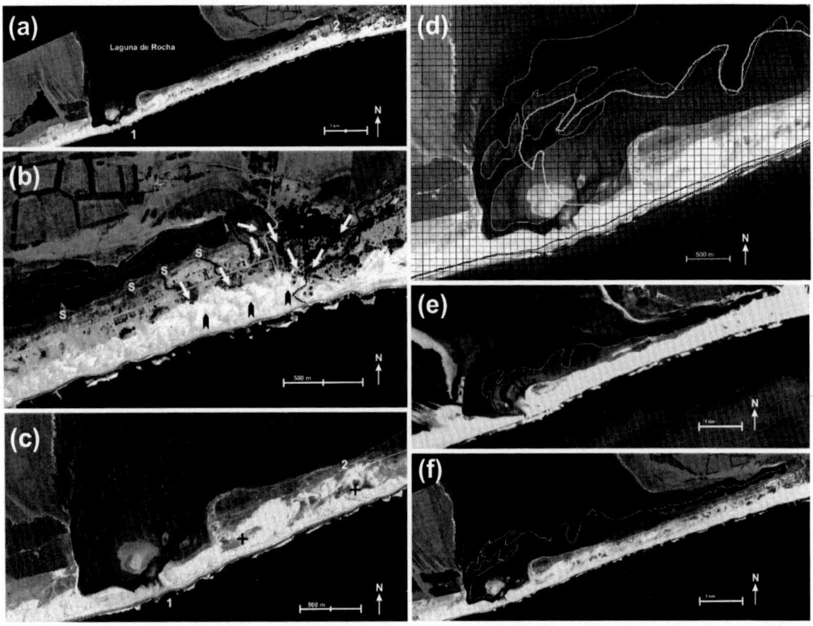

FIGURE 13.7 Details of the Laguna de Rocha sandbar system, based on satellite image WordView II (15/11/2011). Dunes with scarce vegetation (i.e., higher kinetic potential) are shown in whitish tones. (a) Numbers indicate different opening areas of the lagoon (1, current opening zone; 2, a relict area ("barra vieja"). (b) White arrows indicate slopes into streams and creeks (black dotted lines) corresponding to some of the last opening areas; black arrows indicate remnant areas with evidence of seawater intrusion (kidney-shaped forms, corresponding to flooded depressions with emerging water table); S indicates sand ridges formed by internal waves in the lagoon. (c) Soils with level similar to sea level (+ symbols), which correlate with relict openings. The first mark is situated ca. 800 m from the present opening zone. The dotted white line inside the lagoon indicates the limit of the silty-sandy bank with depth less than 1 m, limiting the deeper channels. (d) Movement of sand banks inside the lagoon toward the west are shown (1966, dotted line; 2011, continuous line). On the marine side of the sandbar, the receding shoreline is also shown (1943, dotted black line; 2011, solid black line). The grid side is 50 m (satellite image WordView II—15/11/2011). (e) A detail of the 1943 image (SGM). (f) A detail of the 2011 image (WordView II).

al., 2011), the material supplied from the sea and the existence of plains generated favorable conditions for the formation of a mobile dune system with different types of evolution, complexity (Barnes, 1980; Carter and Woodroffe, 1997), and conditions for confining water bodies. These coastal lagoons originated through this process, by damming caused by dunes, which were mobilized by wind action according to the direction of the prevailing winds. In Uruguay, the estimated age for the origin of the mobile dune systems is between 4000 and 4500 years ago (Piñeiro and Panario, 1993). For Laguna de Rocha sandbar, both the IROSL and TL analysis as well as the sediments surveyed indicate the evolution of a system of mobile bark-haan dunes across the current sandbar, generated from sands of marine origin.

Simultaneously to the formation of the sand dune system during the last period of the Holocene epoch, there also occurred transport and deposition of sands in submarine banks and beach berms by longshore drift currents and waves. Until the early 1900s (after the "bonariensis little ice age," with prevailing winds from the west), the net transport probably had a northeastern direction, filling up with sand the area of "barra vieja" and making less frequent the possibility of opening in that zone. This can be revealed by direct testimonies of a higher opening frequency during the past century in the western sector (opposite to "barra vieja") but still keeping an area of less frequent opening in the northeastern sector. Over the long term, sediments carried by waves took a predominant direction westward, as evidenced by the sandbar structure (Figure 13.7), the trend of sand grain size, and the east-to-west evolution of a dissipative beach toward an intermediate reflective one. The presence until the present of a parallel channel in the interior of the lagoon along the whole sandbar indicates that this area has the potential of opening if extreme marine conditions and high hydraulic pressure from the lagoon prevail simultaneously. Marine sand spits in the interior of the lagoon associated with semipermanent small water bodies and flooded depressions on the sandbar, close to the present sea level, correlate well with local people's knowledge of marine ingressions by waves during high-energy events (overwash) until the end of the 1930s and early 1940s.

The hydraulic pressure of the lagoon water body on the sandbar, besides its relevant role in the opening mechanism, is also a significant process concerning the water soil saturation and the elevation of the water table in that zone. Water saturation of the sandy sediments of the bar reduces the dissipative capacity against sea wave action by decreasing the permeability of the substrate, facilitating the penetration of sea waves in low-lying areas, especially in depressions between dunes. Furthermore, in this latitude extreme rainfall events are associated with an increase of the sea level (i.e., storm surges), strong southern winds with extended fetch (~50 km), and high wave energy, adding up to produce a greater effect on the sandbar structure.

Impact of the Artificial Opening on the Dynamics of the Sandbar

Diverse disturbances caused by human activity act on the sandbar of Laguna de Rocha, with enormous implications for management (e.g., construction of roads and houses, afforestation with exotic species, off-road traffic). However, the deliberate artificial opening of the sandbar stands out as one of the most severe physical and ecological threats. Our results demonstrate that this disturbance has introduced during the last decades a new dynamic equilibrium to the system and has defined new attractors that could generate nonreturn points to predisturbance conditions. One of the key changes was the gradual change of opening mechanism, where the traditional manual opening (using shovels to dig a channel) done by fishermen and other local people has been replaced by excavating machinery, the current opening method. This technological change has enabled openings in conditions that would not be possible by more traditional procedures. For example, mechanized openings have been performed in inappropriate places and beyond optimum conditions, like low hydraulic potential of the lagoon, which does not allow water and sediment to flow out to the sea, favoring the development of internal banks.

There has been an extensive growth of internal banks and the gradual filling up of the lagoon channels, especially close to the opening area (Figure 13.7(c)–(f)). In the past 10 years, this area has been progressively transformed into a tidal delta type structure, where the marine energy prevails as well as the inward flow of marine sands. Since 2003, these banks have generated a threatening positive feedback that makes more difficult the evacuation of sediments due to the decreased depth of the lagoon. Probably, the long-term result is the evolution of the lagoon to a coastal wetland type of environment, as occurred elsewhere (Carter and Woodroffe, 1997). Images change detection analysis showed a correlation between growth and translation of internal banks with the regression of the coastline (Figure 13.7(d)), which reached over 80 m in the opening area between 1966 and 2011. This type of regression has been also reported in other estuaries of the Uruguayan coast (Panario and Gutiérrez, 2005). Therefore, the information on changes in the dynamics of the natural opening and the reported impacts from the 1900s to the present suggest an increase of the marine energy relative to the hydraulic energy of the lagoon, mainly attributable to the fact that over the past 30 years the lagoon has been mechanically open far from the optimal natural conditions.

A CONSENSUAL DECISION MODEL FOR THE ARTIFICIAL OPENING OF THE SANDBAR

The outcomes of this study prompted us to develop a series of technical and management recommendations, hierarchically ranging from (1) a complete ban of the artificial opening, to (2) a partial ban of the practice, and finally to (3) a multidimensional decision tree for the artificial opening.

Primarily, considering all available information on the sandbar functioning, collected or derived from this study, our research team suggested IDR and DINAMA authorities and the CAE give a central recommendation consisting of fully avoiding artificial openings of the sandbar of Laguna de Rocha, given the character of the protected area. This advice is based on the potential long-term impacts that this practice would have on the hydrology and on ecological and ecosystem processes of the lagoon (Rodríguez-Gallego et al., 2010, 2013), as well as on the international experience on this matter (see Section Introduction). In fact, we basically recommend the reestablishment of the natural connection of the system with the ocean, with the conviction that the natural hydrology is the optimal controller of the ecosystem functioning and quality (e.g., by discharging the excess of nutrients, pollutants, and sediments). It has to be emphasized that this general suggestion is not novel, since several research and conservationist groups operating in these coastal lagoons for more than two decades have been permanently informing authorities and local stakeholders about the threat that the practice represents to these highly vulnerable ecosystems (Conde and Rodriguez-Gallego, 2002).

If a complete ban of the artificial opening is not possible to implement in the short term, mainly due to the social pressure experienced by the protected area administrators, our team alternatively recommended maintaining a three-year period without artificial openings. To our understanding, this period would probably allow (through the occurrence of several natural openings) the reconstruction of the principal discharge channel along the main axis of the lagoon, the recovery of an adequate depth in the channel for the water and sediments to be released, and the sand dunes of the bar to be recovered, reestablishing the optimal conditions for a natural connection in the site where the discharge can take place with the highest energy. This partial ban would also make it possible to monitor the functioning and the evolution of the system without artificial manipulation, which has never been done until now. Moreover, researchers will be able to compare both management modes and make sounder recommendations on the best options for the artificial manipulation.

Although we do not consider the artificial opening a sustainable environmental practice, since technical advice is not always politically realistic to be implemented, in case the previous recommendations cannot be applied we have developed a multidimensional decision-making model concerning the artificial opening. The practical purpose of this model is to help those who have to make the decision about when, where, and how this can be done in the best way possible, based on the current decision procedure (the four marks indication; see Section Historical Strategies of Sandbar Use and Management) and upon relevant information obtained during this research.

The uniqueness of the new model is that it is not merely based on the water level criteria (as it has been until now), but also on geomorphological criteria as well as on weather conditions. The final aim of the model is to reduce to a minimum the number of times the lagoon is artificially open along an annual cycle, including the possibility of not opening at all if criteria are not met. Several dimensions and criteria covering various relevant aspects of the opening mechanism guarantee that the new decision system is highly restrictive in comparison to the previous one. The rationale behind the proposed model is that there is evidence of severe impacts if indiscriminate openings take place in this protected and vulnerable site; therefore they must be kept to a minimum, and approved and performed only if diverse technical criteria are fully met.

Here we only present the decision system developed to decide when an artificial opening can be authorized, given that this is the most critical and challenging aspect of the whole problem. This decision tree, and a protocol of good practices, were consulted and adjusted several times with the protected area Director and the rangers, with managers from IDR and SNAP/DINAMA, as well as with the CAE and other stakeholders. The consensus accomplished was only possible due to the already stated tradition of stakeholders of addressing environmental local issues through active and egalitarian participation.

The new model (Figure 13.8) is basically a simple decision tree designed to be easily understood by administrators, rangers, and by the local people as

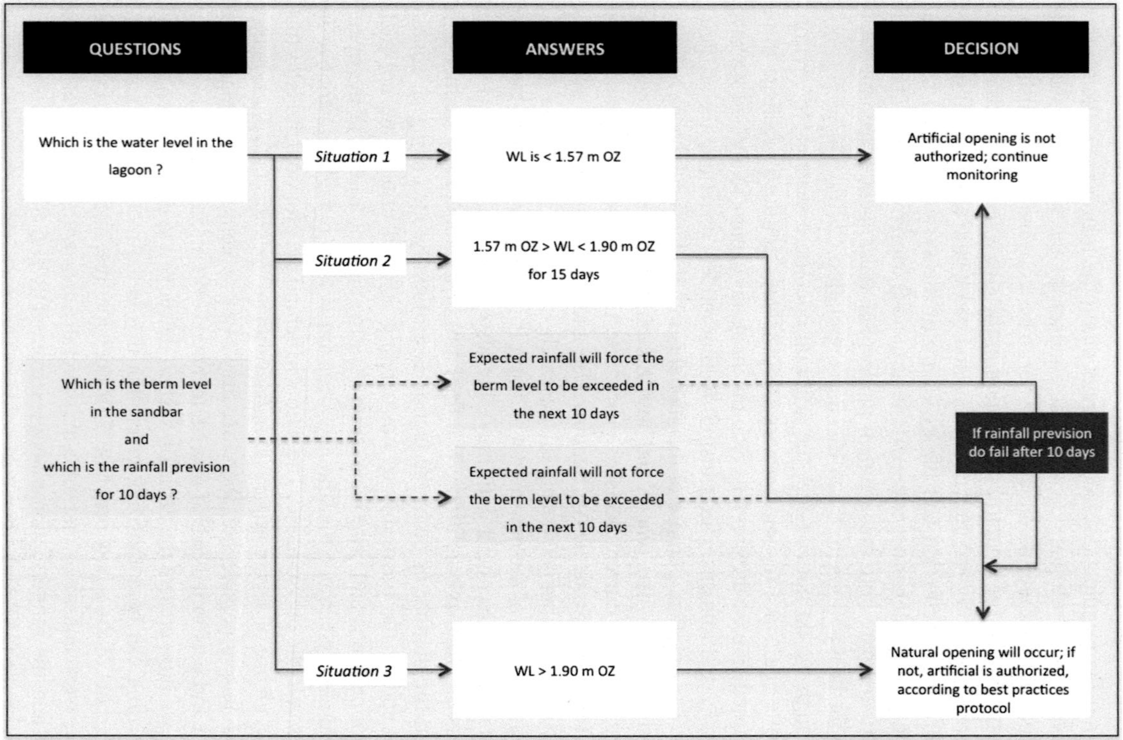

FIGURE 13.8 Proposed decision-making model for the artificial opening of Laguna de Rocha sandbar (see text in Section A Consensual Decision Model for the Artificial Opening of the Sandbar for a detailed explanation). WL, water level.

well. The tree is guided through a series of intuitive and straightforward questions, starting by assuming the lagoon is not connected with the ocean, but a progressive increase of the water level is taking place. Specifically, the first and leading question refers to the water level of the lagoon, which triggers the following steps and questions in the protocol, depending on the flooding situation observed.

Situation 1

If the water level has not attained the four marks already agreed by local stakeholders (which approximately indicate the legal delimitation of the lagoon, i.e., 1.57 m OZ), then no action is taken, because it is considered that the water level is not enough to allow an artificial opening, but mainly because the volume of the lagoon is still well below the limit of private lands. Also, under this situation, the perception of people is that no flooding is still occurring. A continuous daily to hourly monitoring has to be performed by rangers, especially when the water level approaches the marks.

Situation 2

The water level of the lagoon is higher than 1.57 m OZ, but lower than 1.90 m OZ, for up to 15 days (on a continuous series or accumulated). The value 1.90 m OZ represents the maximum water level observed in one of the artificial openings in 2013 (June). Although stakeholders perceived that this artificial opening was performed with a high water level, it did not significantly affect houses or other infrastructure, and did not generate major conflicts in the decision-making process, although being a water level very close to the water level required for a natural opening.

Nevertheless, this flooding situation (range from 1.57–1.90 m OZ) can still produce persistent flooding of cattle-raising fields, some interferences with housing in the floodplain of the lagoon, and a generalized social perception of the need of an artificial opening of the lagoon, thus prompting pressure on authorities and managers to artificially manage the sandbar. Positively, this water level range will trigger an alert to managers, so they can start preparing conditions to perform the artificial opening if the water level continues to rise or keeps more than 15 days in the reported range.

Once this period of time is achieved under the same flooding situation, the second and third questions come in, concerning the rain forecast and the height of the berm. The motive to include these criteria is that if a rainfall event is expected to occur in the basin in the next 10 days, and the water level, due to the rainfall effect, is estimated to be exceeding the height of the berm in the sandbar, then there is no reason to make an anticipated artificial opening, because it is highly probable that the natural one will occur anyhow, once the berm is crowned. In contrast, if the rainfall prevision is low (or if the berm is too high), then the artificial opening makes sense, but only if the other criteria (water level higher than 1.57 m for 15 days, plus 10 days of rainfall prevision) were already achieved. A specific protocol has been designed to estimate, for a given expected rainfall in the watershed, the expected water level in the lagoon, to be able to know if the berm will be overtopped (not shown here).

If after a period of 10 days, estimations that the berm will be exceeded fail, the opening can also be performed. It is assumed that the beneficial ecological processes associated with the flooding period can be properly achieved during the potential accumulated time (from 15 to 25 days).

Situation 3

If water level in the lagoon reaches 1.9 m OZ it is assumed that the sandbar will open naturally, so no action except monitoring has to be taken.

Nevertheless, if for any reason the natural connection does not occur after a prudent time once the 1.9 m OZ is achieved (e.g., 1–2 extra days), the lagoon can be artificially opened. In this case, a rapid response has to be taken, to avoid extreme flooding of fishermen's houses or other areas, which can occur in few hours if rains are persistent and the lagoon is closed. This situation was observed during the second artificial opening conducted in 2013 (in August), where water level increased from 1.9 to 2.17 m during the night (less than 12 h). Even if no risks for human safety were observed, fishing gear and other houses located in the sandbar were partially damaged. Although the recurrent period of flooding achieving a water level of 2.17 m OZ is 5–10 years, it is suggested that correct land planning in this lagoon should strongly promote the migration of littoral houses to safer places or to promote a modification to different building typologies (e.g., pile-dwelling).

In all cases, the artificial opening is to be performed by applying a best practice protocol specifically, which includes a precisely explanation of all variables and steps to be taken for the artificial opening procedure. This includes how and where the water level of the lagoon as well as the berm height must be measured, and an explicit methodology to determine if an expected rainfall can exceed the berm or not. The protocol also determines the precise site to conduct the artificial breaching and the orientation of the excavation, how to determine the moment of the optimal wind to avoid a fast refilling of the channel with littoral sand, and the ideal steps to involve the fishermen into the process, among others features. Details on these aspects are not included in this chapter.

CONCLUSIONS AND PERSPECTIVES

From a general perspective, research is still needed to fully understand the impacts of artificial openings of coastal lagoons and intermittent estuaries on their biological, ecological, and systemic processes (Santangelo et al., 2007; Bertotti Crippa et al., 2013). There are contrasting evidences about the benefits and costs of manipulating sandbars, although results from tropical lagoons (where a major part of the research has been done) cannot be straightforwardly extrapolated to mid-latitude lagoons (Esteves et al., 2008), where less unpredictable rainfall regimes prevail. Comparative results between artificially intervened systems and undisturbed ones, as well as long-term comparisons of management with and without human interventions, may be adequate research questions to address. These studies should ideally be complemented with long-term data series, commonly not available; experimental research simulating different hydrological conditions of the lagoon dynamics and their effects on species and processes; and a modeling approach to capture and simplify the essentials of the problem.

A central characteristic of coastal lagoons and intermittent estuaries, mainly in regions of unpredictable climate path, is their permanent change, making these systems highly dynamic and complex (Haines and Thom, 2007). These features are very attractive to researchers but at the same time pose a high degree of complexity. How changes in rainfall and wind patterns, sea level rise, and temperature changes may further enhance anthropogenic effects and promote disruption and deregulation of the natural driving processes, are some of the main questions to be answered. Local information on these aspects will allow adapting new models and protocols of sustainable management of sandbars in these coastal lagoons.

To fully understand the functioning and evolution of the sandbar system in Laguna de Rocha, new variables associated with climate change should be incorporated (i.e., sea level and wave energy increase, and changes in rainfall and winds patterns). Considering these aspects, it can be projected that risk is even higher than expected. According to local prediction, future scenarios can be more complex and the consequences will significantly depend both on the response of the natural system and on the management actions to be implemented. Alterations in the hydrological regime of the lagoon, as well as the reduction of the hydraulic pressure by the development of internal banks, suggest the possibility of scenarios where ancient areas of openings could be reactivated. Nagy (2012) specifically mentions the effects of climate change for this coast, suggesting an increase in sea level between 12 and 18 cm for 2030, without considering the further increases caused by wind. This would indicate an increasing erosive impact, especially when coupled with wind tides of ca. 2 m above sea level. The regression observed in the coastline of the current opening area and the presence of many depressions in "barra vieja" close to sea level indicate a high vulnerability of the sandbar system considering the time scale of climate change. Taking into account these climatic caveats, it seems indispensable to completely avoid any urban development across the sandbar of the lagoon.

The main outcomes of this study range from ideal technical recommendations to preserve the environment (i.e., permanently banning the artificial opening) to more practical and applicable suggestions (e.g., a temporal ban or a consensus multidimensional decision tree). All of the recommendations emerged from the study are believed to be a substantial improvement concerning the decision making on the artificial opening of the sandbar of the protected area Laguna de Rocha, which could also be generalized for other neighboring systems, if proven to be successful. Suggested courses of action are considered to be adaptive schemes that need to be permanently monitored, revised, adapted, and improved, according both to the results obtained and to the natural dynamics of the local socioecological system. The effective implementation of the new decision system, the monitoring of its beneficial effects on the ecosystem processes, its readjustment following evaluation, and its social acceptance or rejection is at the same time an approach to management, a participatory process, and a research challenge.

ACKNOWLEDGMENTS

The authors thank Hugo Inda for technical support, and Carolina Cabrera and Lucía Nogueira for helping with the references search. Special thanks to Unidad de Cambio Climático and Sistema Nacional de Área Protegidas (DINAMA), Dirección Nacional de Aguas-DINAGUA, Fundación de Amigos de las Lagunas Costeras e Intendencia de Rocha, for supporting the research. Also, to fishermen and other members of the CAE-Laguna de Rocha, for providing valuable information.

REFERENCES

Aldabe, J., Rocca, P., Claramunt, S., 2009. Uruguay. Bird life conservation series no. 16. In: Devenish, C., et al. (Eds.), Important Bird Areas Americas, Priority Sites for Biodiversity Conservation. Bird Life Internacional, Quito, pp. 383–392.

Alonso, R., 2012. Evolución del potencialundimotriz de Uruguay. Tesis de Maestría en Mecánica de los FluidosAplicada, Facultad de Ingeniería de la Universidad de la República, Uruguay.

Armitage, D.R., Plummer, R., Berkes, F., Arthur, R.I., Charles, A.T., Davidson-Hunt, I.J., Diduck, A.P., Doubleday, N., Johnson, D.S., Marschke, M., McConney, P., Pinkerton, E., Wollenberg, E., 2009. Adaptive co-management for social–ecological complexity. Front. Ecol. Environ. 6, 95–102.

Aubriot, L., Conde, D., Bonilla, S., Hein, V., Britos, A., 2005. Vulnerabilidad de unalagunacostera en unaReserva de Biosfera: indiciosrecientes de eutrofización. In: Vila, I., Pizarro, J. (Eds.), Eutrofización de Lagos y Embalses, CYTEDXVIIB. Printers Patagonia, Chile, pp. 65–85.

Bally, R., 1987. Conservation problems and management options in estuaries: the Bot River estuary, South Africa, as a case-history for management of closed estuaries. Environ. Conserv. 14 (1), 45–51.

Barbosa, F.A.R., Scarano, F.R., Sabará, M.G., Esteves, F.A., 2004. Brazilian LTER: ecosystem and biodiversity of Brazilian LTER. Environ. Monit. Assess. 90 (1–3), 121–133.

Barnes, R.S.K., 1980. Coastal Lagoons: The Natural History of a Neglected Habitat. Cambridge University Press, Cambridge.

Berkes, F., 2009. Evolution of co-management: role of knowledge generation, bridging organizations and social learning. J. Environ. Manage. 90, 1692–1702.

Bertotti Crippa, L., Stenert, C., Maltchik, L., 2013. Does the management of sandbar openings influence the macroinvertebrate communities in southern Brazil wetlands? A case study at Lagoa do Peixe National Park Ramsar site. Ocean Coastal Manage. 71, 26–32.

Bonilla, S., Conde, D., Aubriot, L., Perez, M.C., 2005. Influence of hydrology and nutrients on phytoplankton species composition and life strategies in a subtropical coastal lagoon. Estuaries 28 (6), 884–895.

Bonilla, S., Conde, D., Aubriot, L., Rodriguez-Gallego, L., Piccini, C., Meerhoff, E., Rodríguez-Graña, L., Calliari, D., Gómez, P., Machado, I., Britos, A., 2006. Procesos estructuradores de las comunidades biológicas en lagunas costeras de Uruguay. In: Menafra, R., et al. (Eds.), Bases para la conservación y el manejo de la costa Uruguaya. Vida Silvestre Uruguay, Montevideo, pp. 611–630.

Bracco, R., del Puerto, L., Inda, H., Panario, D., Castiñeira, C., García-Rodríguez, F., 2011. The relationship between emergence of mound builders in SE Uruguay and climate change inferred from opal phytolith records. Quat. Int. 245, 62–73.

Buizer, M., Arts, B., Kok, K., 2011. Governance, scale, and the environment: the importance of recognizing knowledge claims in transdisciplinary arenas. Ecol. Soc. 16 (1), 21. www.ecologyandsociety.org/vol16/iss1/art21.

Cañedo-Argüelles, M., Rieradevall, M., Farrés-Corell, R., Newton, A., 2012. Annual characterisation of four Mediterranean coastal lagoons subjected to intense human activity. Estuarine, Coastal Shelf Sci. 114, 59–69.

Carter, R.W.G., Woodroffe, C.D., 1997. Coastal Evolution: Late Quaternary Shoreline Morphodynamics. Cambridge University Press, Cambridge.

Chow, V.T., Maidment, D., Mays, L., 1994. Hidrología Aplicada. McGraw-Hill Interamericana, Buenos Aires.

Christensen, L., Krogman, N., 2012. Social thresholds and their translation into social-ecological management practices. Ecol. Soc. 17 (1), 5. http://dx.doi.org/10.5751/ES-04499-170105.

Conde, D., Aubriot, L., Bonilla, S., Sommaruga, R., 2002. Marine intrusions in a coastal lagoon enhances the effects of UV radiation on the phytoplankton photosynthetic rate. Mar. Ecol. Prog. Ser. 240, 57–70.

Conde, D., Aubriot, L., Sommaruga, R., 2000. Changes in UV penetration associated with marine intrusions and fresh-water discharge in a shallow coastal lagoon of the Southern Atlantic ocean. Mar. Ecol. Prog. Ser. 207, 19–31.

Conde, D., Bonilla, S., Aubriot, L., De León, R., Pintos, W., 1999. Comparison of the areal amount of chlorophyll a of planktonic and attached microalgae in a shallow coastal lagoon. Hydrobiologia 408 (9), 285–291.

Conde, D., Rodríguez-Gallego, L., 2002. Problemática ambiental y gestión de laslagunas costeras atlánticas de Uruguay. In: Dominguez, A., Prieto, R. (Eds.), Perfil ambiental del Uruguay. Nordan-Comunidad, Montevideo.

Dean, R.G., Dalrymple, R.A., 2002. Coastal Processes with Engineering Applications. Cambridge University Press, Cambridge.

Duarte, P., Bernardo, J.M., Costa, A.M., Macedo, F., Calado, G., Cancela da Fonseca, L., 2002. Analysis of coastal lagoon metabolism as a basis for management. Aquat. Ecol. 36, 3–19.

Duck, R.W., da Silva, J.F., 2012. Coastal lagoons and their evolution: a hydromorphological perspective. Estuarine, Coastal Shelf Sci. 110, 2–14.

Espinoza-Tenorio, A., Wolff, M., Espejel, I., Montaño-Moctezuma, G., 2013. Using traditional ecological knowledge to improve holistic fisheries management: transdisciplinary modeling of a lagoon ecosystem of southern Mexico. Ecol. Soc. 18 (2), 6. http://dx.doi.org/10.5751/ES-05369-180206.

Esteves, F.A., Caliman, A., Santangelo, J.M., Guariento, R.D., Farjalla, V.F., Bozelli, R.L., 2008. Neotropical coastal lagoons: an appraisal of their biodiversity, functioning, threats and conservation management. Braz. J. Biol. 68 (4, Suppl.), 967–981.

Fabiano, G., Santana, O., 2006. Las pesquerias en laslagunassalobres de Uruguay. In: Menafra, R., et al. (Eds.), Bases para la Conservación y el Manejo de la Costa Uruguaya. Vida Silvestre Uruguay, Montevideo, pp. 557–565.

Gale, E., Pattiaratchi, C., Ranasinghe, R., 2007. Processes driving circulation, exchange and flushing within intermittently closing and opening lakes and lagoons. Mar. Freshwater Res. 58, 709–719.

Gönenç, E., Wolflin, J.P., 2004. Coastal Lagoons: Ecosystem Processes and Modeling for Sustainable Use and Development. CRC Press, New York.

Griffiths, S.P., 1999. Consequences of artificially opening coastal lagoons on their fish assemblages. Int. J. Salt Lake Res. 8, 307–327.

Griffiths, S.P., West, R.J., 1999. Preliminary assessment of shallow water fish in three small intermittently open estuaries in southeastern Australia. Fish. Manage. Ecol. 6 (4), 311–321.

Haines, P.E., Thom, B.G., 2007. Climate change impacts on entrance processes of intermittently open/closed coastal lagoons in new South Wales, Australia. J. Coastal Res. SI 50, 242–246.

Heinzen, W., Velozo, C., Carrión, R., Cardozo, L., Mandracho, H., Massa, E., 1986. Carta Hidrogeológica del Uruguay, escala 1/2.000.000 y texto explicativo. DINAMIGE, Montevideo. Uruguay.

Hopkins, T.S., Bailly, D., Elmgren, R., Glegg, G., Sandberg, A., Støttrup, J.G., 2012. A systems approach framework for the transition to sustainable development: potential value based on coastal experiments. Ecol. Soc. 17 (3), 39.

Horikawa, K., Maruo, H., 2011. Nonlinear Water Waves. Springer-Verlag, Berlin.

Isacch, J.P., 2008. Implementing the biosphere reserve concept: the case of Parque Atlántico Mar Chiquita biosphere reserve from Argentina. Biodiversity Conserv. 17, 1799–1804.

Kjerfve, B., 1994. Coastal Lagoons Processes. Elsevier Oceanography Series (60). Elsevier Science Publishers, Amsterdam.

Knoppers, B., 1994. Aquatic primary production in coastal lagoons. Elsevier oceanography series (60). In: Kjerfve, B. (Ed.), Coastal Lagoons Processes. Elsevier, Amsterdam, pp. 243–275.

Larson, K., Edsall, R., 2010. The impact of visual information on perceptions of water resource problems and management alternatives. J. Environ. Plann. Manage. 53 (3), 335–352.

Nagy, G.J., 2012. Reporte PACCC Escenarios climáticos y Diagnóstico ambiental para la adaptación en el sitio Piloto Laguna de Rocha y Adyacencias, Proyecto "Implementación de medidas piloto de adaptación al cambio climático en áreas costeras del Uruguay, URU/07/G32.

Norbis, W., Galli, O., 2004. Feeding habits of the flounder *Paralichthys orbignyanus* (Valenciennes, 1842) in a shallow coastal lagoon of the southern Atlantic Ocean: Rocha, Uruguay. Cienc. Mar. 30 (4), 619–626.

Palma-Silva, C., Albertoni, F., Esteves, F.A., 2000. *Eleocharismutata* (L.) Roem. Et Schult. subject to drawdowns in a tropical coastal lagoon, State of Rio de Janeiro, Brazil. Plant Ecol. 148 (2), 157–164.

Panario, D., Gutiérrez, O., 2005. La vegetación en la evolución de playas arenosas. El caso de la costa uruguaya. Ecosistemas 14 (2), 150–161.

Pauly, D., Yañez-Arancibia, A., 1994. Fisheries in coastal lagoons. Elsevier oceanography series (60). In: Kjerfve, B. (Ed.), Coastal Lagoons Processes. Elsevier, Amsterdam, pp. 377–396.

Piccini, C., Conde, D., Alonso, C., Sommaruga, R., Pernthaler, J., 2006. Blooms of single bacterial species in a coastal lagoon of the Southwestern Atlantic ocean. Appl. Environ. Microbiol. 72 (10), 6560–6568.

Piñeiro, G., Panario, D., 1993. Dinámicasedimentaria y geomorfológica de dunas y playas en Cabo Polonio, Rocha. Informe UNCIEP, Facultad de Ciencias, Uruguay.

Pintos, W., Conde, D., De Leon, R., Cardezo, M.J., Jorcin, A., Sommaruga, R., 1991. Some limnological characteristics of Laguna de Rocha. Revista de Biología Brasilera 51 (1), 79–84.

Pollard, D.A., 1994. Opening regimes and salinity characteristics of intermittently opening and permanently open coastal lagoons on the south coast of New South Wales. Wetlands 13, 16–35.

Pomeroy, R., Douvere, F., 2008. The engagement of stakeholders in the marine spatial planning process. Mar. Policy 32, 816–822.

Preciozzi, F., Spoturno, J., Heinzen, W., Rossi, P., 1985. MemoriaExplicativa de la CartaGeológica del Uruguay a la escala 1:500.000. DINAMIGE, Montevideo, Uruguay.

Rabassa, J., Brandani, A., Boninsegna, J.A., Cobos, D.R., 1984. Cronología de la "PequeñaEdad del Hielo" en los glaciares Río Manso y Castafio Overo, Cerro Tronador, Provincia de Rio Negro. 9no. Congreso Geológico Argentino, Actas, vol. 3, pp. 624–639.

Reese, M.M., Stunz, G.W., Bushon, A.M., 2008. Recruitment of estuarine-dependent nekton through a new tidal inlet: the opening of Packery channel in Corpus Christi, TX, USA. Estuaries Coasts 31, 1143–1157.

Rodríguez-Gallego, L., 2010. Eutrofización de laslagunascosteras de Uruguay: impacto y optimización de los usos del suelo, Tesis de doctorado, PEDECIBA, Facultad de Ciencias, Universidad de la Republica, Montevideo.

Rodríguez-Gallego, L., Meerhoff, E., Clemente, J.M., Conde, D., 2010. Can ephemeral proliferations of submerged macrophytes influence zoobenthos and water quality in coastal lagoons? Hydrobiologia 646 (1), 253–269.

Rodríguez-Gallego, L., Sabaj, V., Masciadri, S., Kruk, C., Arocena, R., Conde, D., 2015. Salinity as a major driver for submersed aquatic vegetation in coastal lagoons: a multi-year analysis in the subtropical Laguna de Rocha. Estuaries Coasts 38, 451–465.

Rodríguez-Gallego, L., Santos, C., Amado, S., Gorfinkiel, D., González, M.N., Neme, C., Tommasino, H., Conde, D., 2013. Interdisciplinary diagnosis and scenario analysis for the implementation of a coastal protected area (Laguna de Rocha, Uruguay). In: Yáñez-Arancibia, A., et al. (Eds.), Ecological Dimension for Sustainable Socio Economic Development. WIT, Southampton, pp. 389–411.

dos Santos, A.M., Amado, A.M., Minello, M., Farjalla, V.F., Esteves, F.A., 2006. Effects of the sandbar breaching on *Typha domingensis* (PERS.) in a tropical coastal lagoon. Hydrobiologia 556, 61–68.

Saad, A.M., Beaumord, A.C., Caramaschi, E.P., 2002. Effects of artificial canal openings on fish community structure of Imboassica coastal lagoon, Rio de Janeiro, Brazil. J. Coastal Res. 36, 634–639.

Santangelo, J.M., Rocha, A., Bozelli, R.L., Carneiro, L.S., Esteves, F., 2007. Zooplankton responses to sandbar opening in a tropical eutrophic coastal lagoon. Estuarine, Coastal Shelf Sci. 71, 657–668.

Sarroca, M., 2008. Relevancia de la Laguna de Rocha (Uruguay) como hábitat para *Cygnus melancoryphus* y *Coscoroba coscoroba*: análisis espacio-temporal de la abundancia y estudio de comportamiento. Tesis de Maestría, PEDECIBA-Biología, UDELAR., Montevideo.

Schallenberg, M., Larned, S.T., Hayward, S., Arbuckle, C., 2010. Contrasting effects of managed opening regimes on water quality in two intermittently closed and open coastal lakes. Estuarine, Coastal Shelf Sci. 86, 587–597.

Sei, S., Rossetti, G., Villa, F., Ferrari, I., 1996. Zooplankton variability related to environmental changes in a eutrophic coastal lagoon in the Po Delta. Hydrobiologia 329 (1–3), 45–55.

Shackeroff, J., Campbell, M.L.M., Crowder, L.B., 2011. Social-ecological guilds: putting people into marine historical ecology. Ecol. Soc. 16 (1), 52. www.ecologyandsociety.org/vol16/iss1/art52.

Smakhtin, V., 2004. Simulating the hydrology and mouth conditions of small temporarily closed/open estuaries. Wetlands 24, 123–132.

Sommaruga, R., Conde, D., 1990. Distribución de materia orgánica en los sedimentos recientes de la Laguna de Rocha (Rocha Uruguay). Atlántica 12, 35–44.

Soutullo, A., Bartesaghi, L., Berazategui, P., Clavijo, C., Díaz, I., Faccio, C., García, M., González, E.M., 2010. Diseño espacial del Sistema Nacional de Áreas Protegidas de Uruguay: sitios a integrar al sistema, prioridades de conservación en esos sitios y aportes a la protección de la biodiversidad fuera de áreas protegidas. In: Proyecto Fortalecimiento del Proceso de Implementación del Sistema Nacional de ÁreasProtegidas, Serie de Informes N° 24, 50 pp.

Suzuki, M.S., Figueiredo, R.O., Castro, S.C., Silva, C.F., Pereira, E.A., Silva, J.A., Aragon, G.T., 2002. Sandbar opening in a coastal lagoon (Iquipari) in the northern region of Rio de Janeiro state: hydrological and hydrochemical changes. Braz. J. Biol. 62 (1), 51–62.

Suzuki, M.S., Ovalle, A.R.C., Pereira, E.A., 1998. Effects of sandbar openings on some limnological variables in a hypertrophic tropical coastal lagoon of Brazil. Hydrobiologia 368, 111–122.

Thomas, C.M., Perissinoto, R., Kibirige, I., 2005. Phytoplankton biomass and size structure in two South African eutrophic, temporarily open/closed estuaries. Estuarine, Coastal Shelf Sci. 65 (1–2), 223–238.

Thompson, D., 2008. Economía e identidad de los pescadores de la Barra de la Laguna de Rocha. In: Anuario de Antropología Social y Cultural. Nordan Comunidad, Facultad de Humanidades y Ciencias de la Educación, Montevideo.

Thorton, E.B., Guza, R.T., 1983. Transformation of wave height distribution. J. Geophys. Res. 88 (C10), 5925–5938.

U.S. Army Corps of Engineers, 2002. Coastal engineering manual. Engineer Manual 1110-2-1100. U.S. Army Corps of Engineers, Washington, DC.

UNESCO, 2006. Evaluación de los recursoshídricos. Elaboración del balance hídrico integrado por cuencas hidrográficas. Documentos técnicos del PHI-LAC N° 4. UNESCO.

Vizziano, D., Forni, F., Saona, G., Norbis, W., 2002. Reproduction of *Micropogonias furnieri* in a shallow temperate coastal lagoon in the southern Atlantic. J. Fish Biol. 61, 196–206.

Whitfield, K., Adams, J.B., Bate, G.C., Bezuidenhout, K., Bornman, T.G., Cowley, P.D., Froneman, P.W., Gama, P.T., James, N.C., Mackenzie, B., Riddin, T., Snow, G.C., Strydom, N.A., Taljaard, S., Terörde, A., Theron, A.K., Turpie, J.K., van Niekerk, L., Vorwerk, P.D., Wooldridge, T.H., 2008. A multidisciplinary study of a small, temporarily open/closed South African estuary, with particular emphasis on the influence of mouth state on the ecology of the system. Afr. J. Mar. Sci. 30, 453–473.

Yáñez-Arancibia, A., Day, J.W., Sánchez-Gil, P., Day, J.N., Lane, R.R., Zárate-Lomelíc, D., Alafita Vásquez, H., Rojas-Galavizc, J.L., Ramírez-Gordillo, J., 2014. Ecosystem functioning: the basis for restoration and management of a tropical coastal lagoon, Pacific coast of Mexico. Ecol. Eng. 65, 88–100.

Young, G.C., Potter, I.C., 2002. Influence of exceptionally high salinities, marked variations in freshwater discharge and opening of estuary mouth on the characteristics of ichthyofauna of a normally-closed estuary. Estuarine, Coastal Shelf Sci. 55, 223–246.

Zagonari, F., 2008. Integrated coastal management: top-down vs. community-based approaches. J. Environ. Manage. 88, 796–804.

Making the Link

By Liette Vasseur

Solutions for Sustainable Coastal Lagoon Management: From Conflict to the Implementation of a Consensual Decision Tree for Artificial Opening	⟺	**Challenges to Evaluating Coastal Management in the Twenty-First Century: Lessons from the Lofoten Archipelago**
D. Conde et al.		*Scott Bremer, Anne Blanchard and Matthias Kaiser*

Conde et al. have discussed in their chapter the complexity of the relationships between humans and coastal lagoons. Lagoons are biodiverse and fragile ecosystems threatened by human activities, one of them being artificial opening (or breaching). Through a case study of the Laguna de Rocha, they have demonstrated the need for research on three main aspects: the current and future changes of the lagoon and its ecosystem, the reasons and the needs for artificial opening for both social and ecological components of the system, and the decision-making process used for opening. They underline again the importance of connecting people to research in order to find sustainable solutions.

The chapter by Conde et al. represents a great example of the need for integrated coastal management and the importance of its evaluation, as explained in the next chapter by Bremer et al. Interestingly, both chapters also stress the need for evaluation and assessment to make sure that actions are sustainable. Understanding the outcomes of management actions can be especially useful if both institutional structures and decision-making processes are measured. This reemphasizes Conde et al.'s statement of the need to connect decision-making to causes and impacts. Bremer et al. underline what is a common theme in this book: the importance of stakeholder involvement to co-construct sustainable solutions for coastal zones. As Conde et al. mention in their previous chapter, consensus at the local level can help resolve conflicts and find more sustainable and long-term ways to manage ecosystems.

In the next chapter, Bremer et al. go further by proposing to examine integrated coastal management and its evaluation from a local to a global level since issues are multiscale in nature, as well as bridging ex ante and ex post evaluation for more effective management. Through a case study in Lofoten, Norway, they show the need for stakeholder involvement and dialog at all levels. There are challenges to link the local to the national and international, where objectives of management may greatly differ. In this case, the issues are the protection of the fisheries at the local scale versus the interest for oil exploration at the national level. This brings a contentious and politically hot topic: how to balance between immediate profit and long-term sustainability. In the end, it's a question of coastal governance, where all of these aspects are allowed to be discussed with all stakeholders present.

Chapter 14

Challenges to Evaluating Coastal Management in the Twenty-First Century: Lessons from the Lofoten Archipelago

Scott Bremer, Anne Blanchard, Matthias Kaiser
Center for the Study of the Sciences and the Humanities, University of Bergen, Bergen, Norway

Chapter Outline

Coastal Zones

INTRODUCTION

For more than 40 years the field of Integrated Coastal Management (ICM) has evolved according to a reflexive relationship between theory and practice. This has seen increasing recognition, among ICM practitioners and scholars alike, of the importance of evaluation for more accountable, transparent, and adaptive coastal management initiatives; witnessed by the emergence of a diverse and sophisticated literature on ICM evaluation frameworks. However, this chapter asserts that these evaluation frameworks remain rooted in the resource management paradigm from which ICM emerged in the 1960s, and are bounded by the limitations of that paradigm. We argue that ICM evaluation in the twenty-first century faces three fundamental challenges that demand a revision in the way ICM initiatives are evaluated, including an appreciation for: (1) a plurality of incommensurable measures of high-quality ICM; (2) evaluation of ICM across local, national, and international scales; and (3) the relationship between *ex ante* and *ex post* evaluation. This chapter expounds on these three challenges, and illustrates them by way of a case study of coastal management in Lofoten, Norway. The chapter finishes by suggesting that finding solutions to these challenges may imply a genuine paradigm-shift toward innovations in coastal governance.

The chapter begins in The Section Reviewing the Evaluation of Coastal Management with a review of the ICM evaluation literature, which it argues to be rooted in a resource management paradigm. The Section Challenges to Evaluating Coastal Management in the Twenty-First Century introduces three fundamental challenges to this tradition of ICM evaluation, before The Section Lofoten's Oil: The Challenges of Evaluating Coastal Management in the Context of Uncertainty and Contentiousness explores the nature of these challenges in practice, via the Lofoten case study. The Section Governance Perspectives Providing New Norms of Evaluation for the Twenty-First Century presents a discussion of possible solutions to these challenges as espoused by a governance paradigm.

REVIEWING THE EVALUATION OF COASTAL MANAGEMENT

The Current Terrain of ICM Evaluation

Evaluation has always been a central concern of ICM. As noted by a number of authors (see, e.g., Chua (1993) and the International Oceanographic Commission (2006)), monitoring and evaluation are integral components of an ICM initiative, for at least three key reasons:

1. Accountability: To justify the political support and expenditure on coastal management.
2. Establish causality: To link ICM initiatives with improved collective decision-making and coastal outcomes, and in parallel, build on the scholarship of ICM.
3. Adaptive management: So that management can continually adapt to the changing issues.

The evaluation of ICM initiatives has, however, typically been a neglected aspect of ICM (Tobey and Volk, 2002), not least because of the political nature of evaluation. Cicin-Sain and Knecht (1998) noted the reluctance of those initiating ICM initiatives to include evaluation mechanisms for fear that they may be politically destabilizing for an initiative in its early stages; that is, an evaluation of poor performance may see an ICM initiative prematurely concluded. This is especially true when evaluating the environmental outcomes on the ground, as these are very often shaped by forces outside of the control of an ICM initiative, and indeed if measured prematurely, may show poor progress (Tobey and Volk, 2002). For these reasons, coastal managers have in the past been pressured into using vague or immeasurable ICM objectives (Olsen et al., 1997). Since then, a number of authors have written on ICM evaluation methods, providing multiple lists of indicators to guide coastal managers toward formulating their own frameworks (Bille, 2007; Burbridge, 1997; Ehler, 2003; Henocque, 2003; International Oceanographic Commission, 2006; Olsen, 2003a; Pickaver et al., 2004).

There has been significant debate on exactly which aspect of an ICM initiative should be evaluated: its means or its ends (Stojanovic et al., 2004). In terms of means, much ICM literature has focused on the evaluation of institutions and their necessary human, financial and technical resources; what could be described in terms of an initiative's institutional quality. There are numerous frameworks evaluating "ideal-type" institutions, which experience has shown are successful in giving effect to ICM principles and progressing coastal communities to their goals (see, e.g., the IOC (2006); Olsen (2003a); and Pickaver et al. (2004)). However, evaluating institutional quality alone offers a limited perspective on an ICM initiative: an initiative may be judged highly according to its own measures, without affecting desired change within a coastal context (Bille, 2007). As such, some ICM authors have promoted the joint evaluation of *ends*, as those ecological–social–economic outcomes (here termed *coastal outcomes*) that a coastal community collectively deemed desirable (see the Logical Framework of Ehler (2003)). Generally, these coastal outcomes are contingent on the specific coastal context, however, some authors offer generic lists of indicators of coastal outcomes, such as for water quality or ecological resilience (Ehler, 2003; IOC, 2006). Linking institutional quality to coastal outcomes is a challenge.

The core of the problem is causality. While it is relatively easy to evaluate institutional quality, it is far more difficult to measure how institutional mechanisms influence coastal outcomes (Stojanovic et al., 2004). The linking of coastal outcomes to an ICM initiative can be nearly impossible because (1) causality is difficult to establish within complex and often poorly understood coastal zones, (2) coastal outcomes emerge over a long time frame, (3) there is often a lack of agreed indicators, with the relevance of indicators changing, (4) there is often a lack of a scientific baseline, and (5) policy is often divorced from science anyway (Bille, 2007; McFadden, 2007; Turner, 2000).

A number of ICM authors have developed evaluation models to demonstrate the causality between ICM institutions and desired coastal outcomes, with Olsen's Four Orders of Outcome framework depicting a sequence of outcomes from the instigation of an ICM initiative (first order), to behavioral change in a coastal context (second order), to the attainment of measureable coastal outcomes (third order), to the realization of sustainable development as the fourth order (Olsen, 2002, 2003a; Olsen et al., 1997, 2009). Alternative models are presented by the Driving Forces-Pressure-State-Impact-Response (DPSIR) model, mapping society's response to changing states in the environment (International Oceanographic Commission, 2006); and the Logical Framework utilized by Ehler (2003), attempting to evaluate causality by assessing (1) the inputs into an ICM initiative, (2) the initiative process, (3) outputs from the initiative, and (4) the coastal outcomes.

One criticism of these models of causality relates to their rigid, linear, and instrumental structuring of means and ends. The interface between ICM institutions and coastal outcomes is blurry at best, characterized by mutual influence in an adaptive and evolving process. Moreover, a unidirectional progression from means to ends does not allow for regress, which Glavovic (2006) has shown can occur within all politically charged governing systems. A second criticism of these causal models (excluding Olsen's Four Orders model) is the leap of faith whereby policy outputs are deemed to produce coastal outcomes without any appreciation for an intermediate effect, and how these institutions influence coastal stakeholders' actions that will affect coastal outcomes. A third criticism, particularly relevant to the DPSIR model, concerns the abstract nature of the framework. For instance, the concept of driving pressure or state can be difficult to define in practice.

Evaluating ICM Institutional Quality

Having made a distinction between evaluating coastal management institutions and coastal outcomes, the majority of ICM evaluation scholarship has focused on institutional quality. Here we discuss institutions in the broad sense, as the social, legal, and technical structures that both enable and constrain behavior of, and interaction between, stakeholders. In this way we can equally think of the formal institution embodied by a local government coastal planning tribunal, alongside the less formal institution of a fishing club, or even informal institutions governing acceptable behavior on the beach; all are framed by different rules, norms, and traditions. Coastal and fisheries governance scholars (Kooiman and Bavinck, 2005) assert institutional quality is a function of institutions' ability to give effect to certain accepted principles of quality, in facilitating collective decision making and action toward the community goals for the coast. ICM scholars assert that one key role of ICM initiatives has been to harmonize this complex network of disparate institutions according to accepted ICM principles, such as inclusiveness, adaptability, or precaution (Stojanovic et al., 2004).

The ICM literature is replete with evaluation frameworks measuring the progress of institutions toward satisfying the principles of ICM, based in the reflexive learning of 40 years of ICM practice. Echoing the "logical framework," most frameworks include indicators of inputs, processes, outputs, and outcomes; thereby including an appreciation for both institutional quality and the social, political, legal, financial, and technical capacities necessary to support institutions (International Oceanographic Commission, 2006). Importantly, institutional indicators are simplifying parameters used to qualify changes in the complex governing system, and communicate them. They are not prescriptive steps so much as they are an indication that the institutions of a governing system may espouse the principles of ICM (International Oceanographic Commission, 2006).

Yao (2008) and Ehler (2003) point out that evaluating institutions encompasses both performance measures and outcome measures. Performance measures evaluate the quality of the decision making *process* (e.g., whether all relevant stakeholders are involved in a decision making process); while outcome measures evaluate institutional *structural* outcomes of an ICM initiative (such as the creation of a law, or an environment court). This calls attention to the two symbiotic components of ICM: process and structural outputs. The two components are like two sides of the same institutional coin, distinguishable yet totally interdependent. Moreover, the mutually influencing nature of processes and structures is well established in the ICM literature (Christie, 2005; Olsen, 2003a; Tobey and Volk, 2002); for example, Olsen (2003a) states that successive decision making process iterations result in a growth in institutional capacity and quality.

There are a number of evaluation frameworks that are structured primarily around ICM process or performance measures. Olsen suggests a policy cycle framework that maps the ICM process as an iterative cycle split into five steps, based on the enduring GESAMP (1996) model, with priority actions at each stage of the process (Olsen, 2003a). On the other hand, there are a number of frameworks evaluating structures or outcome measures. Pickaver et al. (2004) list institutional outcomes as a scorecard to be filled out at the end of each process cycle, with outcomes grouped according to growing phases. The Intergovernmental Oceanographic Commission produced a handbook for evaluating ICM initiatives (International Oceanographic Commission, 2006), which included both lists of institutional outcome indicators and environmental outcome indicators. Since their release, the IOC indicators have been utilized in a number of efforts to evaluate ICM (Glavovic, 2008b; Hoffmann and Löser, 2008; Yao, 2008).

Both the International Oceanographic Commission (2006) and the LOICZ report (Olsen et al., 2009) recommend that it is better to use a combination of methods for evaluating ICM institutions rather than one framework in isolation. For instance, if ICM institutional quality is defined by institutional structures and decision making processes, then this necessitates measures of both; to omit one is to evaluate half of the institutional "coin." Moreover, though there are multiple generic evaluation frameworks, it is roundly accepted that these have limited utility compared to indicators developed by stakeholders for objectives in a specific context. The above generic frameworks are just that; frameworks to act as a starting point from which to structure more case specific indicators (Fontalvo-Herazo et al., 2007; Hoffmann and Löser, 2008; International Oceanographic Commission, 2006).

ICM Evaluation within a Resource Management Paradigm

Having traversed the ICM evaluation literature, it can be argued that the evaluation of ICM initiatives remains deeply rooted in the resource management paradigms from which ICM emerged in the 1960s. By way of illustration, many frameworks begin from an idea of some communally agreed "ends" as the focus of an ICM initiative, with an implicit assumption that it is possible to arrive at an aggregated vision of the public good. Another common theme is that most frameworks map ICM as a linear, instrumentally rational process defined in terms of means and ends. Indeed, even where the decision making process is depicted as cyclic and iterative, as for Olsen et al. (1997), for example, the decision making cycles and institutions are depicted in terms of ever-increasing sophistication over time. This embodies linear notions of progress, and ignores the possibility of ICM initiatives stagnating or regressing in sophistication. A third common theme is the idea of transforming certain human, financial, or technical inputs into policy outputs, which has been a guiding assumption of the management paradigm. This is explicit in Olsen's Four Orders framework, where funding and political support is determined to create the conditions whereby behavior is changed and coastal outcomes achieved. A fourth theme is an emphasis on centralized decision making, typically facilitated by coastal managers. All frameworks reviewed include new state-centered institutions and decision making processes, and stakeholder participation is generally described in terms of consultation.

By mapping the evolution of evaluation frameworks for coastal management we can recognize their increasing diversity and sophistication within the broader resource management paradigm. However, while the strengths of these frameworks must be acknowledged, some coastal management practitioners and scholars are coming to recognize that evaluation approaches in the twenty-first century are facing fundamental challenges (Bille, 2007; Bremer, 2011; Olsen et al., 2009; Stojanovic and Barker, 2008). Beyond the widely recognized challenges of "causality" and evaluating both procedural and structural sides of the institutional coin, this paper introduces three further challenges, discussed here in terms of *bridging*: (1) a plurality of incommensurable measures of quality; (2) local, national, and international coastal governance; and (3) ex *post* evaluation of past coastal management efforts and *ex ante* evaluations of possible futures for the coast. Arguably, these challenges demand a genuine paradigm shift toward fundamentally different forms of evaluation.

CHALLENGES TO EVALUATING COASTAL MANAGEMENT IN THE TWENTY-FIRST CENTURY

Bridging a Plurality of Incommensurable Measures of Quality

It is already widely acknowledged in the ICM literature that any evaluation effort ought to be contingent to its particular coastal context. This chapter argues that a central challenge to constructing any contingent evaluation framework is to recognize and bring together the plurality of measures of high-quality coastal management espoused by different groups of stakeholders in a coastal context. That is, for coastal management evaluation to be contingent to a context, it must comprise indicators that are meaningful for stakeholders in that context, implying the collaborative construction of evaluation frameworks engaging all groups of stakeholders. However, with the exception of some commentators (Hoffmann and Löser, 2008; International Oceanographic Commission, 2006), there remains a strong bias in the ICM literature toward centralized and science-based evaluation. This sees the top-down construction of evaluation frameworks by coastal managers drawing on scientifically quantifiable measures to the exclusion of other stakeholder groups and their knowledge systems.

There are at least three broad arguments against a closed reliance on centralized and science-based evaluation alone (see e.g., Fiorino (1990)). First, it is very rare for there to be a complete scientific[1] understanding of a coastal context and all of its issues; most important coastal issues are characterized by *significant uncertainties*, including about what is meaningful to measure as indicators of institutional quality or desirable coastal outcomes. Such scientific uncertainty legitimates drawing on other knowledge systems as well, for better-informed ICM evaluation frameworks. Second, by appealing to scientific objectivity, a centralized and science-based evaluation approach may ignore the *highly political nature of evaluation.* The construction of an evaluation framework requires that choices be made about which indicators are the most important to measure, and by extension, may valorize one form of knowledge (such as scientific knowledge) over other knowledge systems. By recognizing the political nature of this social choice, it can be argued that these choices should be made in a more inclusive forum. Third, an ICM evaluation framework that does not reflect perspectives from the local community may be rejected by that community as illegitimate.

Acknowledging these challenges to centralized, science-based evaluation frameworks, we can argue for more collaborative approaches that bring together different coastal stakeholders espousing different knowledge systems, to negotiate which indicators are important to measure. Such approaches recognize the political nature of evaluation, and attempt to arrive at frameworks of indicators deemed salient, credible, and legitimate to all. However, this

1. Here "science" is an umbrella term used to denote the diversity of applied scientific disciplines drawn on for coastal management, from across the natural and social sciences, and humanities.

inclusive approach is itself not without challenges, primarily relating to how different knowledge perspectives can be integrated, or brought together into a common evaluation framework. O'Connor (1999), for instance, discusses a trade-off between a "Cartesian" and a "dialogic" approach to this integration. The Cartesian approach sees one dominant knowledge system subsume other knowledge perspectives, such that all different indicators are reconciled within a single internally consistent framework—usually scientific. We can consider a measure put forward by the local community, such as local identity, being "scientized" by the scientific community to be scientifically quantifiable. The dialogic approach seeks to bridge incommensurable perspectives through dialog, in a way that allows for their coexistence.

Bridging Local, National, and International Coastal Management

Coastal issues change in character when we look at them at different scales, with implications for how we evaluate the success of their management. This presents a challenge for evaluating coastal management to have regard for the cross-scale nature of many coastal issues, associated with global environmental change and trends to globalization. Up until now, ICM has more often promoted and evaluated vertical integration according to a hierarchal top-down approach, with local ICM initiatives evaluated for how well they gave effect to a national ICM strategy, for example. However, this hierarchal approach arguably fails to account for the multiscale nature of coastal issues. A nested approach to coastal governance demands that we evaluate ICM initiatives not as isolated efforts but as complimentary responses to global issues in a globalized world. This implies some regard both for the coastal outcomes of an initiative beyond its formal boundaries (territorial or administrative boundaries), and for how harmonized an initiative's institutions and coastal outcomes are with other initiatives at other scales. While highly demanding, the alternative may be local, national, and international initiatives that work at odds with each other.

Climate change presents an important example of global environmental change that is threatening coastal communities globally. We understand and manage the impacts of climate change differently at different scales, and this has implications for evaluating coastal management. At the international scale, climate change is largely the focus of scientific enquiry led by the Intergovernmental Panel on Climate Change (IPCC), which dictates the international response to climate change and its effects on the coast, ranging from the Kyoto Agreement to the recently signed "Future We Want" document. At the local scale, the impacts of climate change are part of the immediate lived experience of a community; shaping the lives of different actors in different ways, and motivating different adaptation strategies. We typically see different groups of coastal stakeholders employ their own knowledge toward their own unique portfolio of management approaches, from beach renourishment to creating marine protected areas. Thus we can simplistically illustrate a spectrum from broad science-based international responses to climate change, to specific local-knowledge-led local responses. It is clear that these different mitigative and adaptive management efforts at the international and local scales are complimentary, but demand very different types of evaluation frameworks. It is also clear that it is unrealistic for a local coastal management initiative to evaluate its success based on measures of global change. However, the challenge remains as an open question: how can we include a regard for the cross-scale nature of climate change? Is it possible to construct evaluation frameworks that equally consider management at all scales, or ought we look to a "nested" approach of international, national, and local evaluation?

Bridging Ex Ante and Ex Post Evaluation

The ICM literature rarely devotes more than a brief discussion to the relationship between (1) *ex ante* evaluation, meaning the speculative evaluation of the desirability of possible futures for guiding decision making and (2) *ex post* evaluation, meaning the reflexive evaluation of how successful an initiative has been in steering a coastal context toward the most desirable future. Bridging forward-looking *ex ante* evaluation and backwards-looking *ex post* evaluation is the third significant challenge facing ICM. Up until now, these two forms of evaluation have been depicted as different stages on a cyclic and iterative model of the policy- or decision making process, borrowed from the planning and policy analysis literature. See, for example, the five-step policy cycle established by GESAMP (1996) and developed by Olsen (2002). As part of a cyclic process, this implies a mutually reinforcing relationship between the two forms of evaluation, but there is a need for a closer look at this relationship. For example, how can the framing of possible futures recommend certain indicators as milestones toward the future deemed most desirable? And how can *ex post* evaluation inform a more sophisticated approach to predicting different futures?

Ex ante evaluation is based on the predictability of different futures, and can be compromised depending on how this predictability is framed—either as largely calculable (as for "risk") or largely incalculable (as for "uncertainty"). Employing *ex ante* evaluation is largely contingent on the ability to predict the totality of possible futures, but where there are significant uncertainties, this opens up an infinite number of possible futures, and *ex ante* evaluation is reduced to a largely arbitrary exercise that deals with a miniscule subset of these futures. This has seen ICM scholars advocate for more incremental and adaptive coastal management, that takes small reversible steps into the unknown, and evaluates their success according to an *ex post* framework. That is, where uncertainty is high, *ex post* evaluation is favored over *ex ante* evaluation. However, paradoxically, it is arguably very difficult to develop *ex post* measures of success without a forward-looking appreciation of those ways in which different futures may be preferable to the status quo. In this way, ex post evaluation is dependent on a clear idea of those coastal elements that are important to measure as an indication of coastal management success, and are therefore equally susceptible to significant uncertainty.

LOFOTEN'S OIL: THE CHALLENGES OF EVALUATING COASTAL MANAGEMENT IN THE CONTEXT OF UNCERTAINTY AND CONTENTIOUSNESS

The three fundamental challenges to evaluating ICM discussed above can be illustrated by the significant challenges facing coastal management in the Lofoten Archipelago, in Northern Norway. Communities in the Lofoten, and Norway more broadly, are facing a difficult social choice about how to balance competing coastal activities in the Lofoten with the recent discovery of potentially significant sub-sea petroleum reservoirs in the area. This paradigmatic case study does not follow the evaluation of a particular ICM initiative in Lofoten. Rather it discusses the challenges of collective decision making for Lofoten's coastal trade-off, characterized by significant uncertainties and high political stakes, and the implications for evaluating the quality of collective decision making institutions and the coastal outcomes of this process. For simplicity, this section discusses the ongoing decision making process around Lofoten's coastal trade-off, and consequent management efforts, under the encompassing label of "coastal management."

Lofoten is an archipelago in northern Norway of 1200 km^2 with a population of 24,500 inhabitants (Figure 14.1). The "Norwegian Integrated Management Plan for the Barents Sea—Lofoten" (NME, 2006) describes this area as very valuable and vulnerable, owing to its ecological diversity and cultural uniqueness. Indeed, the area is a key spawning ground and nursery for Northeast Arctic cod, Northeast Arctic haddock, and herring (Olsen et al., 2010). These fish stocks have, since the Viking Age, supported a successful traditional fishing industry (Jentoft and Kristoffersen, 1989), and today Norway is the third largest exporter of fish worldwide (Hjermann et al., 2007). Beyond fish, Lofoten also hosts a very high density of migratory seabirds (Barrett et al., 2006), more than 20 species of marine mammals (Larsen et al., 2001), and the world's largest deep-sea coral reef, of about 100 km^2 (Forsgren et al., 2009). In addition, Lofoten is known for its unspoiled landscapes that are nominated for being on the UNESCO list for the Protection of the World Natural and Cultural Heritage. This natural diversity supports a lucrative tourism sector (NME, 2006).

Against this background, exploratory drillings in Lofoten have indicated the presence of important petroleum reservoirs potentially amounting to 1300 billion barrels of oil equivalent. This brings the prospect of socioeconomic development, with new infrastructure and employment opportunities locally. However, due to its ecological value, Lofoten has not been opened yet to petroleum production. This context, characterized by high stakes and a plurality of interests, is also colored by uncertainties when it comes to potential developments of petroleum activity in Lofoten. These uncertainties concern both the "everyday" petroleum activities, and the hypothetical "worst-case scenario" oil spill (for a discussion of these uncertainties, see Blanchard et al. (2014) and Hauge et al. (2014)).

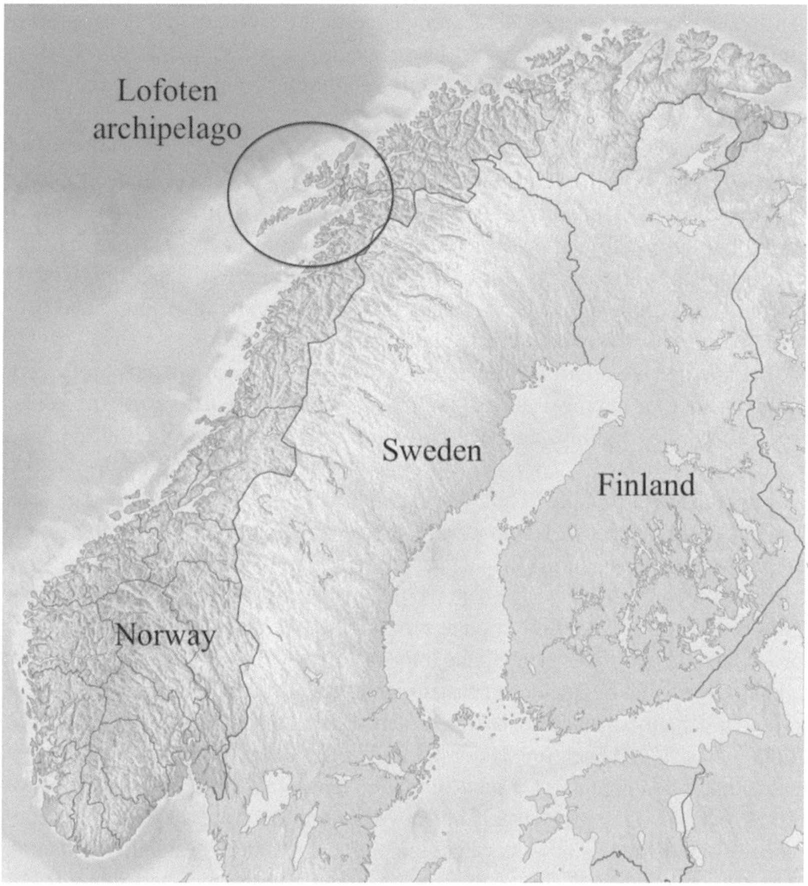

FIGURE 14.1 Lofoten in northern Norway. *From Wikimedia Commons.*

In such a complex, uncertain, and contentious trade-off, what constitutes high-quality coastal management? And how can we evaluate its success? As argued in The Section Challenges to Evaluating Coastal Management in the Twenty-First Century, bridging (1) a plurality of different measures; (2) the local, national, and global scales; and (3) *ex ante* and *ex post* evaluation, are very relevant challenges, and the experience in Lofoten offers insights.

Experts, Decision Makers, Locals: Who Sets the Evaluation Framework?

Around the Lofoten coastal trade-off, there is an expectation that good political decisions should be taken on the basis of sound science. However, even expert assessments are carrying value-biases and choices in the way research questions are posed and framed (Blanchard et al., 2014; Hauge et al., 2014). In addition, even if they constitute a pertinent knowledge base, expert assessments present a limited view of the issue. They focus on specific aspects, such as the environmental impacts of petroleum activity, or the economic gains and losses resulting from this activity.

However, there are other important aspects, such as identity, traditions, or nature, which cannot be reduced to simple monetary measures. Indeed, recent study in Lofoten (Blanchard et al., 2013) showed first that local citizens emphasize global sustainability, with a focus on renewable energies (such as tidal power plants or wave energy, for example). Second, there was a strong wish to preserve the cultural and historical identity of Lofoten, and thus nature and fisheries; not only for the benefit of the region or Norway, but also because Lofoten represents a heritage that belongs to the world. Third, local citizens argued for the importance of having a gender-balanced society, with the petroleum sector argued to employ mainly men, which will further promote a male-dominated community.

In appraising the quality of both the *institutions* and *coastal outcomes* of coastal management in Lofoten, it is important that an evaluation framework comprise the different values and cultural perspectives of the community; for instance, to what extent does local coastal management include a regard for the community's contribution to global sustainability through renewables (see also The Section Lofoten, Norway, and the World: Evaluating the Impacts of Coastal Management at Different Scales)? Or to what degree does coastal management preserve cultural identity or promote a gender-balanced society? Lofoten's stakeholders have been calling for a more inclusive approach to this coastal trade-off, through a coastal management process that is contingent to the particular cultural, social, and economic context of Lofoten; not according to ostensibly objective and universal science-based decision-making. They demand a process where the political nature of evaluation is central, arguing this may promote coastal management that is better informed, more democratic, and with an increased legitimacy (Fiorino, 1990).

However, bridging a plurality of different measures presents numerous challenges in practice. This bridging implies coastal management institutions that support and facilitate participation and inclusiveness, recognizing that participation should not be seen as a panacea, and should be used with caution. With reference to the fieldwork of Blanchard et al., including non-experts was seen at times to result in a simplified discussion of the technical aspects of petroleum developments in Lofoten, and a simplified understanding of the complex and uncertain mechanisms linking these causes and effects. This led to propositions grounded only on suppositions and hypotheses. This experience implies that any effort to include stakeholders in formulating evaluation frameworks for Lofoten's coastal management ought to include scientific and technical experts as well, and be accompanied by accessible information on the technical aspects of the trade-off (even though the debate should not only be limited to these). Moreover, it is important to realize that expert knowledge and local knowledge are complementary. Both knowledge sources focus on different aspects that should all be taken into account to support coastal management decisions and to evaluate the process, institutions, and coastal outcomes.

Lofoten, Norway, and the World: Evaluating the Impacts of Coastal Management at Different Scales

The Lofoten coastal trade-off is seen as having consequences at the local, national, and global scale. At the local scale, local citizens argue that allowing petroleum activity would impact the core identity of Lofoten, putting at risk a thousand-year-old tradition in fisheries, unspoiled landscapes, and rich and unique wildlife (Blanchard et al., 2013). The prospect of increased employment opportunities locally is offset by assertions that petroleum activity is a short-term project, whereas today's tourism sector and the fisheries offer long-term and sustainable livelihoods.

At the national scale, the national government puts forward the economic benefits that could benefit Norway. Petroleum resources nationally are today worth more than fisheries and tourism together, and have brought 2800 billion NOK to the Norwegian state since 1970. The Norwegian welfare state is, to a large extent, built on these resources. However, the unique nature of Lofoten is also argued as selling Norway (as a whole) to the world, with the attractive landscapes of Lofoten attracting many tourists. In addition, Northeast Arctic cod is an economically important fish stock for Norway, which has exported 340,000 metric tons in 2011, with a value of about 4 billion NOK (NDF, 2011).

At the global scale, extracting petroleum in Lofoten can be argued to respond to a growing energy demand worldwide. However, it can be equally argued that the oil and gas produced in Norway will cumulatively contribute to producing greenhouse gas emissions that promote anthropocentric climate change. Moreover, arguments of global sustainability are often put forward by local citizens, who argue that it is their moral responsibility to protect their unique nature, as a heritage that belongs to the world (Blanchard et al., 2013). They further argue that Lofoten should be seen in its global context, noting the interconnectedness of the world's oceans such that in the case of an oil spill, the consequences are unlikely to be localized, and Norway would be responsible.

The cross-scale nature of Lofoten's coastal trade-offs illustrates the complexities for evaluating the institutional quality of coastal management, and its coastal outcomes. By regarding Lofoten's coastal trade-off at different scales, this recommends very different considerations for evaluating coastal management, and conflicting measures of success. While measures related to traditions, identity, and wildlife may be regarded as important measures of successful coastal management at the local scale, at the national scale there is a stronger emphasis on economic growth. Moreover, there is often conflict over what constitutes good performance relative to these different measures, with Lofoten stakeholders emphasizing the trade-off between the national economic growth associated with fossil fuels, and international climate change resulting from the emissions from burning fossil fuels. Ought Lofoten's coastal management to be judged successful according to the economic prosperity it brings Norway, or for foregoing petroleum activities that may contribute to global environmental change? Finally, the debate also highlights the impacts of Lofoten's coastal management on international waters, whereby the adverse effects of an oil spill would not be limited to Norwegian waters. Together, these factors illustrate how important it is for coastal management evaluation to have a regard for the different character of coastal issues at different scales.

Bridging these different scales is very challenging, as the different stakeholders at the different scales (even if the boundaries between the scales are gray and stakeholders move between these scales) have very different agendas, priorities, and interests, and their interaction is characterized by strong power relations. Petroleum industry actors are very powerful in Norwegian political debates, particularly at the national scale. Illustrative of this is the fact that a majority of the Norwegian Parliament is in favor of oil drilling in Lofoten, while the majority of their Norwegian constituents are against it: "Opinion polls [show] a resistance against petroleum activity [among the general public] locally and nationally. In today's parliament, there is an overwhelming majority of pro-petroleum representatives" (Blanchard et al., 2014).

Immediate Profit or Long-Term Sustainability? Considering Risk and Uncertainty Relative to Ex Ante and Ex Post Evaluation

Bridging forward-looking and backward-looking evaluation of coastal management for the Lofoten issue is very challenging as various stakeholders frame the Lofoten trade-off in different ways, with emphasis either on risk or on uncertainty, to promote their interests. Competing concepts of risk and uncertainty are central to the challenge of understanding the relationship between *ex ante* and *ex post* evaluation.

Stakeholders promoting oil drilling in Lofoten frame the future impacts of these activities in terms of risk, implying that oil drilling is controllable and manageable, and in so doing, lending weight to *ex ante* evaluations. These *ex ante* evaluations are promoted for both steering coastal management institutions and evaluating the quality of coastal outcomes. Allowing for oil drilling in Lofoten brings the prospect of socioeconomic development, with new infrastructure and the creation of between 720 and 1340 jobs, depending on the future scenario chosen (Dale, 2011). However, petroleum activity has been asserted to be a short-term project, which would provide immediate profit that is not sustainable in the long term. The actors who are in favor of this immediate profit (such as the petroleum industry and some Norwegian political parties) frame the debates around Lofoten in terms of risk, which allows them to describe petroleum activities as profitable, safe, and compatible with fisheries. For instance, the 2012 report from the Ministry of Petroleum and Energy (OED, 2012) uses methods from risk analysis to conclude: "The assessments show that the environmental risk related to petroleum activity is consistently low, because the probability for significant or severe environmental damage is low" (p. 13; translation by the authors). This report refers to the SINTEF offshore database of 2011, which takes 2×10^{-4} as a typical probability of a blowout per year of production, or in other words, one blowout every 5000 years of production. In sum, even if the potential environmental harm is significant, since the probability of an oil spill is low, the environmental risk is framed as low and acceptable.

On the other hand, actors who wish to see the Lofoten coastal trade-off from a long-term perspective emphasize the uncertainties around the potential impacts of petroleum activity on the marine environment, and put forward the vulnerability of Lofoten ecosystems to pollution from oil. This is the case with the 2009 report from the Norwegian Institute for Nature Research (Forsgren et al., 2009), which underlines the significant uncertainties around impacts of petroleum activity: "We currently have limited knowledge of [oceanographic and geological] factors, and in addition, they show a large degree of uncertainty and stochasticity. Hence, predicting the impact of an oil spill on fish populations is extremely hard" (p. 24). Highlighting uncertainties, as in the NINA report, rather prevents political action; while downplaying uncertainties, like in the OED report, rather promotes immediate action. In this way, an inflated account of uncertainties can be argued to promote small incremental changes over wide-ranging profound ones, and by extension, promote *ex post* evaluation of the success of existing coastal uses.

The Lofoten case study highlights the uncertainties and politics that characterize the relationship between *ex ante* and *ex post* evaluation. From the discussion above it would appear clear that both forms of evaluation are necessary for coastal management, but that balancing such forward-looking or backward-looking perspectives implies an appreciation for the predictability of the future, as embodied in notions of risk and uncertainty.

GOVERNANCE PERSPECTIVES PROVIDING NEW NORMS OF EVALUATION FOR THE TWENTY-FIRST CENTURY

This chapter has argued that coastal management faces fundamental challenges to its evaluation as we enter into the twenty-first century, providing impetus for a genuine paradigm shift. In this section it is proposed that models of coastal governance present a plausible focus. We do not present an exhaustive solution as much as outline the broad characteristics of governance approaches as an area for greater attention in the twenty-first century.

In recognition of complex, plural, and uncertain coastal contexts, ICM authors have increasingly promoted models of a similarly complex and plural governance response (Bille, 2007, 2008; Christie, 2005; Chua, 1993; McFadden, 2007; McKenna and Cooper, 2006; Olsen, 2003b; Stojanovic and Barker, 2008; Tobey and Volk, 2002). In this regard, a number of governance perspectives have been put forward for framing coastal management, including models of "adaptive governance" (Adger, 2000; Gibbs, 2009), "governance of the commons" (Dietz et al., 2003), "good governance" (McKenna et al., 2008), and "interactive governance" (Glavovic, 2008a). Despite this diversity of governance perspectives, authors like Kooiman (1999) argue that they do share a number of common themes and motivations, notably that: (1) the state is not the only crucial actor in addressing major coastal issues; (2) there is a need for both traditional and new government–society interactions; (3) governance arrangements differ for levels of society and vary sector by sector; and (4) most issues are complex, leading to interdependency across societal actors. To take the definition of "interactive governance" (Kooiman and Bavinck, 2005), coastal governance can be thought of as "the whole of public as well as private interactions taken to solve societal problems and create societal opportunities. It includes the formulation and application of principles guiding those interactions and care for institutions that enable them" (emphasis added).

Within the field of ICM, models of governance have been promoted as more participatory and collaborative "governance-beyond-government"; embodying a growing humility among ICM authors (see e.g., Bille (2007, 2008), McFadden (2007), Olsen et al. (2009), Stojanovic and Barker (2008)). This implies recognizing that ICM initiatives represent one small intervention within a complex de facto governing system, shaped by a long history of socioeconomic and political influences. The myth of the omnipotent coastal manager (Bille, 2008) has been replaced with an idea of coastal management constituting a complex network of interactions and partnerships between plural stakeholders in a "governing system" (Stojanovic and Barker, 2008), many of which are informal and beyond the scope of an ICM initiative. Moves toward coastal governance reflect a shift from a unilateral, state-centric model of government, to a more polycentric, multilevel, and multilateral mode of governance. Through coastal governance we can conceive of collective decision making and action as the joint responsibility of state and non-state actors, including the private sector, civil society, and the scientific community. One significant paradigm shift has been the move from strongly centralized government according to a vertical hierarchy, to a devolution of power to the local scale, with an increasingly horizontal focus on "nested" governance networks (Pierre and Peters, 2000).

With the growing popularity of a governance perspective, and acceptance of its ontological, epistemological, and methodological foundations, there is a demand for new norms of evaluation. For instance, by accepting the irreducible plurality within coastal communities, this reduces the likelihood of arriving at any communally agreed "ends" to guide an evaluation. Ideas of linear and instrumental progress toward communally defined ends lose their meaning under governance frameworks; the ends are multiple and shifting, and indeed there are dissenting views on the progress. Furthermore, governance models take a different perspective on what constitutes a rational decision making process. It departs from mechanistic ideas of input-process-output, where decisions seek instrumental goal-attainment, toward accommodating multiple different rationalities, negotiated in a "soup" of stakeholder interactions across multiple institutional settings. In this way, stakeholders employ other types of rationality, not least the rationality of power. Finally, governance models have abandoned ideas of centralized collective decision making around one imagined table, favoring instead a picture of devolved decision making across multiple institutions. So framed, the focus of evaluation zooms out from a narrow focus on one government-led initiative in ICM, to encompass the broader coastal governing system, comprising the multitude of coastal actors interacting within and across different coastal institutions.

However, as the review demonstrated, there has been relatively little attention to governance-based evaluation mechanisms for ICM, and no frameworks that give full effect to a governance model. Indeed, it appears that while ICM has embraced a governance perspective, this perspective has struggled to pervade norms of evaluation. Some frameworks, such as that proposed by the International Oceanographic Commission (2006), have their framework under the heading of "governance indicators," but in reality draw on the same indicators used in management-focused frameworks. Other ICM authors reflect only a limited appreciation for the wider governing system in their discussion of evaluation, such as through evaluating the quality of stakeholder interactions. Stojanovic and Barker (2008) argued that the objective of ICM should be evaluated according to its ability to nurture quality partnerships and interaction measured according to (1) an increased awareness among stakeholders about coastal issues, (2) an altered attitude to their management, and (3) improved communication about roles and responsibilities, leading to coordinated action.

How then might the wider governance literature be drawn on to put together evaluation frameworks for coastal governance? Unfortunately the different governance literature drawn on by ICM scholars does not present a very sophisticated discussion of evaluation either; certainly not a very concrete discussion of specific, measureable, attainable, relevant, and timely indicators. Authors espousing interactive governance have tended to broad statements on the governability of the coast, with Jentoft (2007), for example, arguing that the diversity, complexity, dynamism, and vulnerability of the coast necessitates a governing system that is "context-based, coordinating, learning and safeguarding." Chuenpagdee et al. (2008) offer a similarly high-level framework for evaluating governance, dividing evaluation of the governing system broadly according to (1) the responsiveness of the modes of governance,[2] (2) the fit of the elements of governance,[3] and (3) the quality of the orders of governance.[4] On the other hand, authors espousing adaptive governance models (Armitage et al., 2007; Folke et al., 2007) evaluate governance in terms of stakeholders' increased ability to collectively learn and reach decisions in the face of uncertainty, termed "adaptive capacity." Some authors have gone further to prescribe

2. Modes of governance are split into self-governance, co-governance, and hierarchal governance.
3. Elements of governance are divided into the images that actors use in steering governance, the instruments they use to give effect to these images, and the action of putting instruments into effect.
4. Orders of interactive governance are implicit within its definition above, as first-order interactions, second-order institutions, and third-order meta-principles.

evaluation frameworks for measuring adaptive capacity as stocks of institutional, human, and social capital (Adger, 2003). In any case, what is clear is that if ICM scholars and practitioners are going to meaningfully entertain new norms of governance in ICM evaluation, then there is a demand for a much more accessible and pragmatic literature on evaluating coastal governance.

REFERENCES

Adger, W.N., 2000. Social and ecological resilience: are they related? Prog. Hum. Geogr. 24 (3), 347–364.

Adger, W.N., 2003. Social capital, collective action, and adaptation to climate change. Econ. Geogr. 79 (4), 387–404.

Armitage, D., Berkes, F., Doubleday, N., 2007. Introduction: moving beyond co-management. In: Armitage, D., Berkes, F., Doubleday, N. (Eds.), Adaptive Co-management: Collaboration, Learning and Multi-level Governance. UBC Press, Vancouver.

Barrett, R.T., Lorentsen, S.H., Anker-Nilssen, T., 2006. The status of breeding seabirds in mainland Norway. Atl. Seabirds 8, 97–126.

Bille, R., 2007. A dual-level framework for evaluating integrated coastal management beyond labels. Ocean Coastal Manage. 50 (10), 706–807.

Bille, R., 2008. Integrated coastal zone management: four entrenched illusions. Surv. Perspect. Integr. Environ. Soc. 1 (2), 75–86.

Blanchard, A., Hauge, K.H., Andersen, G., Fosså, J.H., Grøsvik, B.E., Handegard, N.O., et al., 2014. Harmful routines? Uncertainty in science and conflicting views on routine petroleum operations in Norway. Mar. Policy 43, 313–320.

Blanchard, A., Kaiser, M., Hauge, K.H., Fjelland, R., 2013. Oil, Fish, or Renewable Energies in Lofoten? Eight Messages from Lofoten Citizens to Their Decision-Makers. Senter for Vitenskapsteori, Bergen, Norway.

Bremer, S., 2011. Exploring a 'Post-normal' Science-policy Interface for Integrated Coastal Management: A Thesis Presented in Partial Fulfillment of the Requirements for the Degree of Doctor of Philosophy in Environmental Governance at Massey University. Massey University, Palmerston North.

Burbridge, P.R., 1997. A generic framework for measuring success in integrated coastal management. Ocean Coastal Manage. 37 (2), 175–189.

Christie, P., 2005. Is integrated coastal management sustainable? Ocean Coastal Manage. 48 (3–6), 208–232.

Chua, T.-E., 1993. Essential elements of integrated coastal zone management. Ocean Coastal Manage. 21 (1–3), 81–108.

Chuenpagdee, R., Kooiman, J., Pullin, R., 2008. Exploring governability in capture fisheries, aquaculture and coastal zones. J. Transdisciplinary Environ. Stud. 7 (1), 1–20.

Cicin-Sain, B., Knecht, R.W., 1998. Integrated Coastal and Ocean Management : Concepts and Practices. Island Press, Washington, D.C.

Dale, B., 2011. Securing a Contingent Future: How Threats, Risks and Identity Matter in the Debate over Petroleum Development in Lofoten, Norway. University of Tromsø (UiT), Tromsø.

Dietz, T., Ostrom, E., Stern, P.C., 2003. The struggle to govern the commons. Science 306 (5652), 1902–1912.

Ehler, C.N., 2003. Indicators to measure governance performance in integrated coastal management. Ocean Coastal Manage. 46, 335–345.

Fiorino, D.J., 1990. Citizen participation and environmental risk: a survey of institutional mechanisms. Science, Technol. Hum. Values 15 (2), 226–243.
Folke, C., Colding, J., Olsson, P., Hahn, T., 2007. Interdependent social-ecological systems and adaptive governance for ecosystem services. In: Pretty, J., Ball, A.S., Benton, T., Guivant, J.S., Lee, D.R., Orr, D., Pfeffer, M.J., Ward, H. (Eds.), The SAGE Handbook of Environment and Society. SAGE Publications Ltd, London.
Fontalvo-Herazo, M.L., Glaser, M., Lobato-Ribeiro, A., 2007. A method for the participatory design of an indicator system as a tool for local coastal management. Ocean Coastal Manage. 50, 779–795.
Forsgren, E., Christensen-Dalsgaard, S., Fauchald, P., Järnegren, J., Næsje, T.F., 2009. Norwegian Marine Ecosystems—Are Northern Ones More Vulnerable to Pollution from Oil than Southern Ones? Norwegian Institute for Nature Research (NINA), Trondheim.
GESAMP (IMO/FAO/UNESCO-IOC/WMO/WHO/IAEA/UN/UNEP Joint Group of Experts on the Scientific Aspects of Marine Environmental Protection), 1996. The Contributions of Science to Integrated Coastal Management, vol. 61, (GESAMP Reports and Studies).
Gibbs, M.T., 2009. Resilience: what is it and what does it mean for marine policymakers? Mar. Policy 33, 322–331.
Glavovic, B.C., 2006. Coastal sustainability–an elusive pursuit?: reflections on South Africa's coastal policy experience. Coastal Manage. 34 (1), 111–132.
Glavovic, B.C., 2008a. Ocean and coastal governance for sustainability: imperatives for integrating ecology and economics. In: Patterson, M., Glavovic, B.C. (Eds.), Ecological Economics of the Oceans and Coasts. Edward Elgar Publishing Ltd, Cheltenham.
Glavovic, B.C., 2008b. Sustainable coastal development in South Africa: bridging the chasm between rhetoric and reality. In: Krishnamurthy, R.R., Glavovic, B.C., Kannen, A., Green, D.R., Ramanathan, A.L., Han, Z., Tinti, S., Agardy, T.S. (Eds.), Integrated Coastal Zone Management. Research Publishing Services, Singapore.
Hauge, K.H., Blanchard, A., Andersen, G., Boland, R., Grøsvik, B.E., Howell, D., et al., 2014. Inadequate risk assessments—a study on worst-case scenarios related to petroleum exploitation in the Lofoten area. Mar. Policy 44, 82–89.
Henocque, Y., 2003. Development of process indicators for coastal zone management assessment in France. Ocean Coastal Manage. 46, 363–379.
Hjermann, D.Ø., Melsom, A., Dingsør, G.E., Durant, J.M., Eikeset, A.M., Røed, L.P., et al., 2007. Fish and oil in the Lofoten–Barents Sea system: synoptic review of the effect of oil spills on fish populations. Mar. Ecol. Prog. Ser. 339, 283–299.
Hoffmann, J., Löser, N., 2008. Indicators for ICZM—overview and lessons learned based on the Oder Estuary Region in the south-western Baltic Sea. In: Krishnamurthy, R.R., Glavovic, B.C., Kannen, A., Green, D.R., Ramanathan, A.L., Han, Z., Tinti, S., Agardy, T.S. (Eds.), Integrated Coastal Zone Management. Research Publishing Services, Singapore.
International Oceanographic Commission, 2006. A Handbook for Measuring the Progress and Outcomes of Integrated Coastal and Ocean Management. UNESCO.
Jentoft, S., 2007. Limits of governability: institutional implications for fisheries and coastal governance. Mar. Policy 31, 360–370.
Jentoft, S., Kristoffersen, T., 1989. Fishermen's co-management: the case of the Lofoten fishery. Hum. Organ. 48 (4), 355–365.
Kooiman, J., 1999. Social-political governance. Public Manage. Rev. 1 (1), 67–92.
Kooiman, J., Bavinck, M., 2005. The governance perspective. In: Kooiman, J., Bavinck, M., Jentoft, S., Pullin, R. (Eds.), Fish for Life: Interactive Governance for Fisheries. Amsterdam University Press, Amsterdam, pp. 11–24.

Larsen, T., Boltunov, A., Denisenko, N., Denisenko, S., Stanislav, D., Gavrilo, M., et al., 2001. The Barents Sea Ecoregion: A Biodiversity Assessment. WWF, Oslo.

McFadden, L., 2007. Governing coastal spaces: the case of disappearing science in integrated coastal zone management. Coastal Manage. 35 (4), 429–443.

McKenna, J., Cooper, A., 2006. Sacred cows in coastal management: the need for a 'cheap and transitory' model. Area 38 (4), 421–431.

McKenna, J., Cooper, A., O'Hagan, A.M., 2008. Managing by principle: a critical analysis of the European principles of Integrated Coastal Zone Management (ICZM). Mar. Policy 32, 941–955.

NDF, 2011. Economic and Biological Figures from Norwegian Fisheries. Norwegian Directorate of Fisheries, Bergen.

NME, 2006. Integrated Management Plan of the Marine Environment of the Barents Sea and the Sea Areas off the Lofoten Islands–Report No. 8 to the Storting. Norwegian Ministry of the Environment, Oslo.

O'Connor, M., 1999. Dialogue and debate in a post-normal practice of science: a reflexion. Futures 31 (7), 671–687.

OED, 2012. Knowledge Acquisition about the Impact of Petroleum Operations in the Northeastern Norwegian Sea. Ministry of Petroleum and Energy [OED: Olje- og Energidepartementet], Oslo.

Olsen, E., Aanes, S., Mehl, S., Holst, J.C., Aglen, A., Gjøsæter, H., 2010. Cod, haddock, saithe, herring, and capelin in the Barents Sea and adjacent waters: a review of the biological value of the area. ICES J. Mar. Sci. 67 (1), 87–101.

Olsen, S.B., 2002. Assessing progress toward the goals of coastal management. Coastal Manage. 30 (4), 325–345.

Olsen, S.B., 2003a. Frameworks and indicators for assessing progress in integrated coastal management initiatives. Ocean Coastal Manage. 46, 347–361.

Olsen, S.B. (Ed.), 2003b. Crafting Coastal Governance in a Changing World. University of Rhode Island Coastal Resources Center, Narragansett.

Olsen, S.B., Page, G.G., Ochoa, E., 2009. The Analysis of Governance Responses to Ecosystem Change: A Handbook for Assembling a Baseline, vol. 34 LOICZ, GKSS Research Centre, Geesthacht.

Olsen, S.B., Tobey, J., Kerr, M., 1997. A common framework for learning from ICM experience. Ocean Coastal Manage. 37 (2), 155–174.

Pickaver, A.H., Gilbert, C., Breton, F., 2004. An indicator set to measure the progress in the implementation of integrated coastal zone management in Europe. Ocean Coastal Manage. 47 (9–10), 449–462.

Pierre, J., Peters, B.G., 2000. Governance, Politics and the State. MacMillan Press Ltd, London.

Stojanovic, T.A., Ballinger, R.C., Lalwani, C.S., 2004. Successful integrated coastal management: measuring it with research and contributing to wise practice. Ocean Coastal Manage. 47 (5–6), 273–298.

Stojanovic, T.A., Barker, N., 2008. Improving governance through local coastal partnerships in the UK. Geogr. J. 174 (4), 344–360.

Tobey, J., Volk, R., 2002. Learning frontiers in the practice of integrated coastal management. Coastal Manage. 30 (4), 285–298.

Turner, R.K., 2000. Integrating natural and socio-economic science in coastal management. J. Mar. Syst. 25 (3–4), 447–460.

Yao, H., 2008. Lessons learned from ICOM initiatives in Canada and China. Coastal Manage. 36 (5), 458–482.

Making the Link

By Jean-Paul Vanderlinden

Challenges to Evaluating Coastal Management in the Twenty-First Century: Lessons from the Lofoten ⟺ **Motivation for the Viability of the Lobster Fishery: Case Study of the Acadian Coast of New Brunswick**

Scott Bremer, Anne Blanchard and Matthias Kaiser — *Omer Chouinard, Liette Vasseur and Steve Plante*

Bremer and colleagues, using high-stakes and extremely multi-scalar case studies, have shown how evaluating coastal zone management is challenging. One of the fundamental challenges identified is "limited by a bounded rationality anchored in its 'resource management' paradigm." They pursue this by assessing current evaluation proposals and practices according to the multiple "bridging challenges." Yet does this demonstration hold when a resource lies at the center of local communities' concerns?

The answer to this question is yes, without a doubt, as exemplified in the next chapter by Chouinard. Chouinard analyzes a situation where a single resource is the central object of interest: lobster and its associated fishery on the Acadian Coast of New Brunswick, Canada. Through a very different pathway, Chouinard reaches conclusions that resonate strongly with the paradigm shift that Bremer and colleagues are calling for. If "bridge building" is to be achieved for assessment purposes, one of the central keys lies in taking into account territorial needs, understating the coasts as socioecological systems as well as local representations. Putting these recommendations, spelled out in Chouinard's case study, in context, sets the stage for the type of governance and evaluative framework called for by Bremer et al.

Chapter 15

Motivation for the Viability of the Lobster Fishery: Case Study of the Acadian Coast of New Brunswick

Omer Chouinard[1], Liette Vasseur[2], Steve Plante[3]
[1]Université de Moncton, Moncton, New Brunswick, Canada; Marine Sciences For Society, www.marine-sciences-for-society.org; [2]UNESCO Chair in Community Sustainability: from local to global, Department of Biological Sciences, Brock University, St. Catharines, Ontario, Canada; [3]Departement Sociétés, Territoires et Développement, Université du Québec à Rimouski, Rimouski, Québec, Canada

Chapter Outline

INTRODUCTION

The world fishery steadily increased from 19.3 million tons in 1950 to about 110 million tons by 2006 (FAO, 2009). However since the mid-1990s, catches have either remained stable or shown a slight increase. This industry represents a very important sector of the economy of many countries. Unfortunately, a large part of the marine fisheries in the world have been labeled as unsustainable, with 19% being overexploited, 8% being depleted, and another 52% considered as fully exploited (FAO, 2009; Ojea and Loureiro, 2010). While some fish stocks are reported as collapsed or in serious decline due to overcapacity of the fishing fleets and environmental degradation, especially closer to the coast, others, such as the American lobster (*Homarus americanus*), are maintained but under strong pressures. Fishery sustainability has become a serious issue not only for fishers but also for decision-makers who have to define strategies to either maintain the industry or to restore fisheries that have collapsed.

The economic approach to fishery has been based for a long time on maintaining the maximum sustainable yield while optimizing fishing efficiency in function of market prices. This approach tends to lead to overexploitation as the prices drive the need to further exploit the resource (Brown, 2001). Indeed, although we understand better than in the past the dynamics of these populations, the failure of many fishery exploitations lies in how stocks are estimated (Ling et al., 2009). The estimation does not take into account the ecosystem in which the fish live and the other components that may influence survival, such as invasive species and changes in reproduction or growth due to pollutants or climate change (Vasseur, 2015, chapter 17).

Some case studies (FAO, 2009; FRCC, 2011) have shown that once a resource has been overexploited, the main strategy is to establish a restoration plan. Most of these plans are simplistic and continue to only focus on the specific species instead of examining the health of the ecosystem. Several authors (FAO, 2009; Kooiman et al., 2005) have argued that the best way to ensure the sustainability of a natural resource is to examine the various components of the ecosystem in which the resource thrives. However, fishery also integrates other components such as social representations of the stakeholders, territorial development, and the global view of the fish market.

Lobster fishery across the world has remained relatively stable, with 256,120 tons of lobsters in 2009 (ftp://ftp.fao.org/FI/STAT/summary/a1d.pdf). While in some regions the fishery is still growing, as in the Gulf of Maine, in other regions of the world (e.g., Australia), lobster stocks are in decline (Plagányi et al., 2011). In Atlantic Canada, lobster fishery has historically remained a strong industry. The American lobster has a wide distribution from the coasts of North Carolina, USA, up to Labrador, Canada. This species is usually found near the coast within a distance of 20 km from the coast, although there is fishery in deep waters of Georges Bank, for example (FRCC, 1995, 2007; DFO, 2002a,b).

The American lobster is characterized by a complex life cycle and ecology. Female lobsters carry their eggs until larvae are capable to swim. Lobster larvae live in a planktonic stage, freely swimming in the water column for a period of 3–10 weeks depending on the environmental conditions of the region where they are found. Temperature appears to be the major factor influencing the time as a planktonic organism. After the molt of stage III, the young lobster settles on the bottom to complete its growth. Through growth and a series of moltings, the lobster reaches a size when it can become reproductive (http://slgo.ca/en/lobster/context/cycle.html). During its life cycle, several environmental factors can affect its capacity to survive and grow (Boudreau et al., 1991). Temperature may be the main factor, but others will include pH, pollutants, predation rate of the larvae, etc. (Boudreau et al., 1991). Severe fluctuations in the environmental conditions due to natural phenomena (e.g., North American Oscillation (NAO), storm surges) or human activities (pollutants, climate change, ocean acidification) may rapidly jeopardize the stocks (see Tluty et al. in Vasseur, 2015, chapter 17).

Lobster fishery in Atlantic Canada is managed by the Department of Fisheries and Oceans (DFO), and the region has been divided into lobster fishing areas (LFAs). Landings of American lobster along the Canadian coast significantly vary between the different LFAs. For example, along the coast of the Bay of Fundy, landings have continued to slightly increase, while in other regions such as in the southern Gulf of St. Lawrence, a decline has been observed, although this is not uniform (FRCC, 2007; DFO, 2002a). The report of FRCC (2007) indicates that the lobster in the southern Gulf is heavily exploited. This may be fine because of its high abundance, but this is not the case everywhere, especially along the Acadian Coast of New Brunswick, in the Northumberland Strait. The report also underlines that fishery is highly dependent on new recruits.

The current fragility of the system has pushed the various stakeholders to critically rethink the actions to be implemented. As the territory targeted is highly fragmented by seasonal fishery and regulations, adoption of strategies may be quite complex. This is the case in the southern Gulf of St. Lawrence, where the region is divided into LFA 23, 24, 25, and 26A and 26B (Figure 15.1). Each zone has different socioeconomic and environmental issues to deal with. For example, Cape Breton LFA 26 A and B, and Bay of Chaleur LFA 23 A and B, have initiated voluntary restrictive conservation measures such as increasing the size of the lobster carapace being caught, and have experienced some success in maintaining the stock. In this chapter, we focus on Acadian Coast LFA

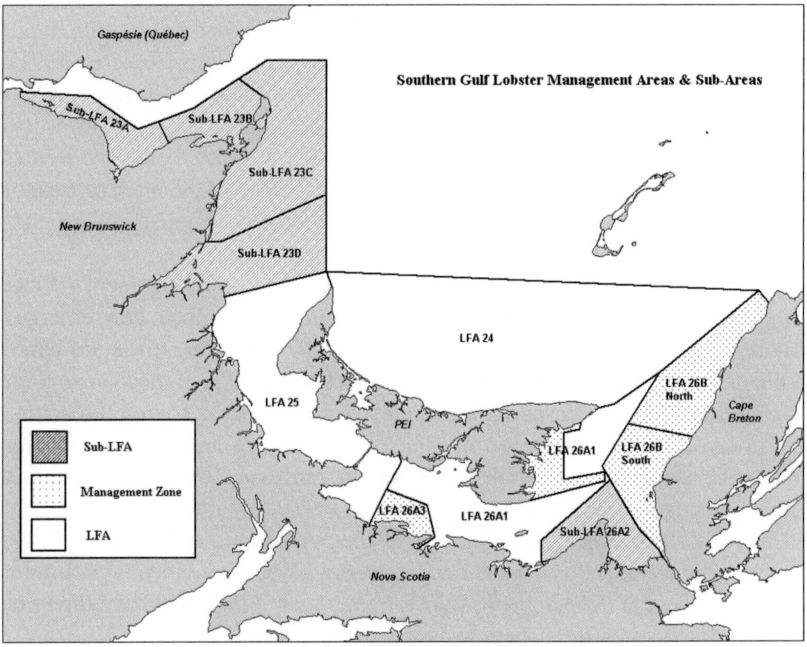

FIGURE 15.1 Map of lobster fishing areas in the Southern Gulf of St. Lawrence. *From Department of Fisheries and Oceans, Canada, 2011, reproduced with permission.*

25, where stocks are uncertain in comparison with LFA 26, where the stocks are healthier. It is worth mentioning that in this region, the lobster industry represents 90% of the revenues of the fishers holding multispecies licenses.

We argue that through a conservation strategy that is socially acceptable by the stakeholders and ecologically sound, the sustainability and resilience of the fishery in this district can be improved, especially in the face of environmental and climate changes. This strategy would have to begin by understanding the motivation of stakeholders, including fishermen, processors, distributors, retailers, and even the consumers, who require greater transparency on the handling of their produce than in the past (David Smith, Sobey's, 2011, presentation at the Atlantic Lobster Sustainability Foundation, Moncton, Crowne Plaza Hotel, Moncton, July 23rd and 24th).

We have divided this chapter in three sections: (1) territorial approach to management and its potential integration in any type of restoration plan; (2) the challenges and opportunities regarding the integration of the concept of the social-ecological system (SES) into territorial development and management; and (3) the integration of social needs to ensure a successful management of the resource. The discussion also integrates some preliminary data from interviews conducted among fishers to determine whether there would be an interest in following such an approach.

TERRITORIAL APPROACH OF MANAGEMENT: THE CHALLENGES ON THE ACADIAN COAST

In the fisheries sector, traditional management tools are based on scientific models that recognize the dynamics of single-species populations (Revéret, 1991). Managers do not usually consider the multitude of species concerned and their interactions in the food web (Dearden and Mitchel, 2009). All these measures underpin a homogeneous character of the resource and space. Folke and Berkes (1995) and the FRCC (2011) view these measures as one of the main causes for the collapse of some of our fisheries, including northern whitefish and groundfish in eastern Canada.

Since the 1990s, several case studies of resource co-management (Wiber et al., 2004; Schumann, 2007) or integrated resource management have emerged (Lal et al., 2001; Belfiore, 2003; Zagonari, 2008). These initiatives are based on the development of individual and collective appropriation mechanisms for resource sustainability based on a certain sharing of access among the actors. Few of these models, however, consider the dynamics among the actors or incorporate social aspects into decision-making processes. The challenge is that from a human point of view, the coastal zone is generally linked to fish habitat, and there is a disconnect with the rest of the ecosystem, especially the land. However, the terrestrial ecosystem encompasses several activities (agriculture, tourism, residential development, transformation, etc.) that may have direct or indirect influences on the fishery (Vasseur, 2015, chapter 17).

We argue here that there is a need to use a territorial approach, as we will see later, integrated with a SES approach, to merge the various issues and their interactions on the resource. Inspired by the community-based management principle (Vasseur and Hart, 2002) and the commons theory, the territorial approach integrates the concepts of customary law, community, social institutions, and the individual and collective strategies. The advantage of this approach is that actions are implemented through a participatory and sustainable mode of governance that promotes sustainability of the communities knowing their interdependence with the environment. We also use the concept of social resilience (Walker et al., 2006; Beatley, 2009; Monasinghe, 2007), where the main components of social capital, such as representations, confidence, social cohesion, sense of belonging, and the identity of the actors, are respected (Reimer et al., 2008; Bouma et al., 2008). These intangible aspects in the decision-making process allow for the valorization of both actors' knowledge (producer, industrial, and consumer) and scientific knowledge.

As a territorial view, the coastal zone not only includes the natural resources but also human activities (fishing, processing, distribution, consumption, etc.) and their place in this territory, and thus there is a need to also discuss SES. This view integrates a multitude of uses and users that are influenced by its geophysical configuration of small bays and estuaries. Discharge of contaminated freshwater from terrestrial activities (agriculture, forestry, etc.), for example, intricately demonstrates the connection between aquatic and terrestrial systems. Under these conditions, it is clear that the lobster fishery applies particularly well to the development of a territorial approach, since the species is rather sedentary. Any changes on land or water will have direct impacts on this crustacean.

The link between the actors and the resource refers to the notion of proximity (Angeon, 2008). Depending on their needs and the economic or physical proximity of the resource, actors may have different representations and conceptualization of their territory, and this can be expressed through institutions. An institution represents an invisible body of formal or informal rules, regulations, and processes that guide decisions (Ostrom, 1990), and where perceptions and social representations of the actors are respected (Keohane and Ostrom, 1994; Baker, 1997). The sharing of management responsibilities among stakeholders depends on the types of relationships (trust, social cohesion, etc.) and social representations that institutions (e.g., regional and national) are based on. In the case of lobster fishery the distribution of responsibilities should be more equitable for communities where fair decision-making mechanisms can occur.

Institutions can be defined as internal (endogenous) and external (exogenous) (Baland and Platteau, 1996). In internal institutions, stakeholders attempt to learn from each other and self-organize their actions in order to improve the resilience of their various components (DaCunha et al., 2012). On the other side, external institutions may influence the decisions made at the internal level and thus positively or negatively affect their resilience (Baland and Platteau, 1996). A territorial approach aims to ensure that actions can adapt to physical and cultural attributes of both external and internal institutions. Interpersonal, social, economic, and political relations play important roles in the establishment and the maintenance of these institutions. Baland and Platteau (1996) consider that the adaptive capacity and the sustainability of these organizations will depend on their collective and constitutional choices (rules). Results can be visible through both collective and individual actions. But for that, capacity building and empowerment are needed (Nunan, 2006).

A territorial approach integrates the concept of resilience and can be adopted if the actors and their social institutions mutually agree on a scale appropriate to protect the resource (Kiousopoulos, 2008). In the case of lobster fishery, territorial scale perceived and experienced by stakeholders would coincide with a local and supra-local perspective, that is, this would include an enlarged coastal zone with linkages between the aquatic and terrestrial ecosystems. This would also mean that adopting a territorial governance approach focusing on mutual construction of the objectives would require a greater place for strategic management and planning, to allow for an improved sense of belonging of all parties through knowledge sharing on their mode of occupation of the territory (Alpizar and Quesada, 2006). The use of this knowledge is similar to a form of adaptation of the process of self-organization of a social system that is in constant feedback relationship with the environment (Hanna et al., 1995; Walker et al., 2009).

SOCIAL-ECOLOGICAL APPROACH: WHAT IS THE CONNECTION?

Ecosystems are composed of a great variety of species, including humans, that interact together and with their surrounding abiotic environment (Vasseur et al., 2002). As we have seen, marine ecosystems are exposed to several risks including overfishing, climate change, invasive species, and pollution (Ling et al., 2009; Vasseur, 2015, chapter 17). Between 2000 and 2010, the concepts of ecosystem health and SES resilience have been proposed to discuss the state of an ecosystem and its capacity to recover after a disturbance (Vasseur et al., 2002). In theory, if an ecosystem is healthy, it should have the capacity to recover by itself either in its original state or in a transformed state (resilience concept). Under this concept, it would be expected that if an ecosystem no longer healthy is disturbed, recovery may not be possible without human intervention. We can define SES management as any human actions or interventions that lead to improve the functionality and productivity, that is, the health and the resilience of an ecosystem. When looking at an ecosystem, humans tend to deal with one species or one factor at a time. Initiatives can focus, for example, on lobster growth, the level of pollution in an estuary, etc. However this approach is limited, and the intricate linkages among ecosystem components must be examined to find more sustainable solutions.

An interdisciplinary approach to SES is considered desirable because it integrates the various components of the system that may be damaged or can become damaged due to some actions (Longstaff, 2009). Decisions should be made in order to better sustain not only the economic activities but also the ecosystem components that are needed and on which humans depend for their economic sustainability (Callicott et al., 1999).

If the aim is to sustain or increase the use of the resource, should we really be talking about ecosystem management, or more specifically restoration and protection for sustained use of the resource? This debate is exemplified by the fact that actions generally target only one aspect of the ecosystem, such as larval growth or habitat of the lobster. If the entire SES were to be considered, actions would have to include changes in policy related to wastewater discharge in the estuaries, combined with new regulations for fishing gear, greener and less polluting vessels, and habitat enhancement. This approach would be controversial, as it would mean removing barriers between issues within a same SES and therefore considering the entire territory instead of the lobster directly. Can coastal communities accept and integrate such concepts and decisions? The first step to accomplish would be to better understand the representations of the stakeholders. In the lobster fishery, we traditionally target the fishers, as they are the closest to the resource. However, an SES approach would require that the other actors directly and indirectly involved in this industry participate in the dialog on sustainability. This would include processors, distributors, restaurant owners, etc. Indirectly, even citizens could be involved, as many of their activities can be impacted by changes in the SES. This would include, for example, farmers located along the shore, whose activities impact the coastal waters due to chemical runoff.

REPRESENTATIONS: WHAT DO WE MEAN BY SUSTAINABILITY OF THE INDUSTRY?

The sharing of values by the social actors involved in the lobster fishery should lead to a common vision of the changes needed in the industry to ensure sustainability. It is from this perspective that here we discuss social representations and their use in collectively defining solutions. According to Gendron (2006) and Jodelet (1989), social representations can be defined as the speech used by stakeholders to develop policies that should link to factors that will trigger social transformations. It represents a form of social knowledge that reflects a common reality and interpretation of a condition or situation and leads to similar actions or decisions (Jodelet, 1989). It is usually anchored in tradition and transmitted from people to people, and can be transmitted from generation to generation in families. Social representations can be dynamic and vary in the function of social, economic, or environmental conditions (Wiber et al., 2012). Understanding these social representations can help understand the barriers for any restoration or conservation strategies proposed by external stakeholders.

In this research, a preliminary analysis of the social representations of fishers was done by conducting semistructured interviews with 15 fishers of districts 25 and in 23, in March 2012. People were selected using a snowball sampling method (Beaud, 2009). In both districts, conservation measures, including the reduction in number of traps, have been implemented. However, only district 23 has implemented the release of large females and the increase in size of the captured lobster.

From the interviews, fishers and governmental agencies have reported concerns regarding loss of ice cover in LFAs 23 and 25 during the winters and the landings of dead lobster after winter storms (http://www.ec.gc.ca/glaces-ice/default.asp?). Will this lead to a collapse of their industry? The question is legitimate, and for most fishers, this leads to questioning the current *status quo* of the fishery and the industry. Fishers are therefore willing to work with other stakeholders of the industry to implement restoration and conservation measures. This can include the protection of large females and small lobsters for a robust fishery. They also understand the need to change the opening date of the fishing season due to consistent changes in ocean temperatures. At the same time, other factors such as pollution coming from the terrestrial or aquatic systems require further discussions with external institutions. This is where a better understanding of the social representations of the other groups can be useful. Dialogs among the various actors should aim to improve synergy by finding common representations and, gradually, solutions. This would lead to a change in governance approach and improve the capacity to adapt to the various needs of one another. We believe this can be achieved through adaptive management that combines both territorial and SES approaches.

INTEGRATING TERRITORIAL AND SES APPROACHES: OPPORTUNITIES?

We have built the conceptual foundation on which it is possible to start discussing the ingredients needed to achieve successful restoration and conservation of the lobster for a sustainable fishery. We acknowledge that first it is crucial to understand the social representations of the stakeholders. But who are they? While it may be easy to target only the fishers, it is clear that external and internal institutional pressures exist. For example, stakeholders interested in lobster include also the transformation operators, the distributors, restaurant operators, etc. Using a territorial approach, it becomes evident that other actors should also be represented in any type of management, including farmers who use chemical pesticides or fertilizers, fishers targeting other fish species, etc. We therefore acknowledge that defining the social institutions on which to build an integrated territorial and ecosystem-based management system will require comprehensive analysis of the interactions and the networks existing in a community and beyond.

In recent times, due to the decline in the lobster landings, fishers of LFA 23 have started experiencing what we can call the concept of community of interest, that is, the understanding of their own representations and the building of their internal institution. This experience was initiated with the realization from the fishers, in conjunction with DFO agents and scientists, of the need to consider a territorial approach to deal with the resource. To this end, LFA 23 has been subdivided into different zones, often called sub-LFAs (Figure 15.1). These zones or territories have been acknowledged by the various stakeholders as being significantly different. This approach seems more appropriate for the restoration of the resource where a sub-LFA is not obliged to have the same management as the rest of the LFA. This allows for a more adaptive and integrated management, which considers the limitation of the SES and the other stresses that occur within each territory.

The FRCC (2011) has recommended that in policies such as the Integrated Fisheries Management Plans (IFMPs), issues such as economic conditions and current environmental impacts on the resource should be considered. Environmental monitoring should not only target the lobster as a single species, but if the SES approach is truly embraced, indicators representative of the system should be defined and used to measure progress in various areas such as water pollution, habitat quality, economic conditions in the communities, etc.

This brings us to the last concept that absolutely needs to be integrated into this management model of territory and SES: adaptive strategies. We agree that as conditions are changing, the premises on which the actions or decisions were first made become absolute. It is essential that all stakeholders understand that in order to remain sustainable, management must be adaptive to changing conditions as well as social representations. We believe that without such an integrated approach that first explores social representations of all the stakeholders and the existing conditions using a territorial and ecosystem approach, restoration and conservation of natural resources such as the American lobster will remain limited or ineffective.

CONCLUSIONS

We have underlined the importance of understanding the social representations in an SES context, meaning that not only the lobster institution must be examined but also the other actors that may directly or indirectly impact the system. There is an urgent need for a more cohesive approach for the sustainability of the industry. We also understand that integrated adaptive management with territorial and SES approaches should be the basis on which actors should be moving toward for a more sustainable SES. While the traditional precautionary principle may be applied in this approach, it is clear that connecting the dots among the elements of the SES and how it is perceived by the actors can enable communities to further develop long-lasting solutions. Restoration measures can be considered, but not in isolation and without the development of other adaptive strategies for the rest of the territory.

Finally, we understand that while this may remain conceptual, integrated adaptive management using territorial and SES approaches may be the only way to stop fixing issues through the eyes of a single institution and a single species. With a better understanding of the social representations of the various stakeholders, synergy can emerge and lead to more comprehensive solutions. There have been a few success stories in this territory of the Acadian Coast of New Brunswick. For example, a local grower group, the Really Local Harvest, partnered with the city of Dieppe in southeast New Brunswick to sell its produce (http://recoltedecheznous.com/fr/), thus making it more sustainable for small farmers. The Coopérative d'énergie renouvelable de Lamèque (cooperative of renewable energy in Lamèque) provides another example of a partnership that has led to the establishment of a wind farm with several stakeholders in northeastern New Brunswick (Chouinard et al., 2014). This integrated approach may be difficult and take a long time to complete, but without this, how can we improve resilience and sustainability? Those elements are more intangible and difficult to measure: they imply values, beliefs, and attitudes.

ACKNOWLEDGMENTS

The authors would like to acknowledge Carolyn Peach Brown for the comments made of the manuscript.

REFERENCES

Alpızar, M.A.Q., 2006. Participation and fisheries management in Costa Rica: from theory to practice. Mar. Policy 30, 641–650.

Angeon, V., 2008. L'explicitation du rôle des relations sociales dans les mécanismes de développement territorial. Rev. Écon. Rég. Urbaine 2, 237–250.

Baker, J.M., 1997. Common property resource theory and the Kuhl irrigation systems of Himachal Pradesh, India. Hum. Organ. 56, 199–208.

Balland, J.M., Platteau, J.P., 1996. Halting Degradation of Natural Resources: Is There a Role for Rural Communities? FAO. Oxford Clarendon Press.

Beatley, T., 2009. Planning for Coastal Resilience: Best Practices for Calamitous Times. Island Press. 200 p.

Beaud, J.-P., 2009. L'échantillonnage, dans Benoît Gauthier, dir., Recherche sociale. De la problématique à la collecte des données, 5è éd. Sillery, PUQ. pp. 251–283.

Belfiore, S., 2003. The growth of integrated coastal management and the role of indicators in integrated coastal management: introduction to the special issue. Ocean Coastal Manage. 46, 225–234.

Boudreau, B., Simard, Y., Bourget, E., 1991. Behavioral responses of the larval stages of the American Lobster, *Homarus americanus*, to thermal gradients and ecological implications. Mar. Ecol. Prog. Ser. 76, 13–23.

Bouma, J., Bulte, E., Soest, D.van., 2008. Trust and cooperation: social capital and community resource management. J. Environ. Econ. Manage. 56, 155–166.

Brown, L.R., 2001. Eco-economy. Building an Economy for the Earth. W.W. Norton Company, New York. 333 p.

Callicott, J.B., Crowder, L.B., Mumford, K., 1999. Current normative concepts in conservation. Conserv. Biol. 13, 22–35.

Chouinard, O., Guillemot, J., Leclerc, A., Rabeniaina, T., 2014. International Summit of Cooperatives, Old Coops and New Coops: The Case of the Coopérative d'énergie renouvelable de Lamèque, Chapter 17, October 6 to 9, 2014, Quebec City. https://www.sommetinter.coop.

Da Cunha, C., Plante, S., Vasseur, L., 2012. Le suivi de la résilience de communautés côtières comme moyen d'évaluation des effets de la recherche action participative. In: Actes du Congrès du GIS Démocratie & Participation, Paris, France. http://www.participation-et-democratie.fr/fr/node/914.

Department of Fisheries and Oceans (DFO), 2002a. Canada's Oceans Strategy: Our Oceans Our Future. Ottawa.

Department of Fisheries and Oceans (DFO), 2002b. Policy and Operational Framework for Integrated Management of Estuarine, Coastal and Marine Environments in Canada, Ottawa.

Dearden, P., Mitchell, B., 2009. Environmental Change and Challenge, third ed. Oxford Press, New York. 624 p.

Folke, C., Berkes, F., 1995. Mechanisms to link property rights to ecological systems. In: Hanna, S., Munasinghe, M. (Eds.), Property Rights and the Environment: Social and Ecological Issues. The World Bank, Washington, DC, pp. 121–137.

Food and Agriculture Organization of The United Nations, 2009. The State of World Fisheries and Aquaculture 2008. FAO, Rome. ISBN: 978-92-5-106029-2. 196 p.

Fisheries Resources Conservation Council (FRCC), 1995. Conservation Framework for Atlantic Lobster, Ottawa, DFO.

FRCC, 2007. Sustainability Framework for Atlantic Lobster, Ottawa, DFO. [Online] http://www.frcc-ccrh.ca/NEWSREL/2007/releaseEP.htm Site visited December 5th, 2011.

FRCC, 2011. Towards Recovered and Sustainable Groundfish Fisheries in Eastern Canada, Ottawa, DFO. [Online] http://www.frcc.ca/2011/FRCC2011.pdf Site visited March 17th, 2012.

Gendron, C., 2006. Le développement durable comme compromis: la modernisation écologique à l'ère de la mondialisation, Québec, PUQ.

Hanna, S., Folke, C., Mäler, K.-G., 1995. Property Rights and Environmental Resources. Beijer International Institute of Ecological Economics and the World Bank, Washington, D.C.

Jodelet, D., 1989. Représentations sociales: un domaine en expansion. In: Jodelet, D. (Ed.), Les Représentations Sociales. Presses universitaires de France, Paris, pp. 31–61.

Keohane, R.O., Ostrom, E., 1994. 1. Introduction. J. Theor. Polit. 6, 403–428.

Kiousopoulos, J., 2008. Methodological approach of coastal areas concerning typology and spatial indicators, in the context of integrated management and environmental assessment. J. Coastal Conserv. 12 (1),19–25. http://dx.doi.org/10.1007/s11852-008-0019-6.

Kooiman, J., Bavinck, M., Jentoft, S., Pullin, R. (Eds.), 2005. Fish for Life: Interactive Governance for Fisheries. Amsterdam University Press, Amsterdam.

Lal, P., Lim-Applegate, H., Scoccimarro, M., 2001. The adaptive decision-making process as a tool for integrated natural resource management: focus, attitudes, and approach. Conserv. Ecol. 5 (2), 11. [online] URL: http://www.consecol.org/vol5/iss2/art11/.

Ling, S.D., Johnson, C.R., Frusher, S.D., Ridgway, K.R., 2009. Overfishing reduces resilience of kelp beds to climate-driven catastrophic phase shift. PNAS 106, 22341–22345.

Longstaff, P.H., 2009. Managing surprises in complex systems: multidisciplinary perspectives on resilience. Ecol. Soc. 14, 49.

Monasinghe, M., 2007. Making Development More Sustainable: Sustainomics Framwork and Practical Applications. MIND Press, Colombo Sri Lanka. 636 p.

Nunan, F., 2006. Empowerment and institutions: managing fisheries in Uganda. World Dev. 34, 1316–1332.

Ojea, E., Loureiro, M.L., 2010. Valuing the recovery of overexploited fish stocks in the context of existence and option values. Mar. Policy 34, 514–521.

Ostrom, E., 1990. Governing the Commons: The Evolution of Institutions for Collective Action. Cambridge University Press, New York. 280 p.

Plagányi, É.E., Weeks, J.S., Skewes, T.D., Gibbs, M.T., Poloczanska, E.S., Norman-López, A., Blamey, L.K., Soares, M., Robinson, W.M.L., 2011. Assessing the adequacy of current fisheries management under changing climate: a southern synopsis. – ICES J. Mar. Sci. 68, 1305–1317.

Reimer, B., Lyons, T., Ferguson, N., Palanco, G., 2008. Social capital as social relations: the contribution of normative structures. Sociol. Rev. 56, 19.

Revéret, J.-P., 1991. La pratique des pêches: Comment gérer une ressource renouvelable. Collection «environnement» L'Harmattan. 198 p.

Schumann, S., 2007. Co-management and "consciousness": fisher's assimilation of management principles in Chili. Marine Policy 31, 101–111.

Tluty, M., Metzler, A., Malkin, E., Goldstein, J., Koneval, M., 2008. Microecological impacts of global warming on crustaceans—temperature induced shifts in the release of larvae from American lobster, *Humarus americanus*, females. J. Shellfish Res. 27, 443–448.

Vasseur, L., 2015. Lobster fishery in Atlantic Canada in the face of the climate and environmental changes: can we talk about sustainability on these coastal communities? (chapter 18). In: Baztan, J., Chouinard, O., Tett, P., Vanderlinden, J.-P., Vasseur, L. (Eds.), Coastal Zones: Developing Solutions for 21st Century. Elsevier.

Vasseur, L., Hart, W., 2002. A basic theoretical framework for community-based conservation management in China and Vietnam. Int. J. Sustainable Dev. World Ecol. 9, 41–47.

Vasseur, L., Rapport, D., Hounsell, J., 2002. Linking ecosystem health to human health: a challenge for this new century (Chapter 9). In: Costanza, B., Jorgensen, S. (Eds.), Integrating Science to Policy—Ecosummit 2000. Elsevier, Cambridge, pp. 167–190.

Walker, B., Salt, D., 2006. Resilience Thinking: Sustaining Ecosystems and People in a Changing World. Island Press, Washington, D.C. 174 p.

Walther, K., Sartoris, F.J., Bock, C., Portner, H.O., 2009. Impact of anthropogenic ocean acidification on thermal tolerance of the spider crab *Hyas araneus*. Biogeosciences 6, 2207–2215.

Wiber, M., Berkes, F., Charles, A., Kearney, J., 2004. Participatory research supporting community-based fishery management. Mar. Policy 28, 459–468.

Wiber, M.G., Young, S., Wilson, L., 2012. Impact of aquaculture on commercial fisheries: fishermen's local ecological knowledge. Hum. Ecol. 40, 29–40.

Zagonari, F., 2008. Integrated coastal management: Top–down vs community-based approaches. J. Environ. Manage. 88, 796–804.

Making the Link

By Paul Tett

Motivation for the Viability of the Lobster Fishery: Case Study of the Acadian Coast of New Brunswick	⟺	Lobster Fisheries in Atlantic Canada in the Face of Climate and Environmental Changes: Can We Talk About Sustainability of These Coastal Communities?
Omer Chouinard, Liette Vasseur and Steve Plante		*Liette Vasseur*

As in many parts of the world, coastal fisheries in eastern Canada have declined as a result of excessive catch coupled with climatic change in the oceans. Typically, fisheries managers have approached this challenge through catch management directed at adjusting pressures on the fish stock. Both of these chapters argue that the solution to these challenges to coastal sustainability lies in understanding the social-ecological system (SES) within which the fishery and its target are located.

The two chapters focus on different aspects of the SES. In moving from the one that you have just read, to the next, it may be helpful to refer to Elinor Ostrom's analysis of the interactions between people and nature that take place when a natural resource is exploited (Figure 1). Simplifying somewhat, in Chouinard's chapter, the "resource system" is the fishing grounds, the "resource units" are the lobsters, the "users" are the fishers, and the "action situations" refer to the choices and decisions made by the fishers in response to the several flows of information coming to them. The challenge for the "governance system" is that of maintaining the sustainability of the lobster fishery and thus the livelihood of the communities that depend on it.

In the next chapter, by Vasseur, the point-of-view pulls back. The resource system is now that of coastal ecosystems. The resource units may be understood as the ecosystem services, including, of course, the lobsters, provided by these systems. The users are now the whole of civil society in these coastal regions of Atlantic Canada, and the long-term issue for the governance system is that of maintaining the resilience of this society to changes in local and external ecological and social systems resulting from, among other things, global warming.

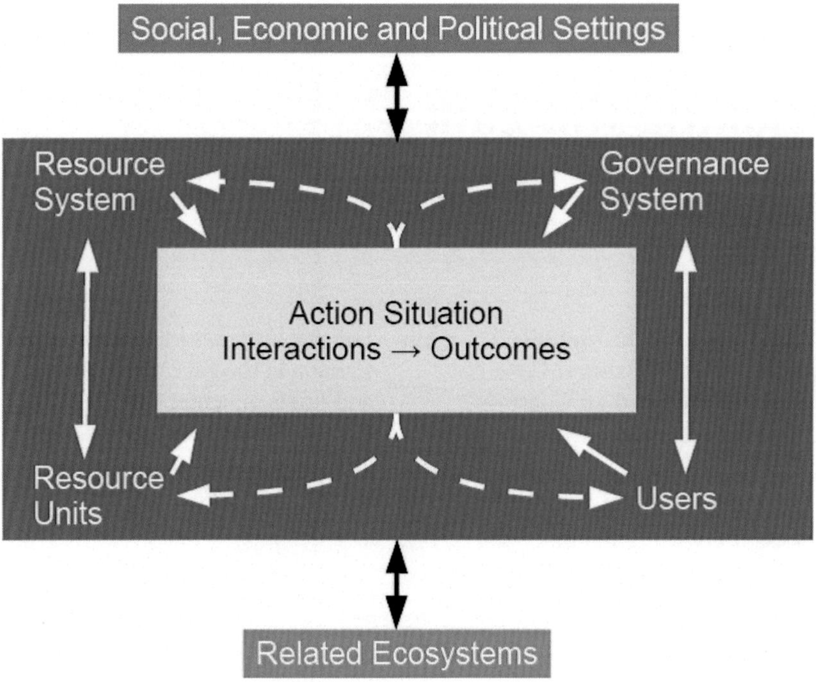

FIGURE 1 Action situations embedded in broader social-ecological systems. *(Based on Ostrom (2009), Figure 6.)*

REFERENCE

Ostrom, E., 2009. Beyond markets and states: polycentric governance of complex economic systems. In: Grandin, K. (Ed.), The Nobel Prizes 2009. Nobel Foundation, Stockholm, pp. 408–444.

Chapter 16

Lobster Fisheries in Atlantic Canada in the Face of Climate and Environmental Changes: Can We Talk About Sustainability of These Coastal Communities?

Liette Vasseur
UNESCO Chair in Community Sustainability: from local to global, Department of Biological Sciences, Brock University, Ontario, Canada

Chapter Outline

INTRODUCTION

Atlantic Canada is often characterized as being a coastal ecosystem because a large part of its population lives along its coasts. These communities since the time of the colonization of Canada have needed the sea for transportation to and from Europe and relied on the sea for food. Canada is one of the major maritime trading nations of the world and is highly dependent on its ports and shipping to maintain its economy. Because of its richness in fish resources, the federal government has historically worked to develop policies that could help protect these resources. However, over time, pressure to further develop coastal communities and their economies has led to the erosion of conservative measures and toward more pro-development policies. While quotas have been placed on most important commercial fisheries, pressures continue to be exerted to increase activities in order to satisfy the demand and the economic needs of these communities. Managers and policymakers have been concerned for a long time about the lack of sustainability of most fisheries (Ojea and Loureiro, 2010).

The lobster fishery is one of the oldest fisheries of this region to be regulated. In Atlantic Canada, over 10,000 harvesters are involved in 45 lobster fisheries. In 2012, landings were 74,790 tons, up from 66,500 tons in 2011; these represent economic values of $662.8 million in 2012 and $619.7 million in 2011 (http://www.dfo-mpo.gc.ca/fm-gp/sustainable-durable/fisheries-peches/lobster-homard-eng.htm). This is mainly a commercial fishery with allowances for subsistence and ceremonies for the First Nations. With increasing pressure on the industry, since the 1990s, conservation efforts have been initiated in order to protect the species and improve the sustainability of the industry. In recent years, the industry has been in crisis. Landing prices remain low since the recession in the United States in 2009.

Until recently, very few efforts have been made to examine and integrate the issues related to climate change into the industry. Climate change has become another issue that most fishers do not want to discuss, as other pressing issues must be dealt with including landing prices, market demands, and quotas. Current observations and projected scenarios are pointing to major changes that can lead to either the displacement of the species or the decline of its stock (Khan et al., 2013). Sea surface temperature (SST) increase may lead to reduced growth and survival, especially for larvae. Ocean acidification can impact molding and survival of adults (Zhang et al., 2012). Combined with other environmental factors, it is unclear as to what can be expected and how these various components interact, possibly exemplifying the stress on some of the ecosystem components. There is a need to view climate change in the overall social-ecological system (SES) that includes many other factors in both human and natural parts and their interactions.

This chapter focuses on the SES of the lobster fishery in Atlantic Canada. I examine the various components that compose the SES, from the biological to the socioeconomic aspects. All these aspects act directly and indirectly on the lobster as a species as well as all the other components that derive from its ecology and industry. It is a complex system that needs to be understood in order to better define solutions to the sustainability challenge that this resource is facing. This chapter briefly describes the fishery and the current environmental and climatic conditions facing this system (refer to Chapter 16 for greater details on the ecology of the lobster). Then, using a simplified interactions diagram of the various components of this SES, I summarize the interactions among the various elements of this SES. I describe some current and potential ways to move forward and ensure sustainability of this coastal ecosystem. By analyzing the various components of this system, I gradually show that climate change represents only one variable in this system, but its impacts can have very important ramifications in both communities and the natural ecosystem. This approach is innovative, as it helps better visualize the various components of the lobster SES that usually are not considered or simply assumed. From this understanding, different solutions and actions are described and proposed to ensure a certain level of sustainability of this species and its industry. Some solutions are already being implemented, such as the actions of the Atlantic Lobster Sustainability Foundation (ALSF), and may be able to support the industry's efforts to maintain its sustainability. Other adaptation strategies, such as changes in fishery seasonality, zoning issues, and ecosystem-based adaptation, will require greater efforts to be implemented in the long term. These solutions are discussed in the light of how such factors may not only help adapt to climate change but also maintain sustainability in the face of other factors such as world market demands.

CLIMATE AND ENVIRONMENTAL CHANGES IN ATLANTIC CANADA

Atlantic Canada's climate has changed on average by 0.3 °C from 1948 to 2005, with increases mainly in the summer months (Galbraith and Larouche, 2013). At the same time, mean annual precipitation has continued to increase, with a change of 10% between 1948 and 1995 (Lines et al., 2003). Vasseur and Catto (2008) report that increases in mean annual temperatures of 2–4 °C and 1.5–6 °C in the winter are projected for 2050. In terms of mean annual precipitation, trends will remain similar to those reported for the past 50 years, with drier summers expected and more precipitation in fall and winter.

Coastal ecosystems are sensitive to environmental (such as water pollution, sedimentation) and climatic changes. Projections suggest that Atlantic Canada will continue to be increasingly affected by storm surges (Forbes et al., 2004). In the last decade, several storm surges have caused substantial damage to coastal infrastructure. With an increased number of hurricanes or tropical storms in the Atlantic Ocean and storm surges exceeding 3.6 m, this ecosystem is quite vulnerable from various perspectives including lobster survival and habitat, fishing fleet, and coastal infrastructure. Sea-level rise also impacts this ecosystem, especially human communities living along the coast (Friesinger and Bernatchez, 2010). The region has been subjected to sea-level rise combined with the postglacial vertical motion of the coast. For Kouchibouguac National Park of Canada, for example, this translates into a sea-level rise of 51 ± 35 cm between 2000 and 2100 with a 7 ± 5 cm vertical motion (Forbes et al., 2006). Considering that the topology of this region is relatively flat, these changes can lead to significant modifications of the coastal infrastructures, including fishing wharfs.

SST is also expected to increase over time, and with this, the area and duration of sea ice along the coast and especially in the Northumberland Strait between New Brunswick and Prince Edward Island. SST variation can also lead to changes in seasonality. The North American Oscillation (NAO) influences the climate of coastal ecosystems on both sides of the Atlantic, mainly between latitudes 40° and 60° N, and its influence can be even greater during the winter months (November–April) (Stenseth et al., 2003). Changes in winds patterns provoked by the NAO (especially positive or negative extremes) influence SST, ice cover, and ocean currents.

Ocean acidification is caused by the accumulation of CO_2 in the oceans (IPCC, 2007). Caldeira and Wickett (2003) have modeled the future scenarios under the current increases of CO_2 in the atmosphere and reported potential sea surface levels of 710 ppm CO_2 by 2100. Ocean current and winds help dissolve CO_2 in the water column, especially in shallow oceanic regions. In addition, cooler waters like frequently found in Atlantic Canada can be a factor that makes the region more vulnerable due to greater solubility (Benoît et al., 2012). Inputs from terrestrial ecosystems can exacerbate the phenomenon by adding nutrient loads. Cai et al. (2011) find that eutrophication may also trigger acidification, since respiration from benthic organisms can increase CO_2 production, thus reducing pH.

Discharges of sewage and agricultural runoff represent other major sources of pollution and gradual eutrophication of the coastal ecosystems (Diaz, 2001). Strong stratification of the water column and proliferation of microbial communities due to enrichment coming from these inputs have led some zones to be seasonally or permanently depleted in dissolved oxygen. This phenomenon, called hypoxia, may be observed more frequently now than in the past, such as in several zones of the southern Gulf of St. Lawrence. For example, Plante and Courtenay (2008) report very low levels of dissolved oxygen in the coastal area of Lamèque, New Brunswick due to the discharge coming from a fish-processing plant. According to Benoît et al. (2012), this phenomenon is becoming common in several small estuaries due to agricultural nutrient inputs. They suggest that continuous monitoring and remedial actions, like reduction of discharge from the fish processing plant at Lamèque (Plante and Courtenay, 2008), may be effective methods to reduce impacts.

LINKING THE LOBSTER TO CLIMATE AND ENVIRONMENTAL CHANGES

When examining the lobster in its SES, many parameters can be included. Figure 16.1 describes the relationships between the lobster and the other trophic levels, the competition among fisheries, the direct and indirect impacts of human activities such as pollution, including climate change effects on the lobster and its habitat, and finally the coastal communities and their livelihood. This schematic representation (Figure 16.1) of the SES does not evaluate variation of biomass or productivity of the lobster like Ecopath and Ecosim by Zhang et al. (2012). It represents the elements that interact and should be considered to make this SES resilient to climate change.

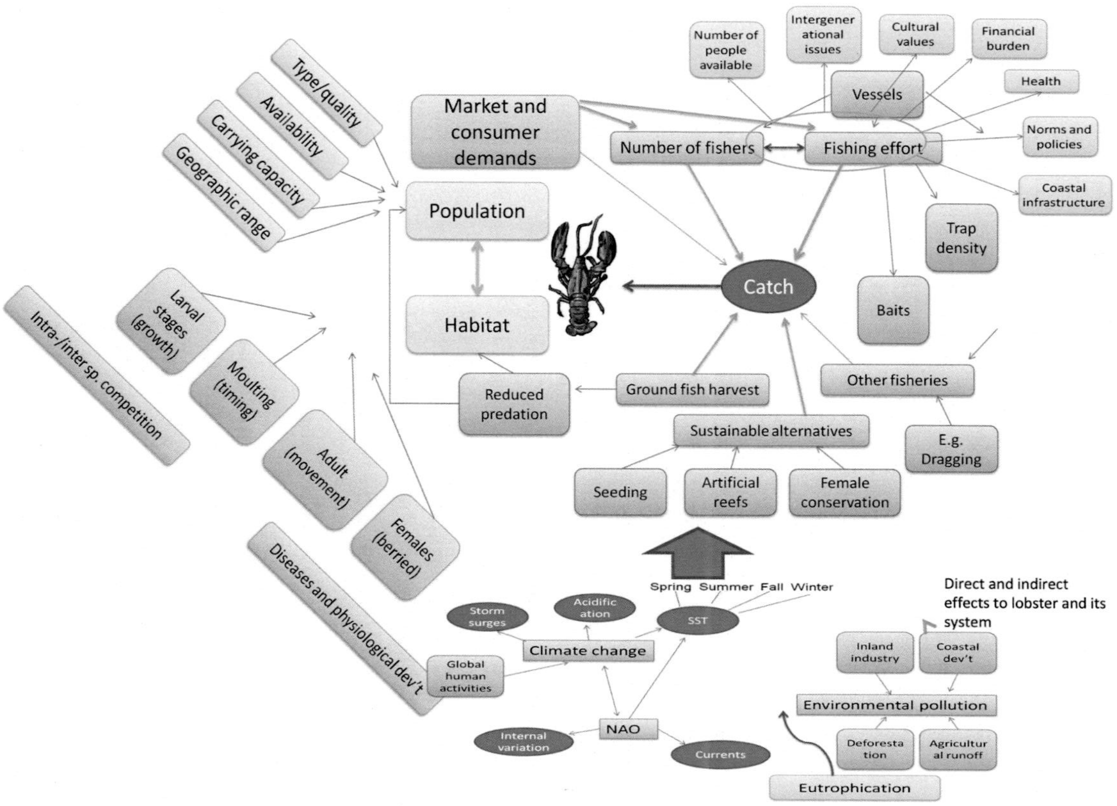

FIGURE 16.1 Schematic diagram describing the links and interactions between the lobster and its social-ecological system. While some of the connections may be positive, others can be negative.

Lobster has a specific temperature range at which its physiology is optimal, and therefore changes in ocean temperatures may have profound impacts on its ecology and distribution. Like the European lobster (*Homarus gammarus*), the development of both embryos and larvae of the American lobster is largely affected by the temperatures to which females and larvae are exposed (Schmalenbach and Franke, 2010). The current temperature regime allows for optimal timing for food availability for larval growth. Dove et al. (2005) suggest that increased temperatures of bottom waters may have had an effect on the lobster physiology, reducing its immune efficiency and survival potential. In the long term, changes in SST and bottom temperatures may reduce the lobster's capacity to respond to stress, becoming more prone to disease and being displaced by other more aggressive species (Tlusty et al., 2008). While SST increases may over the short term lead to increased abundance of lobster, if its thermal tolerance were reached, it is predicted that populations can collapse or migrate to cooler and/or deeper waters (Khan et al., 2013).

Increased water temperatures can also lead to the introduction of exotic and invasive species such as the green crab (*Carcinus maenas*). Some of these species may negatively impact the ecology of the lobster in various ways, such as displacing lobster due to competition for the same habitat or prey, or increasing predation on the lobster. Green crab can feed on various prey including gastropods, bivalves, and algae, many of them also being prey for lobster, thus increasing competition for food among species (Williams et al., 2006). Klassen and Locke (2007) report that while the habitat of green crab and lobster may be different in terms of substrate, fishers from the eastern shore of Nova Scotia have found green crab in their lobster traps. They also mentioned that adult lobster can predate on green crab. However, Lynch and Rochette (2009) suggest that green crab has no influence on the food and shelter of juvenile lobster. But Haarr and Rochette (2012) show that in fact, at a young age, green crab can use lobster as prey. Interestingly they also report that interactions varied depending on the geographic origin of the green crab. There has been discussion to even use green crab as bait for lobster (Atlantic Lobster Sustainability Foundation, personal communication).

Lobsters are known to have few natural pathogens. Most are from holding facilities and include, for example, gaffkemia (bacterium), burnt spot shell disease (various lesion forms with bacteria of genus *Vibrio* and fungi), epizootic shell disease (complex with fungi, bacteria, and protists and linked to environmental conditions), fungal infection, and bumper car disease (scuticociliate *Anophryoides haemophilia*) (Cawthorn, 2011). Increased water temperatures have been suggested, at least in the Gulf of Maine, to be related to the proliferation of epizootic shell disease (Gomez-Chiarri and Cobb, 2012). These authors note that other factors may also play a role in the level of infection including hypoxia, pollutant loads, etc. The main result is a decline in metabolic health of the animals leading to greater susceptibility to other infection.

Water temperature is not only affected by climate change but also by larger oceanic and global systems such as the NAO (González Herraiz et al., 2009).

While few studies have examined its potential impact on American lobster, the Norway lobster, *Nephrops norvegicus*, and its ecosystem have been shown to be greatly affected by changes in the NAO (González Herraiz et al., 2009). In this study, they report that a positive NAO bringing winds and winter storms lead to a decrease in biomass at the lower trophic levels, which serve as food for the lobster. While they admit that other population dynamics mechanisms may influence the abundance of the species, the NAO (e.g., changes in wind intensity and direction in March and April) has as a consequence to reduce larval survival and settlement, thus affecting future recruitment in local populations (González Herraiz et al., 2009). Edwards (2011) reports that seasonality shifts can lead to variation in the predator–prey relationships. Khan et al. (2013) also suggest that, although the responses of fish may be quite complex, some species ranges may also change with SST increase and shifts in seasonality. It is therefore possible that similar changes can occur for the American lobster, leading to changes in its distribution and viability.

Wind intensity has been shown to affect the settlement success of lobster (Hudon and Fradette, 1988, 1993; Wahle and Incze, 1997). Ocean acidification may affect survival and distribution as most aquatic species, including lobster, have physiological limits in terms of pH. For example, changes in pH with increasing temperatures led the spider crab to reach its physiological levels of tolerance faster than with temperature increase alone (Walther et al., 2009). All of these combined effects can impact aquatic communities and the interactions among trophic levels. Zhang et al. (2012) have hypothesized that the change in lobster stock may be related to a trophic cascade in which mesopredator groundfish like cod may have affected larvae and young adult lobster while at the same time, large adult lobster control predation of their prey such as urchins, crab, molluscs, etc. They argue that "understanding the dynamics of ecosystem is important in developing ecosystem-based fisheries management" (Zhang et al., 2012, p. 453).

OTHER ENVIRONMENTAL IMPACTS

Coastal ecosystems may be more impacted by human activities than open sea due to pollutants discharged in estuaries and along coastal shores from terrestrial activities (Collie et al., 2008). These pollutants not only include nutrients but also substances such as endocrine disruptors and other complex chemicals that can affect the health of aquatic organisms. The organochlorine pesticide endosulfan is one of such chemicals that may have a significant effect on lobster. It has been used, especially in Prince Edward Island, for agricultural purposes, where soils are susceptible to erosion and run off. Although it is in the process of being phased out (Health Canada, 2011), as a persistent organic pollutant, it can enter the aquatic environment where concentrations can be higher than the Canadian Council of Ministers of the Environment short exposure levels (Dunn et al., 2010: concentrations of 0.19–0.3 μg/L after a 10-m buffer zone instead of a limit of 0.09 μg/L for the protection of aquatic life). Bauer et al. (2013) show that endosulfan significantly increases mortality, and impacts development and gene expression.

Hypoxic conditions in shallow coastal waters may have more important impacts on sessile fauna than pelagic fish. Benoît et al. (2012, p. 14) state "no published study of the impact of seasonal hypoxia on the fauna of small estuaries in the southern Gulf of St. Lawrence is available". Other studies have found that American lobster like some other crustaceans may be quite sensitive to hypoxia (Miller et al., 2002; Vaquer-Sunyer and Duarte, 2008). Although research is still early in this region regarding the impacts of ocean acidification and hypoxia on lobster, other studies have found that invertebrates including molluscs, crustaceans, and corals can all be greatly affected (e.g., Kleypas and Yates, 2009; Walther et al., 2009; Miller et al., 2009; Whiteley, 2011). The interactions among environmental conditions may lead to more complex responses and changes in the tolerance threshold of most species. Considering the challenges of hypoxia and eutrophication in some regions of Atlantic Canada (e.g., the southern Gulf of St. Lawrence), it is expected that lobster facing acidification, warmer temperatures, and hypoxia may become more vulnerable in the long term, especially during molt.

Fisheries Pressures

The social and economic pressures also exacerbate the current and potential impacts on the ecological system, further affecting lobster sustainability. As several fisheries operate in the same region, direct and indirect impacts on resources can occur. Benoît et al. (2012) state "fishing impacts can be direct, such as the reduction of targeted and incidentally captured populations and truncations in their age and size composition, or indirect via the alteration of foodweb structure." The indirect impacts not only include bycatch and balance of predators–prey but also all the other impacts such as habitat degradation and lost gear. Combined with previously described climate and environmental changes, we can expect that the ecosystem including commercial species may be at risk of extirpation or may migrate in order to survive.

Fisheries have historically targeted not only lobster but mainly groundfish and some pelagic fish (Benoît et al., 2012). Some fisheries such as cod, redfish, and white hake have gradually collapsed. Low numbers of these species have an impact on the food web, and can directly affect the lobster: reduced predation and competition, leading to increased survival and population growth. Decline in the groundfish fishery has led to an increase in landings of numerous high-value invertebrate fisheries such as snow crab (*Chionoecetes opilio*), northern shrimp (*Pandalus borealis*), and lobster (Benoît et al., 2012).

The lobster fishery has not changed in terms of gear over time, except for some minor changes to avoid catching larger females. The issue of baits, however, has been increasingly discussed in fisheries management due to the lack of bait (Brzeski, 2012). They usually consisted of mackerel, herring, capelin, etc.

and were caught when needed. In the past decade, baits have declined due to additional pressures on those fisheries leading to greater prices and having to purchase them in other regions including frozen mackerel from Japan. The economic decline of 2008 to 2009 has brought this fishery to a low level in terms of market prices. Lower values have led fishers to try to catch more lobsters, thus increasing their fishing efforts and further exacerbating the impacts on baits. The consequences are greater pressures on the resources as well as on the float. In the section on solutions, we will discuss the compromise that fishers may have to make in order to maintain the sustainability of this SES.

Coastal Communities

The collapse of the cod fisheries in the 1990s had profound impacts on coastal fishing communities (Miller and Breen, 2010). Most of them relied heavily on this fishery to survive, and communities did not find the need to diversify their economy. This situation has demonstrated the high level of vulnerability of such communities and the lack of, or limited, resilience when conditions are drastically changing. It is important to understand that like in any ecosystem where diversity is considered an essential component for its health, similarly economic diversity should help communities to buffer changes when such a collapse happens. The lobster fishery has become the main, if not the only, economic activity creating precarious conditions for this SES. According to the past experiences in fisheries, we may want to be cautious, and although no one would wish this, it can still happen.

Coastal communities as part of the lobster SES are vulnerable for various reasons. First, the infrastructure is aging and requires constant maintenance (DFO, 2012). These infrastructures include, for example, wharf, float, gear, storage, and transformation plants. With lower prices for lobster, over the years, the financial burden of fishers has certainly increased. In order to compensate, as previously mentioned, most have increased their lobster landings with the challenge of having to sell them at the best price. Unfortunately with market saturation, the prices have yet to increase. In their synopsis, Plaganyi et al. (2011) have demonstrated that for better adaptive management of the fisheries, there is a need for integrating the various components from individuals to ecosystems including the socioeconomic aspects. A challenge that the fishery will most likely face as rapid short-term changes can happen, as they have shown in their case studies on sardine and sea urchins populations.

STRATEGIES AND ACTIONS

Many factors are acting on the lobster and its ecosystem. Zhang et al. (2012) mentioned the challenge of dealing with uncertainty in trying to capture the complete picture of this complex system. These factors must include not only climate change but also other pressure coming from the fishery industry, market demand, and indirectly from other fisheries. It is therefore important to not lose sight of the complexity of the system when actions and strategies are being developed to enhance the sustainability of the industry, which in fact means the sustainability of the lobster as an ecological component of the coastal ecosystem. Over the past decades, most fisheries have been shown to be unsustainable in large part because of the lack of understanding of the complexity of the ecosystem in which they are embedded. As Branch et al. (2006) mention, the current failure in fisheries management relates to a disconnect among disciplines, such as economics, sociology, and ecology, which can have the potential to link the fish to the fisher. Rosenberg (2007) and Pannozzo (2013) have also argued that most sustainable solutions are not implemented due to political reasons and lobbying by the fishing industry and other stakeholders. In this section, we examine some of the possible actions or strategies that can be integrated in order to help enhance resilience and sustainability. Some of them have been or are being implemented already, but several are only at an early stage.

Looking at sustainable fisheries means, as stated by Pilling and Payne (2008, p. 1), "...the maintenance of the quality, diversity and availability of fishery resources in sufficient quantities for present and future generations." To achieve this goal, the fishery would have to consider human needs now and in the future, natural and human-induced changes, and all other components of the ecosystems as I have argued in the previous section. In fact, solutions have been proposed and include development of ecosystem-based co-management plans, establishment of various forms of marine reserves or protected areas, total allowable catch and individual transferable quotas, and reductions in fishing fleet capacity or modified gear regulations. Steinback et al. (2008) suggest that examining various options, reduction of traps not only reduces pressure on the species but also can help enhance employment through diversification. Most of these strategies should integrate some basic elements based on the precautionary principles as proposed by Garcia (1994) during the work on the Code of Conduct for Responsible Fisheries (FAO, 1995):

- Adopt the sustainable development principle.
- Adopt the principle of precautionary management.
- Use the best scientific evidence available.
- Adopt a broad range of management benchmarks and reference points.
- Develop criteria for use when assessing the impacts of development.
- Take a risk-averse stand.
- Agree on acceptable levels of impacts and risk.
- Take a holistic view of resources within their environment.
- Speed up management response time.
- Allow for greater participation by nonfishery users in management bodies.
- Improve decision-making procedures.
- Introduce prior consulting procedures.
- Strengthen monitoring, control, and surveillance.

A very important ingredient that has often been missing in discussions is the understanding of who should be at the table. Fishers and the industry (including processor and suppliers) have been for a long time the main group considered by governments which apply regulations and promote certain types of management. Pauly and Chuenpagdee (2007) and Jentoft and McCay (1995) have argued that to be more effective, management should be more bottom-up, decentralized and community-based in which fishers and their communities have more stake in what is being developed. By trying to meet the current top-down approach with a bottom-up action, more realistic solutions may be adopted (see ecosystem-based adaptation below). While a better representation of all stakeholders may first seem complex, their involvement may increase engagement in decision making and protection of the resources. Under such a scenario, everyone would have to understand the consequences of various actions such as maintaining high landings despite low prices. The question of short-term gain versus long-term sustained gain would have to be discussed. Everyone's involvement in developing community-based "co-management" systems (since decision-makers and communities are both "managers" of the resource) would certainly enhance their role and responsibility in the face of the resource and its SES. Kirkwood and Agnew (2004) argue that illegal, unreported, and unregulated fishing may remain one of the stumbling blocks for developing such a model. It refers to the tragedy of the commons and the current gritty mentality of people.

Currently the use of quotas, limitation in the number of licenses, minimum and maximum size limits, and returning berried females to the ocean have been the most important strategies to try to maintain constant landings. However, in several locations, especially in the Northumberland Strait, the stocks are still declining. Can these solutions be implemented? And how? One organization may have the potential to help enhance the capacity of fishers and their communities to take a larger role in controlling their own industry: the ALSF.

Atlantic Lobster Sustainability Foundation

The ALSF was created in 2011 as a nonprofit organization that aims to support efforts and science research to enhance the sustainability of the lobster industry (www.lobstersustainability.ca). The foundation has been supported by private and public funds that have been given through a competitive process to organizations working on finding solutions. This model has had several advantages, among them having the capacity to be a neutral ground for discussion among researchers, governmental agencies, industry, and fishers regarding the sustainability of the lobster SES. Since 2013, the ALSF has joined efforts with the Lobster Institute in Maine for the annual workshop and to further expand the actions to find solutions.

Pilling and Payne (2008) have pointed out the importance of involving all actors of an SES as an ingredient for success. Cooperation among these actors can help reduce conflicts through dialog and effective partnership. The ALSF exactly reflects this approach. Sharing experiences and knowledge among stakeholders enhances the capacity of all of them to understand the challenges and find solutions that can be more sustainable. Solutions may take various forms including policies, ways to do business, and integrating new best practices.

Ecosystem-Based Adaptation and Resilience

With the lobster SES facing greater uncertainties and impacts due to climate change, new approaches must also be introduced at this level. One approach promoted by the International Union for Conservation of Nature and several other organizations is ecosystem-based adaptation. This toolkit (as it can be composed of several tools) is based on two pylons: SES and resilience (Andrade Pérez et al., 2010). While the number of case studies targeting coastal communities remains low (Rizvi, 2014), the results are promising and demonstrate the need to further develop tools to add to this approach. In this approach, the resilience of the SES is examined and solutions are based on good governance and inclusion of various stakeholders. Ecosystem-based adaptation therefore can effectively complement actions done by organizations such as the ALSF.

CONCLUSIONS

In its State of the Fisheries Report 2011, FAO (2011) states that of the 395 stocks that were assessed in 2009, 57.4% were fully exploited, 29.9% were overexploited, and only 12.7% were not fully exploited. With this current state, it is clear that we need to reflect on sustainability of the ocean SES. Climate change will only exacerbate the current conditions by increasing uncertainties in the conservation of the resources. There is a need to better understand the system in order to adapt to changes.

Although several organizations have been advocating for decades the need to use precautionary principles and sustainable practices in fisheries, stock declines remain an issue mainly because of the lack of understanding at several levels of the interrelationships between the ecology of the lobster and its environment and the human populations, as well as the pressures of the social and economic needs and demands. Decision-makers are constantly being lobbied for more landings in order for fishers to make a decent living, but at what expense? The lack of consistent data on the biology and the stock of lobster populations, and the current and future impacts of other factors such as environmental and climate changes, limits the capacity to predict the long-term health of the lobster SES. Various solutions have been proposed in this chapter. In general, they demonstrate the urgent need to move forward on the interconnectedness among the components of the system. But at the same time, like most other fisheries, it may be time for all levels of the chain (form fishers to consumers) to start rethinking the way fisheries have been perceived in the past as a commodity and focus on how in the long term, it can evolve in a resource that is valued for its ecosystem service.

As Pilling and Payne (2008, p. 8) underline, "There is no magic bullet, however. Specific fisheries require specific approaches to solve the issues inherent in the resources they exploit." For the Atlantic Canada lobster, steps toward solutions have been positive and must be maintained in the right direction. It is important to consider the culture of the fishery, the nature of these coastal communities and their socioeconomic circumstances, but especially the lobster in its natural ecosystem and how human activities are impacting its various components. With a clear understanding of the various elements that make this SES, reduced vulnerability and higher resilience is possible. Coastal communities have been relying for a long time on natural resources to survive. By linking the various actors of coastal communities through discussion and dialog, a greater understanding of these interactions among all the components of the SES will become clearer. This can encourage people to find solutions that may be innovative and more integrated to their livelihoods and the ecology of the species. Such steps are becoming more and more crucial as pure scientific or policy management and decisions may not encompass all new impacts that can occur under scenarios of climate change.

REFERENCES

Andrade Pérez, A., Herrera Fernández, B., Cazzolla Gatti, R. (Eds.), 2010. Climate Change and Ecosystems: Case-studies of Impacts and Ecosystem-Based Adaptation. IUCN Commission on Ecosystem Management (CEM). Climate Change Adaptation Group.

Bauer, M., Greenwood, S.J., Clark, K.F., Jackman, P., Fairchild, W., 2013. Analysis of gene expression in *Homarus americanus* larvae exposed to sublethal. Comp. Biochem. Physiol. Part D 8, 300–308.

Benoît, H.P., Gagné, J.A., Savenkoff, C., Ouellet, P., Bourassa, M.-N. (Eds.), 2012. State-of -the-Ocean Report for the Gulf of St. Lawrence Integrated Management (GOSLIM) Area. Can. Manuscr. Rep. Fish. Aquat. Sci. 2986: viii + 73 pp.

Branch, T.A., Hilborn, R., Haynie, A.C., Fay, G., Flynn, L., Griffith, J., Marshall, K.N., Randall, J.K., Scheuerell, J.M., Ward, E.J., Young, M., 2006. Fleet dynamics and fishermen behaviour: lessons for fisheries managers. Can. J. Fish. Aquat. Sci. 63, 1647–1668.

Brzeski, V., 2012. Adapting Atlantic Canadian Fisheries to Climate Change. Ecology Action Centre. 8 p.

Cai, W.-J., Hu, X., Huang, W.-J., Murrell, M.C., Lehrter, J.C., Lohrenz, S.E., Chou, W.-C., Zhai, W., Hollibaugh, J.T., Wang, Y., Zhao, P., Guo, X., Gundersen, K., Dai, M., Gong, G.-C., 2011. Acidification of subsurface coastal waters enhanced by eutrophication. Nat. Geosci. 4, 766–770.

Caldeira, K., Wickett, M.E., 2003. Anthropogenic carbon and ocean pH. Nature 425, 365.

Cawthorn, R.J., 2011. Diseases of American lobsters (*Homarus americanus*): a review. J. Invertebr. Pathol. 106, 71–78.

Collie, J.S., Wood, A.D., Jeffries, H.P., 2008. Long-term shifts in the species composition of a coastal fish community. Can. J. Fish. Aquat. Sci. 65, 1352–1365.

DFO (Department of Fisheries and Oceans), 2012. Risk-based Assessment of Climate Change Impacts and Risks on the Biological Systems and Infrastructure within Fisheries and Oceans Canada's Mandate – Atlantic Large Aquatic Basin. 40 p.

Diaz, R.J., 2001. Overview of hypoxia around the world. J. Environ. Qual. 30, 275–281.

Dove, A.D., Allam, B., Powers, J.J., Sokolowski, M.S., 2005. A prolonger thermal stress experiment on the American lobster, *Homarus americanus*. J. Shellfish Res. 24, 761–765.

Dunn, A., Julien, G., Ernst, W., Cook, A., Doe, K., Jackman, P., 2010. Evaluation of buffer zone effectiveness in mitigating the risks associated with agricultural runoff in Prince Edward Island. Sci. Total. Environ. 409, 868–882.

Edwards, M., 2011. Change at the community level. Nat. Clim. Change 1, 391–398.

FAO, 1995. Code of Conduct for Responsible Fisheries. FAO Fisheries Department, Rome.

FAO, 2011. State of the World Fisheries and Aquaculture. 2012. Fisheries and Aquaculture Department, Rome, www.fao.org/docrep/016/i2727e.pdf.

Forbes, D.L., Parkes, G.S., Manson, G.S., Ketch, L.A., 2004. Storms and shoreline retreat in the southern Gulf of St. Lawrence. Mar. Geol. 126, 63–85.

Forbes, D.L., Parkes, G.S., Ketch, L.A., 2006. Sea-level rise and regional subsidence. Chapter 4. In: Daigle, R., Forbes, D., Parkes, G., Ritchie, H., Webster, T., Bérubé, D., Hanson, A., DeBaie, L., Nichols, S., Vasseur, L. (Eds.), Impacts of Sea Level Rise and Climate Change on the Coastal Zone of Southeastern New Brunswick; Environment Canada 613 p.

Friesinger, S., Bernatchez, P., 2010. Perceptions of Gulf of St. Lawrence coastal communities confronting environmental change: hazards and adaptation, Québec, Canada. Ocean Coastal Manage. 53 (11), 669–678.

Galbraith, P.S., Larouche, P., 2013. Trends and variability in air and sea surface temperatures in eastern Canada. In: Loder, J.W., Han, G., Galbraith, P.S., Chassé, J., van der Baaren, A. (Eds.), Aspects of Climate Change in the Northwest Atlantic off Canada, pp. 1–18 Can. Tech. Rep. Fish. Aquat. Sci. 3045, x + 190 p.

Garcia, S.M., 1994. The precautionary principle: its implications in capture fisheries management. Ocean Coastal Manage. 22, 99–125.

Gomez-Chiarri, M., Cobb, J.S., 2012. Shell disease in the American lobster, *Homarus americanus*: a synthesis of research from the New England lobster research initiative: lobster shell disease. J. Shellfish Res. 31, 583–590.

González Herraiz, I., Torres, M.A., Farina, A.C., Freire, J., Cancelo, J.R., 2009. The NAO index and the long-term variability of *Nephrops norvegicus* population and fishery off West of Ireland. Fish. Res. 98, 1–7.

Haarr, M.L., Rochette, R., 2012. The effect of geographic origin on interactions between adult invasive green crabs *Carcinus maenas* and juvenile American lobster *Homarus americanus* in Atlantic Canada. J. Exp. Mar. Biol. Ecol. 422-423, 88–100.

Health Canada, 2011. Discontinuation of endosulfan. Pest Management Regulatory Agency, Health Canada, Ottawa. 8 p.

Hudon, C., Fradette, P., 1988. Planktonic growth of larval lobster (*Homarus americanus*) off Îles de la Madeleine (Québec), Gulf of St. Lawrence. Can. J. Fish. Aquat. Sci. 45, 868–878.

Hudon, C., Fradette, P., 1993. Wind-induced advection of larval decapods into Baie de Plaisance (Îles de la Madeleine, Québec). Can. J. Fish. Aquat. Sci. 50, 1422–1434.

IPCC, 2007. Climate change 2007: mitigation of climate change. In: Metz, B., Davidson, O.R., Bosch, P.R., Dave, R., Meyers, L.A. (Eds.), Contribution of Working Group III to the Fourth Assessment Report of the Intergovernmental Panel on Climate Change. Cambridge University Press, Cambridge.

Jentoft, S., McCay, B.J., 1995. User participation in fisheries management: lessons drawn from international experiences. Mar. Policy 19, 227–246.

Khan, A.H., Levac, E., Chmura, G.L., 2013. Future sea surface temperatures in large marine ecosystems of the Northwest Atlantic. ICES J. Mar. Sci. 70, 915–921.

Kirkwood, G.P., Agnew, D.J., 2004. Deterring IUU fishing. In: Payne, A.I.L., O'Brien, C.M., Rogers, S.I. (Eds.), Management of Shared Fish Stocks. Blackwell, Oxford, pp. 1–22.

Klassen, G., Locke, A., 2007. A biological synopsis of the European green crab, *Carcinus maenas*. Can. Manuscr. Rep. Fish. Aquat. Sci. 2818, vii + 75 pp.

Kleypas, J.A., Yates, K.K., 2009. Coral reefs and ocean acidification. Oceanography 22, 108–117.

Lines, G., Pancura, M., Landeer, C., 2003. Building climate change scenarios of temperature and precipitation in Atlantic Canada using statistical downscaling model (SDSM). In: 14th Symposium on Global Change and Climate Variations, American Meteorological Society Annual Meeting, Long Beach, California, pp. 1–25.

Lynch, B.R., Rochette, R., 2009. Spatial overlap and biotic interactions between subadult American lobsters, *Homarus americanus*, and the invasive European green crab *Carcinus maenas*. J. Exp. Mar. Biol. Ecol. 369, 127–135.

Miller, A.W., Reynolds, A.C., Sobrino, C., Riedel, G.F., 2009. Shellfish face uncertain future in high CO_2 world: influence of acidification on oyster larvae calcification and growth in estuaries. PLoS One 4, e5661.

Miller, D., Poucher, S., Coiro, L., 2002. Determination of lethal dissolved oxygen levels for selected marine and estuarine fishes, crustaceans, and a bivalve. Mar. Biol. 140, 287–296.

Miller, R.J., Breen, P.A., 2010. Are lobster fisheries being managed effectively? Examples from New Zealand and Nova Scotia. Fish. Manage. Ecol. 17, 394–403.

Ojea, M., Loureiro, M.L., 2010. Valuing the recovery of overexploited fish stocks in the context of existence and option values. Mar. Policy 34, 514–521.

Pannozzo, L., 2013. The Devil and the Deep Blue Sea. An Investigation into the Scapegoating of Canada's Grey Seal. Fernwood Publishing, Halifax. 175 p.

Pauly, D., Chuenpagdee, R., 2007. Fisheries and coastal ecosystems: the need for integrated management. In: Nemetz, P. (Ed.), Sustainable Resource Management. Edward Elgar Publ, Camberley, Surry, UK, pp. 171–185. Chapter 6.

Pilling, G.M., Payne, A.I.L., 2008. Sustainability and present-day approaches to fisheries management—are the two concepts irreconcilable? Afr. J. Mar. Sci. 30, 1–10.

Plaganyi, E.E., Weeks, S.J., Skewes, T.D., Gibbs, M.T., Poloczanska, E.S., Norman-Lopez, A., Blamey, L.K., Soares, M., Robinson, W.M.L., 2011. Assessing the adequacy of current fisheries management under changing climate: a southern synopsis. ICES J. Mar. Sci. 68 (6), 1305–1317.

Plante, F., Courtenay, S.C., 2008. Increased oxygenation of sediment in Lamèque Bay (New Brunswick) following removal of algae and reduction of nutrient inputs from a seafood processing plant. Can. Tech. Rep. Fish. Aquat. Sci. 2805, v + 36 pp.

Rizvi, A.R., 2014. Nature Based Solutions for Human Resilience. A Mapping Analysis of IUCN's Ecosystem Based Adaptation Projects, IUCN Report: 50.

Rosenberg, A.A., 2007. Fishing for certainty. Nature 449, 989.

Schmalenbach, I., Franke, H.-D., 2010. Potential impact of climate warming on the recruitment of an economically and ecologically important species, the European lobster (*Homarus gammarus*) at Helgoland, North Sea. Mar. Biol. 157, 1127–1135.

Steinback, S.C., Allen, R.B., Thunberg, E., 2008. The benefits of rationalization: the case of the American lobster fishery. Mar. Resour. Econ. 23, 37–63.

Stenseth, N.C., Ottersen, G., Hurrell, J.W., Mysterud, A., Lima, M., Chan, K.-S., Yoccoz, N.G., Ådlandsvik, B., 2003. Review article. Studying climate effects on ecology through the use of climate indices: the North Atlantic Oscillation, El Niño Southern oscillation and beyond. Proc. R. Soc. Lond. B 270, 2087–2096.

Tlusty, M., Metzler, A., Malkin, E., Goldstein, J., Koneval, M., 2008. Microecological impacts of global warming on crustaceans – temperature induced shifts in the release of larvae from American lobster, *Homarus americanus*, females. J. Shellfish Res. 27 (2), 443–448.

Vaquer-Sunyer, R., Duarte, C.M., 2008. Thresholds of hypoxia for marine biodiversity. Proc. Nat. Acad. Sci. 105, 15452–15457.

Vasseur, L., Catto, N., 2008. Chapter 4 – Atlantic Region. In: Lemmen, D.S., Warren, F.J., Lacroix, J., Bush, E. (Eds.), From Impacts to Adaptation: Canada in a Changing Climate 2007. Government of Canada, Ottawa, ON. pp. 119–170.

Wahle, R.A., Incze, L.S., 1997. Pre- and post-settlement processes in recruitment of the American lobster. J. Exp. Mar. Biol. Ecol. 217, 179–207.

Walther, K., Sartoris, F.J., Bock, C., Portner, H.O., 2009. Impact of anthropogenic ocean acidification on thermal tolerance of the spider crab *Hyas araneus*. Biogeosciences 6, 2207–2215.

Whiteley, N.M., 2011. Physiological and ecological responses of crustaceans to ocean acidification. Mar. Ecol. Prog. Ser. 430, 257–271.

Williams, P.J., Floyd, T.A., Rossong, M.A., 2006. Agonistic interactions between invasive green crabs, *Carcinus maenas* (Linnaeus) and sub-adult American lobsters, *Homarus americanus* (Milne Edwards). J. Exp. Mar. Biol. Ecol. 329, 66–74.

Zhang, Y., Li, Y., Chen, Y., 2012. Modeling the dynamics of ecosystem for the American lobster in Gulf of Maine. Aquat. Ecol. 46, 451–464.

Making the Link

By Bethany Jorgensen

Lobster Fisheries in Atlantic Canada in the Face of Climate and Environmental Changes: Can We Talk About Sustainability of These Coastal Communities?	⟺	**Universities as Solutions to Twenty-First Century Coastal Challenges: Lessons from Cheikh Anta Diop Dakar University**
Liette Vasseur		*Alioune Kane et al.*

Between two clearly related chapters about community adaptation approaches in Canada, we find one that seems, at first blush, tangentially related at best. So why is it here?

Set in the West African country of Senegal, Kane et al.'s chapter fits here for precisely this reason—it shakes things up. Offering a different perspective geographically and content-wise, we hope the juxtaposition of this work with that of Vasseur and Plante et al. reminds us that though work confronting coastal zone challenges takes many forms in many places, even the most varied projects can still inform one another.

Vasseur's chapter examines the question of sustainability in the social-ecological system of the lobster fishery in Atlantic Canada. She provides us with a thorough map of the interrelated elements operating within this system, and from there offers current as well as potential actions to be taken to maintain the sustainability of lobster, its industry, and the communities that rely on it.

But an underlying assumption in Vasseur's chapter is that it is within an academic's purview to do research that connects back to community concerns, and for universities to facilitate this kind of work.

Kane et al.'s chapter reminds us that in order for universities to play this role, they must have structures in place to support it. The university featured in Kane et al.'s work did not initially have an adequate framework for research related to coastal zones, but it recognized the need to change. This next chapter tells the story of that change process, and how the university has now become an integral part of national research and development systems related to West Africa's coastal areas.

Chapter 17

Universities as Solutions to Twenty-First Century Coastal Challenges: Lessons from Cheikh Anta Diop Dakar University

Alioune Kane[1], Jacques Quensière[2], Alioune Ba[1], Anastasie Beye Mendy[1], Awa Fall Niang[1], Ndickou Gaye[1], Aichetou Seck[1], Diatou Thiaw[1]

[1]Departement de Geographie, Université Cheikh Anta Diop, master GIDEL, Boulevard Martin Luther King, Dakar, Senegal; [2]IRD, UMI RESILIENCES Bondy Cedex, France

Chapter Outline

INTRODUCTION

In West Africa, coastal areas have gone through radical changes over the last 30 years. Population concentration, economic activities, intensification, and diversification have turned coastal zones into the main locus of economic and social development.

This has had a profound effect. The sudden business growth fundamentally changed coastal dynamics, now dominated by strong intersectoral competition, widespread overexploitation of natural resources, and also unequalled urban growth.

The resulting environmental degradation is all the more dangerous as it amplifies the threats associated with climate change, whose first effects are already visible. It thus considerably undermines the human populations that are concentrated in coastal areas. These are precisely the areas most endangered by environmental and climate risks.

To help the authorities control the situation, and to facilitate the sustainable management of coasts and coastal areas, Senegalese universities had to profoundly reconsider their analytical frameworks and teaching. Their goals shifted toward developing better understandings of complex social–ecological systems and rehabilitating the notion of territorial management as the best approach to sustainable development. The main steps of this long process are presented here.

THE TRANSFORMATION OF COASTAL AREAS IN WEST AFRICA

Despite their limited area and high vulnerability, coastlines and coastal environments have a richness and abundance of natural resources, which explains their strong appeal to human populations (Cicin-Sain et al., 2002). In the early 1990s, 60% of the world's human population was already living within 50 km of the coast. This figure is expected to grow to 75% by 2025, an increase of roughly 1.5% per year, which in absolute terms represents more than one billion people (UNEP, 2007).

In Senegal, as well as in all West African countries, the main economic drivers are now based on the exploitation of inshore and marine natural resources. This is the result of the combined effects of different trends—political, economic, social, and environmental—for which it would be presumptuous to determine the respective influences, as they are intimately intertwined. For instance, the 1970s drought generated massive out-migrations from inland rural areas (Sabrié and L'Hôte, 2003). This climatic deterioration, combined with declining markets, the resulting Senegalese economic crisis, the structural adjustment process in 1979, and the gradual withdrawal of the state through its New Agricultural Policy (Faye and Ndiaye, 1998), has fostered a growing vulnerability for agricultural production, especially groundnuts, as well as the increased impoverishment of rural areas (Meyer, 2009). A large number of farming households went in search of better living conditions in cities, or turned to the exploitation of open access to abundant fishery resources (Gascuel et al., 2002). The development of tourism and the real estate market favored the installation of populations on the coasts and in the major urban centers of the country. These urban centers are all situated on the coast, starting with Dakar, whose population tripled in 30 years (Le Roux, 2005) to around three million today.

Such a concentration of activities and population on the coast has not occurred without causing serious environmental damage. The extreme vulnerability of West African coastal systems has been identified by UNEP (2007), which considers more than 30% of human developments on the coast of Senegal, as well as several thousands of species, threatened with extinction in the short term. The high densification of human populations within coastal areas (Cour and Snrech, 1998; UNEP, 2007) generated a significant weakening of ecosystems through soil degradation (PNUE, 2006), massive urbanization (UNEP, 2007), and the over-exploitation of marine species and ecosystems (UNEP, 2007), forests, soil resources, and water resources (FAO, 2003). Finally, pollution and poisoning of the environment, such as soil and groundwater salinization, also contribute to destabilizing the natural systems and to a reduction of resources.

The most important consequence of the population concentration on the coast is major shifts in dominant drivers. Whereas previously a scarce population exerted a little pressure on the resources that kept, year after year, a natural dynamic marked by alternating tropical seasons, the massive colonization of coastal borders has encouraged the emergence of social–ecological systems governed by strong interaction between the different sectors involved. The complexity of the resulting dynamic now eludes sectorial approaches and requires much more delicate analysis and refined management methods.

COASTAL ZONES AS COMPLEX SYSTEMS

Over the last 30 years, an increasing number of studies have shown that the human and biophysical systems that coexist in the same space are bound by dynamic, reciprocal interactions (Berkes and Folke, 1998; Holling, 2001), generating hybrid systems whose sustainability is the result of a fluctuating balance between usages of natural wealth (resources, services) by humans and ecological constraints (Haberl et al., 2004). Therefore it is not enough to merely study the changes of ecosystem states, but one needs to also analyze the social pressures, which define the level of sustainability (Turner et al., 2003). Nature-society interactions are now considered a dynamic process of interaction between *autopoietic* systems (Fischer-Kowalski and Amann, 2001; Haberl et al., 2006); these interactions may no longer be considered as disturbances, by humans, of the natural balance (Gallopin, 2006). This new science of sustainability relies heavily on theories of complexity (Prigogine, 1976; Prigogine and Stengers, 1986) to deal with a new object: the socioecologicalsystem (Walker et al., 2002; Ostrom et al., 2007).

Thus, far from being a paradox or an oxymoron, as asserted by some economists (Beckerman, 1974; Hardwick, 1977; Solow, 1993), the concept of sustainability provides a unique framework that dovetails economic development with environmental sustainability and social justice (Brundtland, 1987). Risks, climate, sea-level rise, or diversity loss cannot be mitigated by understanding solely natural processes. One must follow the recommendations of the Earth System Science Partnership, the Millennium Ecosystem Assessment, the Resilience Alliance, and similar organizations striving to understand the coevolutionary dynamics of systems social and natural, i.e., socioecologicalsystems.

Under these conditions, collaboration between experts in various fields of research is necessary for the acquisition of relevant information and for a proper understanding of the mechanisms at work (Redman et al., 2004; Liu et al., 2007). This collaborative approach, which is rooted in the respect of disciplinary knowledge, generates interdisciplinary dynamics through the identification of common research objects (Kates et al., 2001; Ostrom et al., 2007; Blanchard and Vanderlinden, 2013).

Taking into account socioecological systems leads to a particular emphasis on the spatial dimension (Riebsame et al., 1994). Land use creates landscapes, and ecological systems, in which changes by humans can be analyzed as changes in the nature of the modalities that ecosystem services use (biomass production, biodiversity, water storage, etc.) (Daily, 1997). Thus, the analysis of nature–society interactions leads to promoting the region as a preferred scale (Rivera-Monroy et al., 2004), and also to integrating the key processes of a wide range of scales, from local to global (Allen and Starr, 1982; Kates et al., 2001), to address questions such as:

- What are the changes that socioeconomic activities cause to natural systems?
- Why do socioeconomic forces cause these changes, and how can we mitigate their impacts?
- What changes in natural systems have an impact on society?
- How can society cope with the changes it has initiated?

THE EVOLUTION OF SCIENCE AT CHEIKH ANTA DIOP UNIVERSITY

Coasts Are Geographical Objects

Before the Rio Summit in 1992, the coastline was not seen as a functional system that should be managed. Very few states were interested in their coastlines. In Senegal, the only coastal activity was fishing. Sectoral and biological models essentially defined coastal resources management. Science focused separately on the components of the coastal region, and was produced by different laboratories according to their specificities: mainly the combination of the Oceanographic Research Center of Dakar Thiaroye and ORSTOM for fisheries (Chauveau, 1984, 1985), regional physical oceanography (Rebert et al., 1976; Gerlotto et al., 1978), coastal agriculture, etc.

Within Dakar University, it is primarily the geography department that contributes to the improvement of marine and coastal knowledge, working closely with the organizations mentioned above and also with UNESCO, which at the time promoted interactions between researchers in the West African region. Senegalese research, in turn, played a leading role in the mobilization of a regional effort to acquire knowledge on coasts and coastal areas through its participation in various international networks.

Lessons from the Rio Summit: The Creation of a UNESCO Chair

During the 1990s, following the Rio Summit, awareness of the critical importance of preserving coastal areas was raised in West Africa through the development of economic activities on the maritime fringes. The intensification of export fisheries, in spite of scientists' recommendations, induced the diversification of activities, especially in the areas of transport and services. At the same time, concerns about global change appeared and highlighted the need for integrated coastal zone management.

Academic work was then undertaken, which, in 1997, led to the creation of a UNESCO Chair dedicated to the "integrated management and sustainable development of coastal areas and small West African islands" (http://www.unesco.org/csi/act/dakar/projec21.htm). The teachings associated with this effort were a first in West Africa in terms of multidisciplinarity and finalization toward natural resource management and uses.

The Science Conducted within and around the GIDEL Master's Program

Along the way, difficulties have arisen, due in part to the special status of the UNESCO Chair and its interdisciplinary practice in teaching and the associated research. In a university structure with strong disciplinarity, interdisciplinary work is less valued, especially as each individual scientist is placed under different hierarchies within his or her discipline. In a similar manner, interdisciplinary teaching lacks the luster of discipline-focused teaching, and thus has less appeal to students. However, the courses, which have been particularly appreciated, and internships have resulted in a very high rate of employment for students coming out of the interdisciplinary program.

The announcement of major higher education reform was therefore a good opportunity to take advantage of the experience gained in order to reinforce the directions taken in 1997. First, the scale was regional, for West Africa countries in consultation with teachers and researchers from other West African countries. The next step was to develop a consensus on content. This work was performed in two stages. A first discussion took place in 2004 on the definition of useful training for integrated and sustainable management of marine and coastal areas. During the workshop, which took place from May 25 to 27, 2004, three elements were confirmed:

- The need for further development of the multidisciplinary approach by having a clearer and more operational identification of objects situated at the nature–society interface, within coastal systems.
- Increased use should be made of the most recent processes in geographic information systems and spatial analysis as tools for efficient IT-based collaborative science.
- The need to federate experience and regional expertise tailored to the needs of countries in order to promote the emergence of regional expertise and knowledge on coastal management.

Based on these elements, a second regional workshop was organized in 2006. The choice of materials and descriptions of lessons were discussed by hundreds of teachers and researchers during the week. The model that resulted was officially approved, and the new master's program began in 2007.

TABLE 17.1 Description of the thematic areas and number of associated theses for the UNESCO Chair and the GIDEL master's programs

Master's Program	Thematic Area	Number of Master's Theses
UNESCO Chair	Landscape and resources on Senegal's northern coast	7
	Quality and use of the Yeumbeul water table	7
	Fisheries dynamics on Senegal's southern coast	6
	Coastal conservation areas and sustainable tourism	8
	Coastal erosion	4
	Coastal agriculture	6
	Natural resources: regional approaches to use and governance	6
	Benefits and vulnerability of mangroves	7
GIDEL	Freshwater resources on the coast	11
	Resilience of coastal communities	7
	Marine protected areas	7
	Coastal erosion under climate change	7
	Land use planning for coastal areas	11

One of the key elements that led to creation of the UNESCO Chair was the need for knowledge on the conflict arising between the needs of the coastal population and the need to preserve coastal systems facing multisource pressures. With the UNESCO Chair, a bridge was built between the university and national and international structures. It allows the university to position itself in the sphere of national research and development systems. The early years of the UNESCO Chair were characterized by exploratory projects and increasingly specific coastal thematic areas. Related fieldwork has taken place on the Cabo Verde Islands and on the southern and northern coasts of Senegal. The thematic areas (see Table 17.1) have been associated with the production of more than a 100 master's theses, some of which were extended into PhD dissertations.

After the higher education reform that led to its creation, the GIDEL master's program, while relying on the work of the UNESCO Chair, expanded its research into new areas such as vulnerability, resilience, and the management of Senegal's new marine protected areas (see Table 17.1).

The openness of the master's program toward West Africa allows it to position itself as a leader in the management of the West African coast. Conventions and cooperative relationships have been established with regional NGOs (PRCM, RAMPAO, FIBA, WETLANDS, WWF, etc.), while programs such as the Intra African Academic Mobility Program (PIMASO) and the International Oceanographic Commission were also created. The expertise of the UNESCO Chair/GIDEL master's program is increasingly sought and it welcomes professionals working in state structures (Ministry of Environment, sanitation, marine fisheries, etc.). GIDEL also relies on international events, such as Scientists' Days on the coast and various forums (Coastal Zone Canada IMPAC3, Africité, Assistant Regional Forum PRCM, EcoCity, etc.).

Nationally, the expertise developed in the GIDEL master's program is applied in various scientific and technical committees that focus on the future of coastal wetlands, on the institutional and legal management of MPAs, and on the implementation of the project "Mission Observing the West African Coast" (Moloa).

Current Thematic Work

The work carried out since the creation of the UNESCO Chair in 1997, has devoted much attention to spatial and temporal variability at different scales. Special attention is given to past and present choices related to the social components of the systems.

They converge toward approaches akin to the ORE, Observatoire Recherche Environnement; and to the SOERE, Systèmes d'Observation et d'Expérimentation pour la Recherche en Environnement; or even to GMES, Global Monitoring for Environment and Security. The central element is to sustain observation in the long run (Walker et al., 2002). Such an approach favors the creation of regional observatories (Haberl et al., 2004, 2006) as structures for dynamically documenting the phenomena and also as archival spaces (Liu et al., 2007).

Furthermore, if the monitoring of socioecological systems provides a better understanding of their operations, it also provides society with a finer picture of its interactions with nature. It contributes to local governance and decision-making (Ostrom et al., 2007). The work undertaken must therefore explicitly match social priorities and choices. The analysis of these choices and priorities should in turn be the focus of research (Folke et al., 2005).

For these reasons, the observation of two regional systems is currently being established. One focus will be on the evolution of the entire coastline of Senegal's southern coast; and the other, closely complementary to the first, will be looking at the evolution of the land. Particular attention will be paid to the economic and environmental changes induced by the process of metropolization of the Cap-Vert peninsula, which is closely associated with the building of a new toll highway. The development of these research tools reflects the evolution of Senegalese research seeking to be increasingly attuned to the development needs of Senegal and the West African region.

REFERENCES

Allen, T.F.H., Starr, T.B., 1982. Hierarchy: Perspectives for Ecological Complexity. University Chicago Press, Chicago.

Beckerman, W., 1974. In Defence of Growth. London Jonathan Cape.

Berkes, F., Folke, C. (Eds.), 1998. Linking Social and Ecological Systems: Management Practices and Social Mechanisms for Building Resilience. Cambridge University Press, Cambridge, UK.

Blanchard, A., Vanderlinden, J.-P., 2013. Prerequisites to interdisciplinary research for climate change: lessons from a participatory action research process in Île-de-France. Int. J. Sustainable Dev. (IJSD) 16, 1–22.

Brundtland, H.H., 1987. Our Common Future, Report of the World Commission on Environment and Development, Oxford University Press, Oxford.

Chauveau, J.-P., 1984. La pêche piroguière sénégalaise: les leçons de l'histoire. Revue mer no Spécial: 10–15.

Chauveau, J.-P., 1985. Histoire de la pêche maritime et politiques de développement de la pêche au Sénégal: représentations et pratiques du dispositif de l'intervention moderniste. Anthropologie maritime. 301–318.

Cicin-Sain, B., Bernal, P., Vandeweerd, V., Belfiore, S., Goldstein, K., 2002. A Guide to Oceans, Coasts, and Islands at the World Summit on Sustainable Development.

Cour, J.M., Snrech, S., 1998. Pour préparer l'avenir de l'Afrique de l'Ouest. Une vision à l'horizon 2020. Etudes des perspectives à long terme en Afrique de l'Ouest. C. d. Sahel, OCDE: 157.

Daily, G.C., 1997. Nature's Services: Societal Dependence on Natural Ecosystems. Island Press, Washington, D.C., US.

FAO, 2003. TERRASTAT Database: Land Resource Potential and Contrainst Statistics at Country and Regional Level.

Faye, J., Ndiaye, O., 1998. Formulation d'un Projet d'Appui Institutionnel aux Organisations Paysannes du Sénégal. Rapport provisoire. Cncr/Cirad-Tera.

Fischer-Kowalski, M., Amann, C., 2001. Beyond IPAT and Kuznets curves: globalization as a vital factor in analysing the environmental impact of socio-economic metabolism. Popul. Environ. 23, 41.

Folke, C., et al., 2005. Adaptive governance of social–ecological systems. Annu. Rev. Environ. Res. 30 (1), 441–473.

Gallopin, G.C., 2006. Linkages between vulnerability, resilience, and adaptive capacity. Glob. Environ. Change 16 (3), 293–303.

Gascuel, D., Laurans, M., Sidibé, A., Barry, M.D., 2002. Diagnostic comparatif de l'état des stocks et évolutions d'abondance des ressources démersales dans les pays de la CSRP. In: Actes du symposium international DAKAR Juin, pp. 205–222.

Gerlotto, F., Marchal, E., Stéquert, B., 1978. Le courant des Canaries: upwelling et ressources vivantes. In: Le Courant des Canaries: Upwelling et Ressources Vivantes: Symposium, 1978/04/11-14, Las Palmas.

Haberl, H., et al., 2004. Progress towards sustainability? what the conceptual framework of material and energy flow accounting (MEFA) can offer. Land Use Policy 21 (3), 199–213.

Haberl, H., et al., 2006. From LTER to LTSER: "Conceptualizing the socioeconomic dimension of long-term socioecological research. Ecol. Soc. 11 (2), 13.

Hardwick, J., 1977. Intergenerational equity and the investisment from ehaustible resources. Am. Econ. Rev. 65 (5), 972–994.

Holling, C.S., 2001. Understanding the complexity of economic, ecological, and social systems. Ecosystems 4 (5), 390–405.

Kates, R.W., et al., 2001. Sustainability science. Science 292 (5517), 641–642.

Le Roux, S., 2005. Pêche et territoire au Sénégal. thèse de géographie. université de Nantes.

Liu, J., et al., 2007. Complexity of coupled human and natural systems. Science 317, 1513–1516.

Meyer, J.M., 2009. Jeune Afrique (14/11/2009).

Ostrom, E., et al., 2007. Going beyond panaceas. Proc. Natl. Acad. Sci. 104 (39), 15176–15178.

PNUE, 2006. Rapport sur l'atelier régional pour l'Afrique consacré à l'adaptation 21–23 September 2006. UNFCCC. Accra (Ghana).

Prigogine, I., 1976. Order through Fluctuations: Self-organisation and Social Systems. Evolution and Consciousness: Human Systems in Transition. Addison-Wesley. Reading. Addison-Wesley. pp. 93–130.

Prigogine, I., Stengers, I., 1986. La nouvelle alliance: métamorphose de la science, Gallimard Paris %@ 2070323242.

Rebert, J.P., Amade, P., Privé, M., 1976. Hydrologie et courantométrie sur le plateau continental sénégalais en période d'alizés: résultats d'observations. Dakar: CRODT, 21 p. multigr.

Redman, C.L., et al., 2004. Integrating social science into the long-term ecological research (LTER) network: social dimensions of ecological change and ecological dimensions of social change. Ecosystems 7 (2) %U.

Riebsame, W.E., et al., 1994. Modeling land use and cover as part of global environmental change. %U Clim. Change 28 (1–2), 45–64.

Rivera-Monroy, V.H., et al., 2004. A conceptual framework to develop long-term ecological research and management objectives in the wider Caribbean region. BioScience 54 (9), 843–856.

Sabrié, M.L., L'hôte, Y., 2003. Sahel: une sécheresse persistante. *Fiches scientifiques* 178. IRD, Paris.

Solow, R., 1993. Intergenerational equity and exhaustible resources. Review of economics studies. In: Special Issue Symposium on the Economics of Exhaustibles Resources.

Turner, B.L.I., et al., 2003. Illustrating the coupled human-environment system for vulnerability analysis: three case studies. Proc. Natl. Acad. Sci. 100, 8080–8085.

UNEP, 2007. GEO-4: Fourth Global Environment Outlook. DEWA, Nairobi, UNEP. Nairobi 00100 Kenya.

Walker, B., et al., 2002. Resilience management in social ecological systems: a working hypothesis for a participatory approach. Conserv. Ecol. 6 (1), 14.

Making the Link

By Bethany Jorgensen

Universities as Solutions to Twenty-First Century Coastal Challenges: Lessons from Cheikh Anta Diop Dakar University ⟺ **Engaging Local Communities for Climate Change Adaptation: A Case Study in Quebec, Canada**

Alioune Kane et al. — *Steve Plante, Liette Vasseur and Charlotte Da Cunha*

In the previous chapter, Kane et al. illustrate the process of a Senegalese university shifting its goals to be more attentive to understanding the complex socioecological systems that come together and affect each other in coastal areas. This call to become attuned to interconnected socioecological systems is a leitmotif reverberating throughout the book, and we encounter it once again in this final chapter from Plante et al.

We close this collection of diverse and far-ranging efforts with a case study from a small town in Quebec, Canada, and the example of Plante et al.'s work in a successful community–university research alliance.

In the following chapter, we clearly see the enormous potential for universities and researchers to work and learn with coastal communities to increase their resilience to climate change and other challenges. This exemplary case study offers important lessons for researchers and practitioners alike through the authors' detailed explanation of the Participatory Action Research approach and other practical tools and techniques for public engagement and resilience building.

Enjoy!

Chapter 18

Engaging Local Communities for Climate Change Adaptation: A Case Study in Quebec, Canada

Steve Plante[1], Liette Vasseur[2], Charlotte Da Cunha[3]
[1]Departement of Societies, Territories and Development, University of Quebec at Rimouski, Rimouski, Quebec G5L 3A1, Canada; [2]UNESCO Chair in Community Sustainability: from local to global, Department of Biological Sciences, Brock University, St. Catharines, Ontario, Canada; [3]Université de Versailles Saint-Quentin-en-Yvelines, OVSQ, CEARC, Guyancourt, France

Chapter Outline

INTRODUCTION

Coastal ecosystems have long been modified by humans, who rely on the ocean as a means of transportation and as a natural resource. Until the middle of the twentieth century, coastal communities were highly integrated into their environment to the point of being a very good example of what a resilient[1] coastal

1. Holling (1973, p. 17) defines resilience as the capacity of a system to maintain these functions in a context of changes and so "determines the persistence of relationships within a system and is a measure of the ability of these systems to absorb changes of state variables, driving variables, and parameters, and still persist."

social-ecological system (SES) could be (Moser et al., 2012). With increasing pressures from development and exploitation of natural resources, coastal SESs have become increasingly vulnerable to coastal hazards. Moreover, climate change has exacerbated the impacts of coastal damage. Therefore, these SESs are no longer resilient to extreme events such as storm surges and hurricanes.

With the need to better protect their communities, decision-makers have been pushing for adaptation solutions in the face of the changing climate. For local actors (citizens, entrepreneurs, NGOs, and even municipal counsellors), however, the jargon used in the climate change adaptation framework has been confusing and abstract. Actors involved in the decision-making process must anticipate the inherent risks of these events due to climate change and adapt to new conditions in a context characterized with uncertainties, high political issues, and an overemphasis on technical aspects (Touili et al., 2014). Several issues in terms of territorial development emerge from trying to protect infrastructure and people from impacts of climate change. In recent years, new practical concepts have been developed but have yet need to be tested to attempt to better integrate communities and SES into the process of adaptation to climate change (Lahsen et al., 2010; Biagini et al., 2014) and adopt a territorial stakeholder perspectives (Raymond et al., 2010).

Through a case study conducted in the municipality of Bonaventure, Quebec, Canada, we describe the process used to help this community adapt to climate change. More precisely, we tested the Method of Evaluation by Group Facilitation (MEGF) as an approach to implement a tool that was developed in the Climate Change Challenges–Community-University Research Alliance (CCC-CURA) project called community resilience capacity building (Vasseur, 2012). This case study is part of a larger project called Coastal Communities Challenges (funded by Social Sciences and Humanities Research Council of Canada, under the Community-University Research Alliance program). This project aims to strengthen the resilience of individuals and communities by using on the researchers' side Participatory Action Research (PAR) and innovative public engagement techniques on the community's side. PAR is a nontraditional approach that considers local actors as partners in research rather than objects or subjects (van Aalst et al., 2008). During the course of this project, several tools have been tested and developed to support communities in their path toward improving resilience. In the next sections, we first describe the community resilience capacity-building tool, the PAR, approach and the MEGF technique. Then we summarize the steps that were accomplished with the community to implement these techniques and tool to finally examine the efficiency of MEGF to enhance public engagement in a community.

COMMUNITY RESILIENCE CAPACITY BUILDING

Adaptation to climate change and resilience can be foreign concepts for most communities. This tool on community resilience capacity building was developed within the CURA-CCC project in order to help the communities that are part of this project to initiate the process of adapting to climate change. The tool itself provides a

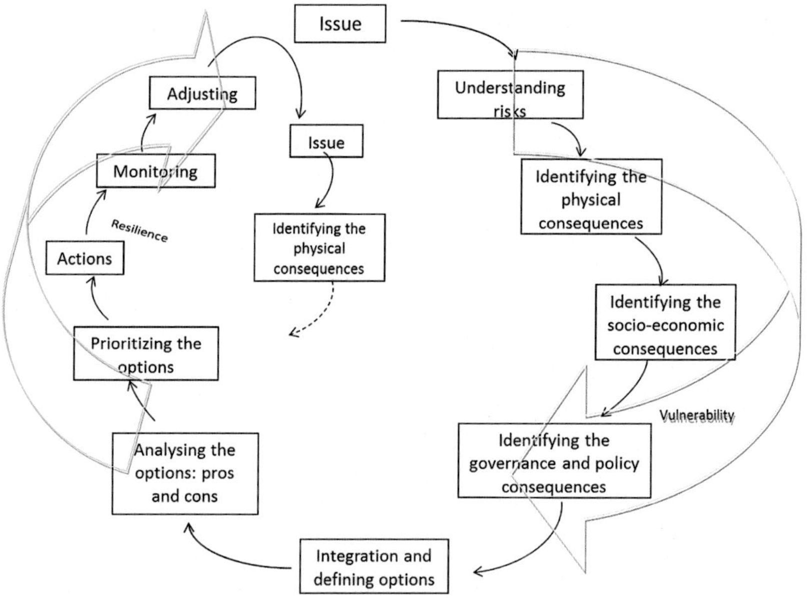

FIGURE 18.1 Community resilience capacity-building tool developed during the CCC-CURA project used for developing a community adaptive resilience plan in Bonaventure. *From Vasseur (2012).*

step-by-step explanation of the process with three sections: the first section explains the process, the second describes how to assess risks and vulnerabilities looking at various aspects of SES, and finally the third examines the way people can define solutions and strategies and how they can prioritize them (Vasseur, 2012). The process itself is composed of 11 steps, which can be done through a workshop of a few days, or—most likely due to limited community availability for a large block of time—through a series of meetings completed over a certain period of time (Figure 18.1).

The advantage of this process is that it is simple, inclusive, and requires integrating the various components of the community such as the natural ecosystem, health, social and cultural aspects, economy, and regulations. Using this social-ecological integrated view, it is easier to understand that each solution defined later may have some consequences for some components of the system. The final advantage of this tool is the possibility to integrate monitoring and evaluation components, which can be defined during the process. This tool is also very conducive to public engagement and the involvement of all actors at the territorial level. This is the main reason for having selecting PAR in this project, and gradually for the start of the process, MEGF, as approaches to better understand how these can be used in communities to help them move forward in adaptation and resilience to climate change.

PARTICIPATORY ACTION RESEARCH

For decision-makers and governments, adaptation is limited to civil protection (civil engineering structure) and policymaking (zoning law, rules, and program), and examined during a crisis or a recovery period after a disaster (Spalding, 2013). Indeed, most adaptation plans have been built using a disaster risk-reduction approach (management plan, land use, and development plans) (Ayyub et al., 2007; Hall et al., 2003; Ikeda et al., 2008; Campbell-Lendrum and Woodruff, 2006). They have been simplified to focus on emergency measures or technical advice for risk areas (breakwater, sea wall, or retreat) (Measham et al., 2011).

This approach may seem appropriate for coastal communities, since they are no strangers to extreme events, especially flooding and coastal erosion due to storm surges. However, is this the only adaptation priority for these communities in the long term? With more frequent extreme event scenarios and greater vulnerability of the SES, there is a need for deliberative approaches to consider the mechanisms of negotiation and apprehension of reality (Armitage et al., 2008, 2009; Chambers, 2003). Indeed, communities—i.e., those living in a coastal SES—should understand the consequences of these changing conditions and develop long-term adaptation strategies that are not only reactions after a storm. The techniques to enhance participation, cooperation, and consensus-building are increasingly used by professionals as an aid to develop adaptation strategies (Raymond et al., 2010). Unfortunately, most professionals, especially in local organizations, still do not have the adequate knowledge or skills to successfully implement such techniques (Poitras et al., 2003).

PAR can be considered as a way to examine such techniques through research that encourage actors to deliberate on issues and collectively find solutions. PAR has been adopted by several researchers with their partner communities and organizations to better reach all actors at the community level, enhance engagement, and legitimize the research results (Bell et al., 2013). The mobilization of local actors with researchers can generate individual and collective learning on current situations and the sharing of existing knowledge, i.e., scientific, local, ecological, cultural, or traditional knowledge (Raymond et al., 2010; Biagini et al., 2014). Social learning is believed to be an initial and essential step to negotiate and initiate changes, a necessary condition for an SES to have the ability to adapt (Folke, 2006). Through social learning, local actors can better understand the issues and take greater ownership. This process leads to an exchange of information and knowledge in which solutions can be developed, instead of only using a top-down approach where a decision-maker, after consulting a researcher or an expert, dictates what should be done. At the earliest stages of the development of solutions, PAR promotes the emergence of partnerships and stimulates dialog (Brydon-Miller et al., 2011). Knowledge and solutions can be co-constructed and co-produced by an alliance between researchers and actors, as adaptation actions can only be sustainable if they are socially and ecologically sound.

PAR goes further than traditional action research by including all actors from the initial development of the research question. These research questions are often more relevant to community members than action research where activities are carried out according to the interests of the researcher. Then through exchange and knowledge-sharing, issues are identified, discussed, and solutions are proposed in function of the capacity of the community and the level of acceptance of various solutions. Such an approach leads to a better degree of implementation of solutions, and using monitoring indicators usually defined by the community, evaluation of projects allows for reiterative processes (Maiter et al., 2008; Measham et al., 2011). By focusing on capacities of renewing, restructuring, and territorial development, we allow actors to better negotiate the changes and adapt appropriately (Beuret, 2010).

METHOD OF EVALUATION BY GROUP FACILITATION

The MEGF is a variant of the Nominal Group Technique (NGT) (Talbot, 1992; Lespinasse and Talbot, 2006). The NGT was developed during the 1960s with the work of de Delbecq and Van de Ven (1974), especially in the studies on individual and collective decision-making processes in the areas of health (Pineault and Daveluy, 1995), social services, and public administration. It is particularly appropriate for setting priorities in an evaluation. However, the use of MEGF should not be carried out systematically as a unique tool. It is inappropriate for activities that deal only with negotiation and confrontation.

The MEGF approach is based on the principle of formative evaluation. It includes five steps: (1) collecting statements or opinions from participants, (2) developing a common understanding of these statements, (3) sorting the statements into groups of common themes, (4) prioritizing the groups of statement, and (5) debating the results and evaluating the process. Indeed, MEGF may be suitable in the context of consensus-building (surrounding the process itself and how to establish the decision), planning (strategic planning and action plan), solution-seeking (problem identification, resolution program, and prioritization), identification of needs and goals (visioning exercise), and evaluation of research or teaching programs. It can also be used to initiate actions such as developing questionnaires and interview schemes to better understand the perceptions of the community (selection of themes, questions, order, and sampling choice).

It is an approach that promotes learning through co-construction and co-production of outputs and dialog among participants as well as the integration of existing and scientific knowledge (ideas, observations, opinions, and concepts). In this sense, a common language must be first developed to ensure a comprehensive understanding about the key components of the debate or the issue. It is also important that the facilitator or the experts do not impose their personal definitions and views and do not define or dominate the type of knowledge provided to the participants. It reduces the possibility that the process is being highjack for political and administrative reasons.

INITIATING COMMUNITY PLANNING FOR RESILIENCE IN BONAVENTURE

The municipality of Bonaventure was first contacted in late summer 2012. Bonaventure is a municipality of about 2800 people with 92% being Francophone and 5% Anglophone (Canada, 2006). Bonaventure covers 109.2 km^2 and is part of the administrative region of regional municipality of Bonaventure. The municipality has been affected by floods since the early 2000s as well as coastal erosion.

Initial Meetings

Prior to initiating the community resilience capacity-building process, people from different sectors of the municipality received open invitations to an introductory meeting in September 2012. This invitation was also published in the municipal newsletter. At the same time, more specific invitations were sent to environmental and social organizations (e.g., ZIP Baie des Chaleurs committee, Conseil régional en environnement-CREGIM, Bioparc de la Gaspésie or CIME Aventure). Additional participants were also targeted using a snowball sampling technique[2] (Figure 18.2).

Three introductory meetings were organized to reach as many actors as possible. Following these three meetings, a committee was created to move ahead with the community resilience planning process. The committee was composed of 10 people representing the main actors of the community. The first mandate of the committee was to identify the main issues regarding climate change adaptation faced by the community. To identify these issues, it was decided to use a participatory mapping technique in which all members were able to contribute their knowledge on where impacts and risks about climate extreme events threaten their community.

To complete this mapping, the committee used the kitchen assembly technique. A kitchen assembly is an activity taking place at the home of a participant, who invites people of the neighborhood to participate in the discussion. Four kitchen assemblies were organized by the committee to target four subdivisions of the municipality vulnerable to flooding and erosion. This exercise was also to be used to enhance awareness, trust in the committee, credibility of the process, and engagement of the community. A total of 25 people from these subdivisions participated in this process.

During the kitchen assemblies, participants asked a multitude of questions, several of which were related to how to define adaptation actions and to become more resilient. Several of them became interested in being involved in future activities. The researcher suggested the possibility of using the tool called community resilience capacity-building (Vasseur, 2012) to collectively co-construct

2. The research team initially inventoried the various groups of people that could be contributing to this process, including public and private sectors, community organizations, and civil societies, and sought to reach as many of them as possible.

FIGURE 18.2 Location of the community of Bonaventure. *From Her Majesty the Queen in Right of Canada, Natural Resources Canada, 2002.*

adaptation strategies. The participants were highly interested by this process. It was therefore proposed to the committee, which agreed to go ahead with this process.

The Use of an MEGF in the Community Resilience Capacity Building Process

At the beginning of the community resilience capacity-building process, the researcher suggested several approaches to accomplish the different steps of the process. The first step was to define the priority issues or components that the community felt more important in terms of risks in function of climate change; the MEGF was selected for this purpose by the committee. It took five hours to define these priority issues, i.e., to accomplish the five steps of the MEGF technique. It is important to note that the committee remained at 10 participants. A few of the initial members left the committee but were immediately replaced by others involved in the kitchen assemblies. To complete the rest of community resilience capacity-building process, seven additional meetings were organized. The average number of participants per meeting was eight, ranging from four to twelve per meeting.

Analysis of MEGF Results

In this section, we analyze how the MEGF technique was used to define the priority issues to be used in the community resilience capacity-building process and the results obtained from the discussion at the five-hour meeting. Prior to starting the MEGF, the committee decided on the question to be used to identify these issues. The question agreed upon was: "In your opinion, what are the key issues and challenges your community will face over the next 15–20 years due to climate change?" Once defined, the participants of the committee agreed to reflect upon and discuss this question with relatives, neighbors, and friends. They allocated two weeks for this reflection before officially starting the MEGF.

At the beginning of the MEGF meeting, the facilitator gave about 10 min to the participants to identify three statements (individually by writing) that seemed most important in order to answer the question previously proposed. Then, the facilitator asked the participants to present one by one their statements. The discussion stopped when no new statement was added. Statements and comments were recorded in an electronic format and projected. To meet the requirements of confidentiality and ethics, during the course of this process, participants were advised that they had the choice to leave the activity at any time if they felt uncomfortable or simply wanted to stop being involved. Discussion followed to ensure that all participants understood the meaning of these statements. It is important to note that understanding the statements does not necessarily mean agreeing with them. This discussion allowed for the sharing of knowledge and expertise and learning from one another. It provided a time to rework the statements, define concepts and representations, and classify the statements into categories to avoid duplication.

From this discussion, the committee identified 31 possible statements, which were grouped into 12 issues that could be integrated into the community resilience capacity-building process. The issues were territorial management and planning, coastal and riparian erosion, awareness and education, governance, emergency measures, drinking water supply, local economy, denial, ecological safeguard, events versus knowledge, zoning, and territorial valorization. For example, the territorial management and planning issue included statements concerning house relocation, flooding zone and occupation, public and private forest resource management, transportation, urban extension, and land use. To be successful in an MEGF, a maximum of 15 issues should be identified (Plante et al., 2009). The advantage of this process was to improve the ownership of the individual statements by the collective.

The final step of the MEGF was to vote on the priority issues through a secret ballot. This strategy allowed depersonalizing the discussion and arriving quickly at an understanding of the current situation. Each participant received a card to prioritize his or her first five most important issues among the twelve. The most important of these issues received five points, while the least important issue received one point. Once the coding was completed, the results were compiled and presented to the group for comments (Table 18.1). In this case, eight participants put at least one score for territorial management and planning, and nine for awareness and education. Six people voted for ecological safeguard, coastal and riparian erosion, and territorial valorization. The recurrence is important, as the same result in terms of points may reflect distinct realities: a strong evaluation (five points) by a small number of participants reveals the importance of this issue for some participants, although low evaluations (one to two points) by all the participants show the consistency of the issue for the entire community. The scores should be interpreted with caution considering that points alone may not reflect the representation of all participants. In term of relative scores, territorial management and planning had 36 points, and awareness and education and ecological safeguard had 24 and 21 points, respectively. It is interesting to note that the participants did not select events versus knowledge, and only once selected denial of the phenomenon.

Post MEGF

Using the results of the MEGF and the prioritization of the issues, the committee was then in a position to go ahead with the community resilience capacity-building-process, as described in Vasseur (2012). During the following months, the researchers acted as facilitators in this process. Since the first step of identifying priority issues was already completed, the next steps were to co-produce information and knowledge for one issue at a time. The committee selected and co-constructed potential adaptive options and solutions that could then be implemented.

More precisely, the second step of the community resilience planning tool required the participants to examine for each issue separately the risks and hazards associated with it. A focus group of two-and-a-half hours was organized to treat the first issue using the concepts of risk and hazard. After 40 min of debate on the meaning of each term, in which participants were able to develop a better understanding, potential options were co-produced. The committee worked on the physical, environmental, social/cultural, political, and economic consequences of these risks.

TABLE 18.1 List of the Twelve Issues Stated during the MEGF with the Three Most Important Issues Listed in Bold for Bonaventure in 2013

	Participants												
Issues	1	2	3	4	5	6	7	8	9	10	Total Points	Number of Votes	Rank
Territorial management and planning	5	3	5	5	5	4	5			4	36	8	**1**
Coastal and riparian erosion		4	1		3			1	5	2	16	6	5
Awareness and education		5	2	4		5	4		4		24	6	**2**
Governance	4							5	2		11	3	6
Emergency measures	1	2	4		1						8	4	8
Drinking water supply					2						2	1	10
Local economy				1		2	1	3	3		10	5	7
Denial											0	0	0
Ecological safeguard	3	1	3	2	4	1	3		1	3	21	9	**3**
Events versus knowledge											0	0	0
Zoning								2		1	3	2	9
Territorial valorization	2			3		3	2	4		5	19	6	4
	15	15	15	15	15	15	15	15	15	15	150		

Once the first issue completed, the second issue was then tackled. During the meetings, researchers were able to observe an increase in ownership of the process by participants as well as a greater sharing of existing knowledge and information. The results of this exercise in which a tool like the MEGF is used showed that actors can gradually take ownership of their conditions and define their own priorities regarding the issues that may affect them individually and collectively.

DISCUSSION

Adaptation or resilience planning at the community level can be achieved through various techniques such as kitchen assemblies, participative mapping, and MEGF. They became strategic in this case study. These techniques were collaboratively chosen by the participants because they provided support (through the presence of experts/researchers), inspiration (with participants' existing knowledge), and a territorial context (which can be modified in time and space in response to participants' reactions). In Bonaventure, this process stimulated the engagement of actors. Most participants were present during the entire process. They used their experience, expertise, and collective representations to develop this vision and gradually designed realistic solutions adapted to their conditions.

The particularity of the MEGF is the possibility to put on the table in objective fashion information relevant for decision making. It is important to underline that the meetings should be structured in such a way that the facilitator can direct the dialog toward a specific objective. The researcher must be able to clearly explain the purpose of the tool to the participants (van Aalst et al., 2008). It is therefore important that all information is disclosed to enhance the knowledge and understanding of all actors. The MEGF allows for the provision of data and information on the various components of the SES in such a way that is better balanced instead of being top down and too technical (van Aalst et al., 2008). Each intervention or comment should be brief and validated as often as possible with a round table discussion to ensure a common understanding of the statements or issues being brought forward to the group. The knowledge-sharing and exchange among actors lead to greater feedback, thus leading to the co-construction of solutions or strategies (Raymond et al., 2010). It is necessary to ensure the right to freedom of speech and expression in such a way that there is no risk that any leader performs pressures (formal or informal) or monopolizes the discussion to reach consensus in its favor (Plante et al., in press).

Through knowledge co-production and the co-construction of solutions, we noted a transformation in the perceptions of the participants regarding climate change adaptation and priority risks. For example, initially most members of the community targeted coastal erosion as the main factor contributing to the loss of land. Following discussion and sharing knowledge, the participants realized that in fact storm surges were the risk that caused coastal erosion. They acknowledged the importance of differentiating what is a risk and a hazard and how they can influence them. The MEGF can also generate a significant number of ideas that can be treated on site and within short periods of time.

MEGF, like many other public engagement techniques, can be flexible enough to adapt to the conditions of local communities in function of their needs and the challenges they face. While the results obtained in one community cannot be generalized to another one, the advantage of these techniques is that they are readily accessible. They do not require advanced technologies or expert firms. Local environmental organizations, such as those involved in the CCC-CURA, can acquire the knowledge and skills necessary to be able to facilitate community adaptation using these techniques. Our experience in this case study reinforces the idea that communities and local environmental organizations must examine which techniques will be the most effective for their own situation, as no approach fits all (Preston et al., 2011).

Through PAR and public engagement approaches, communities can improve their knowledge regarding the impacts of climate change and better understand how their actions now and in the future can influence their livelihoods and their interdependence with the SES. Denying such interrelationships may lead to unsustainable communities, as they increasingly face risks linked with maladaptation or nonadaptation (Ribot, 2011). In coastal communities, lack of knowledge and denial of the relationship between risks and hazards push people to make decisions, such as the construction of protection walls that are considered, in many cases, maladaptations, especially in the long term. This short-term view of reaction to change may have important consequences on the SES, not only for the present generation, but also for the future.

CONCLUSIONS

The Intergovernmental Panel on Climate Change (IPCC), United Nations Framework Convention on Climate Change (UNFCCC), and many countries have promoted mitigation efforts in coastal communities for many years. However, it is clear that adaptation measures are critical to maintain their sustainability and improve their resilience in the face of climate change. Having selected PAR in this CURA project has been beneficial for both researchers and communities. For researchers, through this approach it was possible to test the community resilience capacity-building tool and some public engagement techniques and validate their capacity to help community in defining their adaptation strategies and improving their resilience to climate change. For the community, this was an opportunity to share information, knowledge, and experience from different sources, improve governance adaptive capacity, and enhance their awareness regarding the various spatial and temporal challenges related to climate change adaptation. It will be interesting to follow whether this process will continue in the future when the researchers will no longer be involved.

ACKNOWLEDGMENTS

The author would like to acknowledge the financial support of the Social Sciences and Humanities Research Council of Canada. They also would like to thank the municipality of Bonaventure for their openness and willingness to work with the CCC-CURA in this project and Nicole Plank for the comments made of the manuscript.

REFERENCES

van Aalst, M.K., Cannon, T., Burton, I., 2008. Community level adaptation to climate change: the potential role of participatory community risk assessment. Glob. Environ. Change 18, 165–179.

Armitage, D., Marschke, M., et al., 2008. Adaptive co-management and the paradox of learning. Glob. Environ. Change 18 (1), 86–98.

Armitage, D.R., Plummer, R., et al., 2009. Adaptive co-management for social–ecological complexity. Front. Ecol. Environ. 7 (2), 95–102.

Ayyub, B.M., McGill, W.L., et al., 2007. Critical asset and portfolio risk analysis: an all-hazards framework. Risk Anal. 27 (4), 789–801.

Bell, S., Correa Pena, A., et al., 2013. Imagine coastal sustainability. Ocean Coastal Manage. 83 (0), 39–51.

Beuret, J.E., 2010. De la négociation conflictuelle à la négociation concertative : un « Point de Passage Transactionnel ». Négociations 13, 43–60.

Biagini, B., Bierbaum, R., Stults, M., Dobardzic, S., McNeeley, S.M., 2014. A typology of adaptation actions: a global look at climate adaptation actions financed through the global environment facility. Glob. Environ. Change 25, 97–108.

Brydon-Miller, M., Kral, M., et al., 2011. Jazz and the Banyan Tree: Roots and Riffs on Participatory Action Research, fourth éd. The Sage Handbook of Qualitative Research. Denzin, N.K, Lincoln, Y.S., California, Sage.

Campbell-Lendrum, D., Woodruff, R., 2006. Comparative risk assessment of the burden of disease from climate change. Environ. Health Perspect. 114 (12), 1935–1941.

Chambers, S., 2003. Deliberative democratic theory. Annu. Rev. Pol. Sci. 6, 307–326.

Delbecq, A., Van de Ven, A., 1974. The effectiveness of nominal, delphi, and interacting group decision making process. Acad. Manage. J. 17 (4), 605–621.

Folke, C., 2006. Resilience: the emergence of a perspective for social–ecological systems analyses. Glob. Environ. Change 16, 235–267.

Hall, J.W., Meadowcroft, I.C., et al., 2003. Integrated flood risk management in England and Wales. Nat. Hazards Rev. 4 (3), 126–135.

Holling, C.S., 1973. Resilience and stability of ecological systems. Annu. Rev. Ecol. Syst. 4, 1–23.

Ikeda, S., Sato, T., et al., 2008. Towards an integrated management framework for emerging disaster risks in Japan. Nat. Hazards 44 (2), 268–280.

Lahsen, M.R., Sanchez-Rodriguez, P., Romero Lankao, P., Leemans, R., Gaffney, O., Mirza, M., Pinho, P., Osman-Elasha, B., Stafford Smith, M., 2010. Impacts, adaptation and vulnerability to global environmental. Curr. Opin. Environ. Sustainability 2, 364–374.

Lespinasse, P., Talbot, R.W., 2006. Méthode d'Évaluation en Animation de Groupe (MEAG) appliquée à la pharmacie hospitalière. Le Pharmacien Hospitalier et Clinicien 41 (164), 43–47.

Maiter, S., Simich, L., et al., 2008. Reciprocity. An ethic for community-based participatory action research. Action Res. 6 (3), 305–325.

Measham, T.G., Preston, B.L., Smith, T.F., Brooke, C., Gorddard, R., Withycombe, G., Morrison, C., 2011. Adapting to climate change through local municipal planning: barriers and challenge. Mitig. Adapt. Strat. Glob. Change 16, 898–909.

Moser, S.C., Williams, J.S., et al., 2012. "Wicked" challenges at land's end: managing coastal vulnerability under climate change. Annu. Rev. Environ. Resour. 37, 51–78.

Pineault, R., Daveluy, C., 1995. La Planification de la santé: concepts, méthodes, stratégies. Montréal, Éditions Nouvelles.

Plante, S., Vasseur, L., DaCunha, C. Adaptation to climate change and participatory action research (PAR): lessons from municipalities in Quebec, Canada. In: Knieling, Jörg (Ed.). Climate Adaptation Governance. Theory, Concepts and Praxis in Cities and Regions, in press.

Plante, S., Boisjoly, J., et al., 2009. Participative governance and integrated coastal management. An experiment of dialogue in an insular community at isle-aux-coudres (Quebec, Canada). J. Coastal Conserv. 13, 175–183.

Poitras, J., Bowen, R., et al., 2003. Challenges to the use of consensus building in integrated coastal management. Ocean Coastal Manage. 46, 391–405.

Preston, B.L., Westaway, R.M., Yuen, E.J., 2011. Climate adaptation planning in practice: an evaluation of adaptation plans from three developed nations. Mitig. Adapt. Strat. Glob. Change 16, 407–438.

Raymond, C.,M., Fazey, I., Reed, M.S., Stringer, L.C., Robinson, G.M., Evely, A.C., 2010. Integrating local and scientific knowledge for environmental management. J. Environ. Manage. 91, 1766–1777.

Ribot, J., 2011. Vulnerability before adaptation: toward transformative climate action. Glob. Environ. Change 21 (4), 1160–1162.

Spalding, M.D., Ruffo, S., et al., 2013. The role of ecosystems in coastal protection: adapting to climate change and coastal hazards. Ocean Coastal Manage. 90 (0), 50–57.

Talbot, R.W., 1992. Méthode d'évaluation pour l'amélioration des performances dans l'enseignement postsecondaire. Revue des sciences de l'Éducation XVIII (2), 217–235.

Touili, N., Baztan, J., et al., 2014. Public perception of engineering-based coastal flooding and erosion risk mitigation options: lessons from three European coastal settings. Coastal Eng. 87, 205–209.

Vasseur, L., 2012. Vers une planification de la Résilience Communautaire. Une trousse pour initier le dialogue sur la planification de la résilience communautaire face aux changements environnementaux et climatiques, Trousse de formation préparée pour la Coalition pour la viabilité du sud du golfe du Saint-Laurent et l'Alliance de recherche universités–communautés - Défis des communautés côtières: 23.

Chapter 19

Conclusion

Liette Vasseur[1], Jean-Paul Vanderlinden[2,3], Paul Tett[4], Bethany Jorgensen[5,3], Omer Chouinard[6,3], Juan Baztan[2,3]

[1]UNESCO Chair in Community Sustainability: from local to global, Department of Biological Sciences, Brock University, St. Catharines, Ontario, Canada; [2]Université de Versailles Saint-Quentin-en-Yvelines, OVSQ, CEARC, Guyancourt, France; [3]Marine Sciences for Society, www.marine-sciences-for-society.org; [4]Scottish Association for Marine Science, Scottish Marine Institute, Oban, Argyll, UK; [5]The University of Maine, Orono, Maine, USA; [6]Université de Moncton, Moncton, New Brunswick, Canada

In our world of inequalities, increasing pollution, decreasing resources, environmental degradation, and global challenges, individual and shared responsibilities and actions are needed now more than ever. While this book calls for a sea change in how we as researchers, community members, and decision-makers approach coastal zone challenges, we know that this is not enough to turn the tide. For this reason, we remain committed to pursuing alternatives and implementation approaches that will help coastal communities move toward a more sustainable future.

The chapters in this book have taken us to many countries, and through them we have gained insight into many challenges facing coastal communities. As we have seen, the studies presented here offer specific examples of what are actually common challenges facing our coastal zones globally. We hope the solutions suggested may become common, too—adapted, expanded upon, and improved by communities around the world to fit their particularities and value their diversity.

As repeatedly depicted in these examples, one of the main challenges is also one of the main solutions for effective ICZM: the need to involve all coastal zone stakeholders—decision-makers, researchers, and all citizens—in the management process. This requires thinking about social-ecological systems holistically, considering temporal, spatial, political, and personal scales of management for people and natural processes in the coastal zone—and the coupling of such scales—to reach truly integrated yet flexible approaches, reflective of the dynamic contexts within which they operate. These scales must account for on-the-ground realities without losing sight of the global picture, and vice versa. It also means working with stakeholders, in accordance with community values, to identify challenges, inform potential solutions, and implement them collaboratively. Since many of the challenges involve the difficult task of striking a balance between immediate profit and long-term sustainability, it is critical that all stakeholders be allowed a voice in the coastal governance process.

Another crucial realization illustrated in these chapters is that developing sustainable solutions requires transdisciplinarity, respectful collaboration between all disciplines and sectors of society, from the social sciences to the arts and humanities to the physical and natural sciences to local and First Nations' knowledge. We know this is possible; often it is "just" a question of personal political will. Solutions to such multifaceted, "wicked" problems like climate change and plastic pollution will only grow from open-mindedness and transparency among actors, drawing from all existing knowledge, including the traditional, cultural, ecological, scientific, and so on. Lessons from the past can and must help us learn for the future.

Additionally, these chapters have shown there is a dearth of baseline data for coastal zone stakeholders to measure against in order to quantify or track the challenges they face, and evaluate the progress they are making in addressing them. Nontechnical as well as computational techniques and other methods are being developed to help communities meet these needs. Local knowledge from community members is invaluable for understanding the social-ecological history of a place. Going forward, there is much to be done to document the changes occurring in coastal zones, and to make sure proposed solutions are sustainable.

As we have mentioned before, the seed for this book was planted in 2011, at a regional meeting in Helsinki. It took root in anticipation of the Rio+20 Summit, when 110 members of the "Coastal Zones: Twenty-First Century Challenges" working group submitted a declaration outlining coastal zone challenges and potential solutions (Appendix I). The synergy from that widespread collaboration for the public good sprouted into the vision for this book, which has been tended by a smaller group of authors while keeping in mind the aims of the former working group.

As we reach the last page of this entwined effort, the editors would like to thank the authors and reviewers who contributed their time and expertise, and made this book possible. We would also like to thank the numerous others who have supported these efforts and continue their diligent work addressing coastal issues. Global and local challenges persist for most coastal communities. While our separate endeavors may seem like insignificant saplings, when taken together these local labors make up a vast forest of knowledge, spanning the globe with examples of communities moving forward and improving the outlook for our coasts.

Thanks to you, also, dear reader, for your attention to the examples we have shared with you here. We hope you nourish any seeds these chapters have planted, and that they may flourish into many fruitful pursuits.

We offer a final note of appreciation for the communities that opened their doors to us and allowed us to learn with and from them, traveling together down the sometimes rocky path of collaborative work to develop liveable, resilient solutions to coastal zone challenges.

Appendix

Input for the Compilation Document from the "Coastal Zones: 21st Century Challenges" Working Group

CONTEXT OF THE DOCUMENT

This document represents the input provided by the authors in order to participate in the construction of the "focused political document" for the Rio + 20 outcomes. As part of the effort to construct and achieve the Rio + 20 goals, the authors' points of view are comprised of contributions from members of the scientific and technological community and NGOs concerned with the sectoral priority of coastal zones.

Coastal zones are the most productive regions in the world, both biologically and economically, but they are also the most populated. They face a harsh future due to greater challenges stemming from hunger, wars, and health-related issues threatening populations and countries' economies. These challenges constitute the core of this document. Will nations work together to save these zones that buffer our world? They must be included in the next ten-year agenda before it is too late.

Following the recommendations provided by the "co-chairs' guidance note," this document has been redacted in the form of focused input. The material the authors submitted was coded, synthesized, and condensed to create this document. While this document communicates the most significant concerns of the authors as a group, it is not a consensus document. This document enables us to (1) establish our input to the Compilation Document, (2) propose adapted material to the Rio + 20 delegations from various governments through each country's correspondent, and (3) publish an extended reference document concerning the sectoral priority of coastal zones from the perspective of twenty-first century challenges.

One-hundred and fifteen authors from the "Coastal Zones: 21st Century Challenges" working group actively participated in the creation of this document. They are from 30 countries, and the following institutions, universities, research centers, and NGOs:

Aristotle University of Thessaloniki, ARUC, Aurecon, Australian National Center for Ocean Resources and Security University of Wollongong, Australian Rivers Institute, Baltic Environmental Forum, Boskalis Offshore, Brock University, CALCH, CEAB/CSIC, Center of Research in Material Sciences of Borj Cedria, Center de Suivi Ecologique, Centro Desarrollo y Pesca Sustentable, CETMEF/DS, Chao Pescao, CNR-INSEAN, Coastal Protection and Restoration Authority-Louisiana, Cooper Ecological Monitoring, Inc., École des Hautes Études en Sciences Sociales, ESA PWA, Environmental Hydrology, European Commission, Joint Research Center, Gaz-system, German Association of Aquaculture, Greenpeace, Hellenic Center for Marine Research, Helzel, IFMGEOMAR, Institut Universitaire Européen de la Mer, Jagiellonian University, Kyushu University, Laboratorio Nacional de Energia e Geologia, Latvijas Universitāte, LittOcean, Marine Sciences for Society, National Research Institute for Rural Engineering Water and Forestry, Nelson Mandela Metropolitan University, Norwegian Directorate of Fisheries, NUI Galway, OANNES, Oceanógrafos Sin Fronteras, OGS National Institute of Oceanography and Experimental Geophysics, Pepperdine University, Queen's University, Regional Ministry of Environment of Andalusia-Consejería de Medio Ambiente/Junta de Andalucía, Scripps Institution of Oceanography, Snowchange Cooperative, the Norwegian University of Life Sciences, the Pomeranian Maritime and Vistula River Basin Cluster Association, the Royal Marine Conservation Society of Jordan, UNESCO-IHE, Universidad Autónoma de Baja California Sur, Universidad de Cádiz, Universidad de La Laguna, Universidad de Las Palmas de Gran Canaria, Centro Interdisciplinario Manejo Costero Integrado, Universidad de la República-Uruguay, Universidad del Magdalena, Universidade do Algarve, Universidade Federal do Rio Grande, Universidad Nacional Autonoma de Mexico, Universidad Politécnica de Madrid, Universitat Autonoma de Barcelona, Universitat Politècnica de Catalunya, Université Bordeaux 1, Université de Moncton, Université de Picardie Jules Verne, Université du Québec à Rimouski, Université de Versailles Saint-Quentin-en-Yvelines, Université du Havre, Université du Québec à Rimouski, University of Bergen, University of Florence, University of Patras, University of Tartu, University of Ulster, University of Western Australia, Waseda University, and Technical University of Delft.

Authors Listed in Alphabetical Order:

Octavio Aburto-Oropeza, Mohammad Al-Tawah, Chistos Anagnostou, Christian Appendini, Matías Asun, Juan M. Barragán-Muñoz, Robert Battalio, Juan Baztan, Anne Blanchard, Camilo-Mateo Botero, Scott Bremer, Laura Cabrera-Vega, Ana Campos, Donata Canu, Maria-Elena Cefalì, Mike Chadwick, Semia Cherif, Omer Chouinard, Katarzyna Chrulska, Daniel Conde, Dan Cooper, Alejandra Cornejo, Susana Costas, Louise Coyle, Charlotte Da Cunha, Debora De Freitas, Irene Delgado-Fernandez, Federico Demaria, Javier

Diez, Patricia Dominguez, Jean-Marc Douguet, Anne Duperret, Matthieu Dutertre, Raimonds Ernstein, Miguel Esteban, Frédérique Eynaud, Elizabeth Figus, Vanessa Finance, Rainer Froese, Noureddine Gaaloul, Jesús Gabaldon, Shari L. Gallop, Marli Geldenhuys, Lea Gracia, Francesco Guercio, Casandra Gutierrez-Gálvez, Birgit Hansen, Linda R. Harris, Yves Henocque, Luis Hernández-Calvento, Ron Hirschi, François Hissel, Alejandro Iglesias-Campos, Claudia Abreu, Frank Jacobsen, Bethany Jorgensen, Aristomenis Karageorgis, Syed Khalil, Matthias Kniephoff, Zoi Konstantinou, Yannis N. Krestenitis, Xavier Lafon, Lisa Levin, Camino Liquete, Merle Looring, Jose-Santos Lopez-Gutierrez, Karen Martin, Loraine McFadden, Amadou M. Dieye, Edwin Mendez, Jan-Olaf Meynecke, Francisco J. Miranda-Avalos, Tero Mustonen, Vicente Negro, João-Luiz Nicolodi, Mélinda Noblet, Julian Orford, André Pacheco, Kaliopi Pagou, Panayotis Panayiotidis, Yves-Marie Paulet, Tiago Pedrosa, Margit Pelzer, Steve Plante, Roshanka Ranasinghe, Nadia Ribas, Gabriel Roa Medvinsky, Elisabet Roca-Bosch, Karyn-Nancy Rodrigues-Henriques, Mireia Roura, Blanca Rubi, Alberto Ruiz, Mauro Salvatore, Agustín Sánchez Arcilla, Felicita Scapini, David Schoeman, Moura Schreiner, Sabine Schultes, Satoquo Seino, T.M. Silveira, Marcos Sommer, Knut-Bjørn Stokke, Céline Suretteve, Josefina Terrera, Evangelos Tzanatos, Jean-Paul Vanderlinden, Liette Vasseur, Kristina Veidemane, Chloé Vlassopoulou, Witold Wacławik-Narbutt.

INTRODUCTION

More than twenty years have passed since the Rio Earth Summit in 1992. Years of efforts to better understand, inform, and improve the relationships between our societies and our planet's coastal zones. These efforts have crystallized into tangible outcomes in the form of improvements in environmental culture and international agreements upheld by over 100 national and transnational coastal zone plans, protocols, and conventions.

While moving forward with these national and international efforts, we realize that the balance between development and stewardship is still broken, and many more efforts are needed to create a harmonious relationship between the use of knowledge in society and our planet's coastal zones. Through the active participation of 115 researchers from 30 countries, the following baseline document has been constructed to highlight the perspectives of academia regarding "Coastal Zones: 21st Century Challenges." Please consider it our input for the RIO + 20 compilation document.

Input for Compilation Document

The majority of our planet's population is concentrated in coastal zones, narrow spaces that amplify the most urgent and emerging questions of sustainability and development. In coastal zones, we clearly see the fragility of the three elements that constitute sustainability: world population growth, economic tenuity, and the increase of environmental degradation. Coastal zones are key in illustrating

(1) the challenges our societies face and (2) the potential solutions, priorities, and views regarding the implementation of practices and policies that build upon previous successes. These two points structure the document.

The Challenges Our Societies Face

Any initiative to truly help society progress sustainably must integrate the limits of the planet and be co-constructed with the affected communities. The consensus is that the challenges we face in coastal zones are mostly anthropogenic or amplified by human activities that clearly transgress ethical limits.

Due to human development on the shoreline and in river basins, along with unsustainable off-shore industrial activities, our challenges are:

Red Flag Challenges Impacting Lives of Coastal Zone Residents:

- Malnutrition, hunger, freshwater availability
- Wars and other violent conflicts
- Lack of education
- Climate change and its consequences
- Over-exploitation of marine living resources
- Toxins in fish and shellfish, and pathogens such as cholera and hepatitis, are threats to human health
- Population growth
- Global economic crises

Challenges in Policy:

- Harmonize the interests of coastal environment users, including local community members, coastal municipalities, regional and/or inter-municipal planning, national, transnational, and international stakeholders, through the continuous improvement of economic-legislative instruments and the elaboration and implementation of coordinated strategies for the use of natural, social, cultural, and institutional resources
- Rethink economic growth and the flows of energy and materials
- Preserve 100% of the areas where the indigenous peoples of the coasts remain, including the Saami, Chukchi, Siberian Yupiaq and many others
- Integrate research and education into the decision making process
- Make information readily and easily accessible to facilitate informed decision-making
- Protect natural and cultural resources at all levels: local, regional, national, and international, while keeping coastal communities safe
- Monitor and control the coastal and littoral maritime traffic, industrial activity, and the related hazards of oil pollution, chemical transport, collision, GIS reduction, and technical failures
- Introduce policies that: make change trends mandatory, protect existing coastal habitats and ecological functions, recover the fishing stocks, and prevent illegal and habitat-destructive fishing

- Prevent over-population in developing regions and minimize damage in already overpopulated areas
- Balance urban growth by using space more efficiently
- Plan sustainable spatial allocation and management for fisheries and aquaculture
- Plan sustainable spatial allocation and management for energy production and supply

Challenges from Pollution and Climate Change:

- Oceanic temperature warming and change in alkalinity
- Decreasing oxygen levels leading to dead zones, species extirpation, and noxious gas emissions
- Seawater intrusion in coastal aquifers
- Coastal and sea pollution caused by wastewaters and solid wastes that have been treated ineffectively or not at all
- Pollution by toxic waste, metals, nutrients, contaminants
- Floods, erosion, and rising sea-levels
- Illegal or little regulated extraction of natural resources to fill increasing demand
- Amplified vulnerability of coastal populations, particularly the economically disadvantaged
- Loss of habitat and loss of biodiversity
- Irreversible ecological destruction
- Toxic blooms due to pollution
- Coral reef bleaching
- Introduction of invasive species
- New diseases among organisms
- Dispel the assumption that the coast is "safe"

Challenges in Research:

- Generate an information baseline of coastal ecological and social processes that researchers can measure against
- Take into account the social and human dimensions of uncertainty
- Study in greater depth the interconnectedness of natural systems to better understand how to sustain coastal and oceanic health
- Conduct research in support of management on multiple spatial and temporal scales
- Develop and establish an integrated oceans monitoring network, and create interoperable open-access databases that can provide reliable data on a user community's defined goal(s)
- Interdisciplinary approaches to solve any challenge
- Integrate all stakeholders in the research process
- Develop innovative techniques for the restoration of ecosystem functions

- Evaluate the success of the integrated costal management political processes and practices on a local to global basis
- Identify and quantify the human-induced stressors acting on coastal ecosystems and populations

Potential Solutions, Priorities, and Views Regarding the Implementation of Practices and Policies that Build upon Successes

Economy and Development Models:

- The model of development based on infinite economic growth needs to be questioned: To what degree do activities on coastal areas facilitate general development and what manner of development is currently needed?
  Can development be based on sustainability and how can the socio-economic structure respond to international competitiveness?
- A trade-off between the economy and the environment exists; destructive industries have to be challenged and held accountable for their social and environmental consequences
- G20 announced the preparation of a charter on "sustainable economics," we must make explicit *how* such a charter should be implemented from a global governance perspective

Governance, Global/Local Articulation:

- The efforts cannot only come from local governments and communities the challenges are global in nature
- UN Ocean should be supplemented by other transgovernmental and nongovernmental networks as additional forms of governance
- Intergovernmental Panel on Maritime Basins (IPMB) should contribute to providing governance systems with common and reliable information and promote coherent responses from these systems
- Build strong connections between transboundary maritime basins related to large marine ecosystems and maritime regions of the world
- Apply a deliberative approach that concentrates on managing emerging challenges and linking all spatial and temporal scales

Collaborative Policy Making:

- The instruments for the implementation of integrated coastal zone management are: an integrated approach to coastal land and marine spatial planning, cross-sectoral and multiregional agreements, public participation, effective cross-border consultation system, monitoring and assessment of socio-economical and ecological changes and trends, comprehensive analysis of sustainable development indicators, financial and legal mechanisms for ICZM implementation, and connected and collaborative decision-making between all administrative levels from global to local

- Move from the theoretical framework into realizing the necessary actions
- Improve the articulation between ICZM and adaptation measures
- Integrate local and traditional knowledge with policy making
- Learn from international experience and practices in integrated coastal management, and adjust lessons to other contexts
- Evaluate the success of the integrated coastal management political processes and practices on a global basis
- Participation of coastal communities is vital, not only to vindicate the legitimacy of strategies, but also to provide them with the opportunity to express their doubts, to rebuild their trust, to learn how to live in a changing environment, and to manage social conflict
- Create respectful partnerships with traditional societies on Earth, as they can provide crucial observations and knowledge regarding emerging challenges
- Make decisions that are compatible with the core values of affected coastal communities and coast-dependent peoples
- Increase interdisciplinary training and cross-collaborations among tertiary programs and teams
- Natural science data must be combined with social science understandings of the places where regulations are to be implemented
- Natural scientists, engineers, economists, lawyers, and social scientists must recognize their responsibility and role in the process and collaborate with each other to achieve common goals

Legislation and Regulation:

- Define coastal zones in both spatial and temporal dimensions, since coastal dynamics cast some legal uncertainty on how coastal zones are determined
- Improve and reinforce legal frameworks controlling coastal activities
- Make good practices mandatory for stakeholders; hold elected politicians accountable for their promises
- Coordinate states and various sectors

Information, Education, and Awareness:

- Improve the competence of and resources for local and regional coastal zone authorities
- Knowledge must be shared, promoted, and used in order to (1) aid society in developing a critical approach, (2) exercise pressure on policy makers, and (3) develop realistic, sustainable, and feasible policies
- Promote public awareness of the socio-ecological values of the coastal resources and ecosystems

Research:

- Encourage the scientific development of new sustainable, useful technologies
- Improve treatment plant performance; increase general use of new biodegradable materials

- Long-term studies that identify past and present evolutionary trends of the coast
- Intelligent and sustainable use of marine resources to enable the development of new, sustainable medical and pharmaceutical products
- Develop and establish integrated monitoring networks and coherent forecasting systems that provide coastal managers and policymakers with critical coastal state indicators in order to ensure the safety of coastal communities while assuring the preservation of the natural coastal dynamics
- Bring together competence and synergy to develop an ecological sustainable aquaculture in order to protect the biodiversity and environment in the ocean, as well as provide a safe and sustainable source for human food
- Combine infrastructures like offshore wind energy facilities and aquaculture installations

CONCLUSION

A general comment to conclude our document is this: the challenges we face in coastal zones are mostly anthropogenic or amplified by human activities that clearly transgress reasonable limits. We insist on the fact that any initiative to truly help society progress sustainably must integrate the limits of the planet and be co-constructed with the affected communities.

REMINDER

This document represents the input provided by the authors in order to participate in the construction of the "focused political document" for the Rio+20 outcomes. As part of the effort to construct and achieve the Rio+20 goals, the authors' points of view are comprised of contributions from members of the scientific and technological community and NGOs concerned with the sectoral priority of coastal zones.

Following the recommendations provided by the "co-chairs' guidance note," this document has been redacted in the form of focused input. The material the authors submitted was coded, synthesized, and condensed to create this document. While this document communicates the most significant concerns of the authors as a group, it is not a consensus document. This document enables us to (1) establish our input to the Compilation Document, (2) propose adapted material to the Rio+20 delegations from various governments through each country's correspondent, and (3) to publish an extended reference document concerning the sectoral priority of coastal zones from the perspective of twenty-first century challenges.

Index

Note: Page numbers followed by "f" and "t" denote figures and tables, respectively.

D

E

J

K

L

M

N

O

P

Q

R

Printed in the United States
By Bookmasters